Lecture Notes in Bioinformatics 7821

Edited by S. Istrail, P. Pevzner, and M. Waterman

Subseries of Lecture Notes in Computer Science

W0234525

Minghua Deng   Rui Jiang   Fengzhu Sun
Xuegong Zhang (Eds.)

Research
in Computational
Molecular Biology

17th Annual International Conference, RECOMB 2013
Beijing, China, April 7-10, 2013
Proceedings

 Springer

Series Editors

Sorin Istrail, Brown University, Providence, RI, USA
Pavel Pevzner, University of California, San Diego, CA, USA
Michael Waterman, University of Southern California, Los Angeles, CA, USA

Volume Editors

Minghua Deng
Peking University, Beijing, China
E-mail: dengmh@math.pku.edu.cn

Rui Jiang
Tsinghua University, Beijing, China
E-mail: ruijiang@tsinghua.edu.cn

Fengzhu Sun
University of Southern California
Los Angeles, CA, USA
E-mail: fsun@usc.edu

Xuegong Zhang
Tsinghua University, Beijing, China
E-mail: zhangxg@tsinghua.edu.cn

ISSN 0302-9743
ISBN 978-3-642-37194-3
DOI 10.1007/978-3-642-37195-0
Springer Heidelberg Dordrecht London New York

e-ISSN 1611-3349
e-ISBN 978-3-642-37195-0

Library of Congress Control Number: 2013933233

CR Subject Classification (1998): J.3, H.2.8, G.3, F.2.2, I.5.2, I.6.3-5

LNCS Sublibrary: SL 8 – Bioinformatics

Typesetting: Camera-ready by author, data conversion by Scientific Publishing Services, Chennai, India

Printed on acid-free paper

Springer is part of Springer Science+Business Media (www.springer.com)

Preface

The RECOMB conference series — with its full name the Annual International Conference on Research in Computational Molecular Biology — was started in 1997 by Sorin Istrail, Pavel Pevzner, and Michael Waterman. The 17th Annual International Conference on Research in Computational Molecular Biology or RECOMB 2013 was held at Tsinghua University, Beijing, China, hosted by the Bioinformatics Division of Tsinghua National Laboratory for Information Science and Technology (TNLIST), Tsinghua University. This volume contains the 32 extended abstracts selected for oral presentation at RECOMB 2013, which were selected by the Program Committee (PC) out of 167 submissions. Each submission was assigned to at least three reviewers and 14 papers received two reviews and the rest received at least three reviews from the PC or external reviewers invited by PC members. Accepted papers were also invited for submission of an edited journal version to a special issue of the *Journal of Computational Biology*.

Besides the selected talks, this year's RECOMB conference also featured six invited keynote talks by leading scientists in life sciences around the world. The keynote speakers were Scott Fraser (University of Southern California, USA), Takashi Gojobori (National Institute of Genetics, Japan), Deborah Nickerson (University of Washington, USA), Nadia A. Rosenthal (Monash University in Melbourne, Australia), Chung-I Wu (Beijing Institute of Genomics, Chinese Academy of Sciences, China), and Xiaoliang Sunney Xie (Harvard University, USA).

Following a tradition begun with RECOMB 2010, RECOMB 2013 also featured a special track for highlights presenting work that had been published in journals during the last 15 months. Five such highlight talks were selected for oral presentation this year.

The success of RECOMB depends on the efforts, dedication, and devotion of many colleagues who spent countless of hours on the organization of the conference. We thank the PC members and the external reviewers for the timely review of the assigned papers despite their busy schedules. We also thank all the authors for submitting their excellent work to RECOMB. The Steering Committee consisting of Vineet Bafna, Serafim Batzoglou, Bonnie Berger, Sorin Istrail, Michal Linial, and Martin Vingron (Chair) gave many excellent suggestions on the organization of the conference. I would like to personally thank the local Organizing Committee members especially the Co-chairs Xuegong Zhang, Minghua Deng, and Rui Jiang, and the local secretary Zhuwei Joan Zhang for their efforts that insured smooth cooperation on the administrative and logistic details. Various organizations including Tsinghua University, TNLIST, the National Science Foundation of China (NSFC), the US National Science Foundation (NSF), the International Society of Computational Biology (ISCB), and all the industry sponsors for their financial support. Mona Singh (Princeton University) helped with the application for the US NSF student support. Finally, we thank the authors of the papers and posters and all the attendees for their enthusiastic participation of the conference.

January 2013   Fengzhu Sun

Conference Organizations

Steering Committee

Vineet Bafna	University of California, San Diego, USA
Serafim Batzoglou	Stanford University, USA
Bonnie Berger	Massachusetts Institute of Technology, USA
Sorin Istrail	Brown University, USA
Michal Linial	The Hebrew University of Jerusalem, Israel
Martin Vingron	Max Planck Institute for Molecular Genetics, Germany (Chair)

Conference Chair

Xuegong Zhang	Tsinghua University, China

Program Chair

Fengzhu Sun	University of Southern California, USA

Program Committee

Tatsuya Akutsu	Kyoto University, Japan
Frank Alber	University of Southern California, USA
Max Alekseyev	University of South Carolina, USA
Kiyoshi Asai	University of Tokyo, Japan
Joel Bader	Johns Hopkins University, USA
Vineet Bafna	University of California, San Diego, USA
Ziv Bar-Joseph	Carnegie Mellon University, USA
Nuno Bandeira	University of California, San Diego, USA
Serafim Batzoglou	Stanford University, USA
Bonnie Berger	Massachusetts Institute of Technology, USA
Sebastian Böcker	Jena University, Demark
Michael Brent	Washington University, USA
Michael Brudno	University of Toronto, Canada
Dongbo Bu	Chinese Academy of Sciences, China
Kun-Mao Chao	National Taiwan University, Taiwan

Brian Chen	Lehigh University, Canada
Luonan Chen	Chinese Academy of Sciences, China
Phoebe Chen	La Trobe University, Australia
Ting Chen	University of Southern California, USA
Francis Chin	The University of Hong Kong, Hong Kong
Minghua Deng	Peking University, China
Nadia El-Mabrouk	University of Montreal, Canada
Mikhail Gelfand	Institute for Information Transmission Problems RAS, Russia
Eran Halperin	Tel Aviv University, Israel
Si-Min He	Chinese Academy of Sciences, China
Wen-Lian Hsu	Academia Sinica, Taiwan
Heng Huang	University of Texas, USA
Daniel Huson	University of Tübingen, Germany
Sorin Istrail	Brown University, USA
Daniel Huson	University of Tübingen, Germany
Rui Jiang	Tsinghua University, China
Simon Kasif	Boston University, USA
Daniel Huson	University of Tübingen, Germany
Jens Lagergren	Royal Institute of Technology, Sweden
Doheon Lee	Korea Advanced Institute of Science and Technology, Korea
Hyunju Lee	Gwangju Institute of Science and Technology, Korea
Thomas Lengauer	Max Planck Institute for Informatics, Germany
Lei M Li	Chinese Academy of Sciences, China
Ming Li	University of Waterloo, Canada
Yixue Li	Shanghai Center for Bioinformation Technology, China
Michal Linial	Hebrew University, Israel
Jinze Liu	University of Kenturky, USA
Stefano Lonardi	University of California, Riverside, USA
Satoru Miyano	Tokyo University, Japan
Bernard Moret	Swiss Federal Institutes of Technology, Switzerland
William Noble	University of Washington, USA
Arlindo Oliveira	INESC-ID, Portugal
Teresa Przytcka	NIH NCBI, USA
Ben J Raphael	Brown University, USA
Knut Reinert	Freie Universität Berlin, Germany
Marie-France Sagot	INRIA, France
Cenk Sahinalp	Simon Fraiser University, Canada
Russell Schwartz	Carnegie Mellon University, USA
Roded Sharan	Tel Aviv University, Israel
Mona Singh	Princeton University, USA

Steven Skiena	SUNY, USA
Andrew Smith	University of Southern California, USA
Peter Stadler	Universität Leipzig, Germany
Yuzhen Ye	Indiana University, USA
Kai Tan	University of Iowa, USA
Chao Tang	Peking University, China
Martin Vingron	Max Planck Institute for Molecular Genetics, Germany
Jerome Waldispuhl	McGill University, Canada
Lusheng Wang	City University of Hong Kong, China
Limsoog Wong	National University of Singapore, Singapore
Xiaohui Xie	UC Irvine, USA
Dong Xu	University of Missouri, USA
Yuzhen Ye	Indiana University, USA
Sungroh Yoon	Seoul National University, Korea
Shibu Yooseph	JCVI, USA
Louxin Zhang	National University of Singapore, Singapore
Michael Q Zhang	UT Dallas, USA and Tsinghua University, China
Xuegong Zhang	Tsinghua University, China

Organizing Committee

Xuegong Zhang	Tsinghua University, China (Co-chair)
Minghua Deng	Peking University, China (Co-chair)
Rui Jiang	Tsinghua University, China (Co-chair)
Jin Gu	Tsinghua University, China
Jingdong Han	PICB, CAS, China
Ruiqiang Li	Peking University, China
Xuan Li	SIBS, CAS, China
Yixue Li	Shanghai Bioinformatics Center, China
Jingchu Luo	Peking University, China
Geng Tian	Tsinghua University, China
Xiujie Wang	Institute of Genetics and Development, CAS, China
Peiheng Zhang	Institute of Computing, CAS, China

External Reviewers

Aguiar, Derek	Antipov, Dmitry	Bagherian, Misagh
Ahrne, Erik	Arndt, Peter	Bahrami, Emad
Aliphanahi, Babak	Askenazi, Manot	Bandyopadhyay,
Andreotti, Sandro	Atias, Nir	Nirmalya

Bankevich, Anton
Baran, Yael
Baudet, Christian
Bazykin, Yegor
Becerra, David
Beglov, Dmitri
Behnam, Ehsan
Bernstein, Laurence
Bienkowska, Jadwiga
Blin, Guillaume
Bozdag, Serdar
Bryant, David
Buske, Orion
Chang, Chia-Jung
Chen, Ching-Tai
Chen, Tiffany
Chen, Yi-Ching
Cheng, Cheng-Wei
Cheng, Chia-Ying
Chiu, Ka Ho
Cho, Dongyeon
Choi, Jeong-Hyeon
Choi, Kwok Pui
Chowdhury, Salim
Chu, An-Chiang
Clevert, Djork-Arné
Costello, James
Daley, Timothy
Daniels, Noah
Dao, Phuong
David, Matei
Dieterich, Christoph
Dondi, Riccardo
Donmez, Nilgun
Doose, Gero
Duma, Denisa
Ermakova, Ekaterina
Eskin, Itamar
Fischer, Martina
Fiser, Andras
Fleischauer, Markus
Frånberg, Mattias
Fu, Yan
Fuentes, Gloria
Gao, Jianjiong

Gao, Xin
Gautheret, Daniel
Gitter, Anthony
Golan, David
Guthals, Adrian
Hach, Faraz
Hajirasouliha, Iman
Halldorsson, Bjarni
Halloran, John
Han, Buhm
Hao, Xiaolin
Harris, Elena
He, Danning
He, Xin
He, Zengyou
He, Zhiquan
Henry, Henry
Hoffman, Michael
Holtby, Dan
Hosur, Raghavendra
Howbert, Jeff
Hu, Jialu
Hu, Yin
Huang, Yan
Hwang, Woochang
Irannia, Zohreh
Jiang, Shuai
Joshi, Trupti
Kaell, Lukas
Kalinina, Olga
Kaplan, Tommy
Katenka, Natallia
Kehr, Birte
Khrameeva, Ekaterina
Kim, Yoo-Ah
Kim, Younghoon
Kirkpatrick, Bonnie
Klau, Gunnar W.
Kochetov, Alex
Korkin, Dmitry
Korobeynikov, Anton
Kozakov, Dima
Kulikov, Alexander
Kyriazopoulou-
    Panagiotopoulou,

Sofia
Lacroix, Vincent
Lam, Henry
Le, Hai-Son
Lee, Sael
Lee, Sejoon
Lee, Sunjae
Lehmann, Kjong
Leiserson, Mark
Lemaitre, Claire
Leung, Henry
Li, Ning
Li, Shuai Cheng
Li, Wei
Li, Weizhong
Liao, Chung-Shou
Libbrecht, Max
Lin, Wei-Yin
Lin, Yen Yi
Liu, Yan
Liu, Yizhou
Love, Michael
Lu, Bingwen
Lynn, Ke-Shiuan
Ma, Bin
Ma, Wenxiu
Ma, Xiaotu
Madhusudhan, M.S.
Mahlab, Shelly
Makeev, Vsevolod
Mammana, Alessandro
Markowetz, Florian
Marti-Renom, Marc
Mazza, Arnon
Medvedev, Paul
Meusel, Marvin
Mezlini, Aziz
Mirebrahim, Seyed
Mironov, Andrey
Misra, Navodit
Molla, Michael
Navlakha, Saket
Ng, Kal Yen Kaow
Nikolenko, Sergey
Numanagic, Ibrahim

Nurk, Sergey
Oesper, Layla
Panchin, Alexander
Parviainen, Pekka
Pasaniuc, Bogdan
Pelossof, Raphael
Peng, Jian
Pfeifer, Nico
Pham, Son
Polishko, Anton
Qu, Jenny
Rahnenführer, Jörg
Rajasekaran,
    Rajalakshmi
Rampasek, Ladislav
Rappoport, Nadav
Ray, Pradipta
Rho, Mina
Richard, Hugues
Ritz, Anna
Rodriguez, Jesse
Roman, Theodore
Rozov, Roye
Sacomoto, Gustavo
Salari, Raheleh
Sayyed, Auwn
Scheubert, Kerstin

Schulz, Marcel
Sheng, Quanhu
Sheridan, Paul
Shibuya, Tetsuo
Shimamura, Teppei
Shiraishi, Yuichi
Silverbush, Dana
Sinaimeri, Blerina
Sindi, Suzanne
Sjöstrand, Joel
Snedecor, June
Souaiaia, Tade
Steffen, Martin
Stegle, Oliver
Subramanian,
    Ayshwarya
Sul, Jae Hoon
Sun, Ruping
Sun, Shiwei
Swenson, Krister
Tannier, Eric
Thomas-Chollier,
    Morgane
Tjong, Harianto
Tofigh, Ali
Tran, Ngoc Hieu
Uren, Philip

Vandin, Fabio
Varoquaux, Nelle
Wan, Lin
Wang, Hung-Lung
Wang, Jian
Wang, Kendric
Wang, Mingxun
Wang, Yi
Wang, Yunfei
Wise, Aaron
Wojtowicz, Damian
Wojtowicz, Danian
Xu, Jinbo
Yamaguchi, Rui
Yan, Xifeng
Yeger-Lotem, Esti
Yu, Zhaoxia
Zeng, Feng
Zhang, Chao
Zhang, Jingfen
Zhang, Jiyang
Zhang, Shihua
Zheng, Jie
Zheng, Yu
Zhong, Shan
Zinman, Guy

The RECOMB Chronology

#	Year	Dates and Location	Hosting Institution	Program Chair	Conference Chair
1	1997	Jan 20-23, Santa Fe, NM, USA	Sandia National Lab	Michael Waterman	Sorin Istrail
2	1998	Mar 22-25, New York, NY, USA	Mt. Sinai School of Medicine	Pavel Pevzner	Gary Benson
3	1999	Apr 22-25, Lyon, France	INRIA	Sorin Istrail	Mireille Regnier
4	2000	Apr 8-11, Tokyo, Japan	University of Tokyo	Ron Shamir	Satoru Miyano
5	2001	Apr 22-25, Montreal, Canada	Université de Montreal	Thomas Lengauer	David Sankoff
6	2002	Apr 18-21, Washington, DC, USA	Celera	Gene Myers	Sridhar Hannehalli
7	2003	Apr 10-13, Berlin, Germany	German Federal Ministry for Education and Research	Webb Miller	Martin Vingron
8	2004	Mar 14-18, San Diego, CA, USA	University of California San Diego	Dan Gusfield	Phillip E. Bourne
9	2005	May 14-18, Boston, MA, USA	Broad Institute of MIT and Harvard	Satoru Miyano	Jill P. Mesirow and Simon Kasif
10	2006	Apr 2-5, Venice, Italy	University of Padova	Alberto Apostolico	Concettina Guerra
11	2007	Apr 21-25, San Francisco, CA, USA	QB3	Terry Speed	Sandrine Dudoit
12	2008	Mar 30- Apr 2, Singapore	National University of Singapore	Martin Vingron	Limsoon Wong
13	2009	May 18-21, Tucson, AZ, USA	University of Arizona	Serafim Batzoglou	John Kececioglu
14	2010	Aug 12-15, Lisbon, Portugal	INESC-ID and Instituto Superior Ténico	Bonnie Berger	Arlindo Oliveira
15	2011	Mar 28-31, Vancouver, Canada	Lab for Computational Biology, Fraser University	Vineet Bafna	S. Cenk Sahinalp
16	2012	Apr 21-24, Barcelona, Spain	Centre for Genomic Regulation (CRG)	Benny Chor	Roderic Guigó
17	2013	Apr 7-10, Beijing, China	Tsinghua University	Fengzhu Sun	Xuegong Zhang

Table of Contents

Reconciliation Revisited: Handling Multiple Optima When Reconciling with Duplication, Transfer, and Loss

Mukul S. Bansal[1], Eric J. Alm[2,3], and Manolis Kellis[1,3]

[1] Computer Science and Artificial Intelligence Laboratory,
Massachusetts Institute of Technology, Cambridge, USA
[2] Dept. of Biological Engineering, Massachusetts Institute of Technology, Cambridge, USA
[3] Broad Institute of MIT and Harvard, Cambridge, USA
mukul@csail.mit.edu, {ejalm,manoli}@mit.edu

Abstract. Phylogenetic tree reconciliation is a powerful approach for inferring evolutionary events like gene duplication, horizontal gene transfer, and gene loss, which are fundamental to our understanding of molecular evolution. While Duplication-Loss (DL) reconciliation leads to a unique maximum-parsimony solution, Duplication-Transfer-Loss (DTL) reconciliation yields a multitude of optimal solutions, making it difficult the infer the true evolutionary history of the gene family.

Here, we present an effective, efficient, and scalable method for dealing with this fundamental problem in DTL reconciliation. Our approach works by sampling the space of optimal reconciliations uniformly at random and aggregating the results. We present an algorithm to efficiently sample the space of optimal reconciliations uniformly at random in $O(mn^2)$ time, where m and n denote the number of genes and species, respectively. We use these samples to understand how different optimal reconciliations vary in their node mapping and event assignments, and to investigate the impact of varying event costs.

Keywords: Gene family evolution, gene-tree/species-tree reconciliation, gene duplication, horizontal gene transfer, host-parasite cophylogeny, phylogenetics.

1 Introduction

The systematic comparison of a gene tree with its species tree under a reconciliation framework is a powerful technique for understanding gene family evolution. Specifically, gene tree/species tree reconciliation shows how the gene tree evolved inside the species tree while accounting for events like gene duplication, gene loss, and horizontal gene transfer, that drive gene family evolution. Thus, gene tree/species tree reconciliation is widely used and has many important applications; e.g., for inferring orthologs, paralogs and xenologs [1–6], reconstructing ancestral gene content and dating gene birth [7, 8], accurate gene tree reconstruction [5, 9], and whole genome species-tree reconstruction [10].

Duplication-Loss (DL) reconciliation, which accounts for only gene duplication and gene loss events, has been widely studied and extensively used [11–15]. However, since it does not account for horizontal gene transfer events, it only applies to multi-cellular eukaryotes, a very small part of the tree of life. An interesting and extremely useful

M. Deng et al. (Eds.): RECOMB 2013, LNBI 7821, pp. 1–13, 2013.

property of DL-reconciliation is that, assuming that loss events have a non-zero positive cost, the most parsimonious reconciliation is always unique [14]. In addition, the most parsimonious reconciliation remains the same irrespective of the chosen event costs for duplication and loss. Given these properties, there is no ambiguity in interpreting the results of DL-reconciliation, making it very easy to use in practice.

The limited applicability of DL reconciliation has led to the formulation of the Duplication-Transfer-Loss (DTL) reconciliation model, which can simultaneously account for duplication, transfer, and loss events and can be applied to species and gene families from across the entire tree of life. Indeed, the DTL-reconciliation model and its variants have been widely studied in the literature [8, 16–22]. In addition, DTL-reconciliation has also been indirectly studied in the context of the host-parasite cophylogeny problem [23–27].

The DTL-reconciliation problem is typically solved in a parsimony framework, where costs are assigned to duplication, transfer, and loss events, and the goal is to find a reconciliation with minimum total cost. DTL-reconciliations can sometimes be *time-inconsistent*; i.e, the inferred transfers may induce contradictory constraints on the dates for the internal nodes of the species tree. The problem of finding an optimal *time-consistent* reconciliation is known to be NP-hard [18, 27]. Thus, in practice, the goal is to find an optimal (but not necessarily time-consistent) DTL-reconciliation. The problem of finding an optimal time-consistent reconciliation does become efficiently solvable [17] if the species tree is fully dated. However, accurately dating the internal nodes of a species tree is a notoriously difficult problem [28], which severely restricts its applicability. Thus, for wider applicability and efficient solvability, in this work, unless otherwise stated, we assume the input species tree is undated and seek an optimal (not necessarily time-consistent) DTL-reconciliation [8, 18, 20, 21]. This problem can be solved very efficiently, with our own algorithm achieving the fastest known time complexity of $O(mn)$ [21], where m and n denote the number of nodes in the gene tree and species tree respectively.

Despite its extensive literature, the DTL-reconciliation problem remains difficult to use in practice for understanding gene family evolution. The first reason for this difficulty is that there are often multiple equally optimal reconciliations for a given gene tree and species tree and for a fixed assignment of event costs. The second reason is that event costs, which can be very difficult to assign confidently, play a much more important role than in DL reconciliation, as varying the costs can result in different optimal reconciliations.

Thus, when applying DTL-reconciliation in practice, it is unclear whether the evolutionary history implied by a particular given optimal solution is meaningful, as many other optimal reconciliations exist with the same minimal reconciliation cost. Moreover, it is unclear whether the properties of an optimal reconciliation are representative of the space of optimal reconciliations, and also how large and diverse this space is. Furthermore, the number of optimal reconciliations is often prohibitively large, as it can grow exponentially in the number of events required for the reconciliation, making even the basic task of enumerating all optimal reconciliations unfeasible for all but the smallest of gene trees [20]. Here, we directly address these problems and seek to make DTL-reconciliation as easy to use as the DL-reconciliation model.

Our Contribution. In this work, we develop the first efficient and scalable approach to explore the space of optimal DTL-reconciliations and show how it can be used to infer the similarities and differences in the different optimal reconciliations for any given input instance. Our approach is based on uniformly random sampling of optimal reconciliations and we demonstrate the utility of our approach by applying it to a biological dataset of approximately 4700 gene trees from 100 (predominantly prokaryotic) taxa [8]. Specifically, our contributions are as follows:

1. We analyze the gene trees in the biological dataset and show that even gene trees with only a few dozen genes often have many millions of optimal reconciliations. This analysis provides the first detailed look into the prevalence of optimal reconciliations in biological datasets.
2. We show how to efficiently sample the space of optimal reconciliations uniformly at random. Our algorithm produces each random sample in $O(mn^2)$ time, where m and n denote the number of nodes in the gene tree and species tree, respectively. This algorithm is fast enough to be applied thousands of times to the same dataset and scalable enough to be applied to datasets with hundreds or thousands of taxa.
3. We use our algorithm for random sampling to explore the space of optimal reconciliations and investigate the similarities and differences between the different optimal reconciliations. We show how to distinguish between the parts of the reconciliation that have high support from those that are more variable across the different multiple optima.
4. We show that even in the presence of multiple optimal solutions, a large amount of shared information can be extracted from the different optimal reconciliations. For instance, we observed that, for fixed event costs, any internal node taken from a gene tree in the biological dataset had a 93.31% chance of having the same event assignment (speciation, duplication, or transfer) and a 73.15% chance of being mapped to the same species tree node, across all (sampled) optimal reconciliations.
5. Our method allows users to compare the space of optimal reconciliations for different event costs and extract the shared aspects of the reconciliation. This makes it possible to study the impact of using different event costs and to meaningfully apply DTL-reconciliation even if one is unsure of the exact event costs to use. We applied our method to the biological dataset using different event costs and observed that large parts of the reconciliation tend to be robust to event cost changes.

Thus, in this work, we introduce the first efficient and scalable method for exploring the space of optimal reconciliations. Our new method allows for the very first large-scale exploration of the space of optimal reconciliations in real biological datasets.

The remainder of the paper is organized as follows: The next section introduces basic definitions and preliminaries. In Section 3 we study the prevalence of multiple optimal reconciliations in biological data. We introduce our sampling based approach and algorithms in Section 4. Section 5 shows the results of our analysis of multiple optimal reconciliations for the biological dataset, and in Section 6 we show how our method can be applied to study the impact of using different reconciliation costs. Concluding remarks appear in Section 7.

2 Definitions and Preliminaries

We follow the basic definitions and notation from [21]. Given a tree T, we denote its node, edge, and leaf sets by $V(T)$, $E(T)$, and $Le(T)$ respectively. If T is rooted, the root node of T is denoted by $rt(T)$, the parent of a node $v \in V(T)$ by $pa_T(v)$, its set of children by $Ch_T(v)$, and the (maximal) subtree of T rooted at v by $T(v)$. If two nodes in T have the same parent, they are called *siblings*. The set of *internal nodes* of T, denoted $I(T)$, is defined to be $V(T) \setminus Le(T)$. We define $\leq_T$ to be the partial order on $V(T)$ where $x \leq_T y$ if y is a node on the path between $rt(T)$ and x. The partial order $\geq_T$ is defined analogously, i.e., $x \geq_T y$ if x is a node on the path between $rt(T)$ and y. We say that v is an *ancestor* of u, or that u is a *descendant* of v, if $u \leq_T v$ (note that, under this definition, every node is a descendant as well as ancestor of itself). We say that x and y are *incomparable* if neither $u \leq_T v$ nor $v \leq_T u$. Given a non-empty subset $L \subseteq Le(T)$, we denote by $lca_T(L)$ the least common ancestor (LCA) of all the leaves in L in tree T; that is, $lca_T(L)$ is the unique smallest upper bound of L under $\leq_T$. Given $x, y \in V(T)$, $x \to_T y$ denotes the unique path from x to y in T. We denote by $d_T(x, y)$ the number of edges on the path $x \to_T y$. Throughout this work, unless otherwise stated, the term tree refers to a rooted binary tree.

We assume that each leaf of the gene trees is labeled with the species from which that gene was sampled. This labeling defines a *leaf-mapping* $\mathcal{L}_{G,S} \colon Le(G) \to Le(S)$ that maps a leaf node $g \in Le(G)$ to that unique leaf node $s \in Le(S)$ which has the same label as g. Note that gene trees may have more than one gene sampled from the same species. Throughout this work, we denote the gene tree and species tree under consideration by G and S respectively and will assume that $\mathcal{L}_{G,S}(g)$ is well defined.

2.1 Reconciliation and DTL-scenarios

Reconciling a gene tree with a species tree involves mapping the gene tree into the species tree. Next, we define what constitutes a valid reconciliation; specifically, we define a Duplication-Transfer-Loss scenario (DTL-scenario) [18, 21] for G and S that characterizes the mappings of G into S that constitute a biologically valid reconciliation. Essentially, DTL-scenarios map each gene tree node to a unique species tree node in a consistent way that respects the immediate temporal constraints implied by the species tree, and designate each gene tree node as representing either a speciation, duplication, or transfer event.

Definition 1 (DTL-scenario). *A DTL-scenario for G and S is a seven-tuple $\langle \mathcal{L}, \mathcal{M}, \Sigma, \Delta, \Theta, \Xi, \tau \rangle$, where $\mathcal{L} \colon Le(G) \to Le(S)$ represents the leaf-mapping from G to S, $\mathcal{M} \colon V(G) \to V(S)$ maps each node of G to a node of S, the sets Σ, Δ, and Θ partition $I(G)$ into speciation, duplication, and transfer nodes respectively, Ξ is a subset of gene tree edges that represent transfer edges, and $\tau \colon \Theta \to V(S)$ specifies the recipient species for each transfer event, subject to the following constraints:*

1. *If $g \in Le(G)$, then $\mathcal{M}(g) = \mathcal{L}(g)$.*
2. *If $g \in I(G)$ and g' and g'' denote the children of g, then,*
   (a) $\mathcal{M}(g) \not\leq_S \mathcal{M}(g')$ and $\mathcal{M}(g) \not\leq_S \mathcal{M}(g'')$,
   (b) At least one of $\mathcal{M}(g')$ and $\mathcal{M}(g'')$ is a descendant of $\mathcal{M}(g)$.

3. *Given any edge* $(g, g') \in E(G)$, $(g, g') \in \Xi$ *if and only if* $\mathcal{M}(g)$ *and* $\mathcal{M}(g')$ *are incomparable.*
4. *If* $g \in I(G)$ *and* g' *and* g'' *denote the children of* g, *then,*
   (a) $g \in \Sigma$ *only if* $\mathcal{M}(g) = lca(\mathcal{M}(g'), \mathcal{M}(g''))$ *and* $\mathcal{M}(g')$ *and* $\mathcal{M}(g'')$ *are incomparable,*
   (b) $g \in \Delta$ *only if* $\mathcal{M}(g) \geq_S lca(\mathcal{M}(g'), \mathcal{M}(g''))$,
   (c) $g \in \Theta$ *if and only if either* $(g, g') \in \Xi$ *or* $(g, g'') \in \Xi$.
   (d) *If* $g \in \Theta$ *and* $(g, g') \in \Xi$, *then* $\mathcal{M}(g)$ *and* $\tau(g)$ *must be incomparable, and* $\mathcal{M}(g')$ *must be a descendant of* $\tau(g)$, *i.e.,* $\mathcal{M}(g') \leq_S \tau(g)$.

Constraint 1 above ensures that the mapping $\mathcal{M}$ is consistent with the leaf-mapping $\mathcal{L}$. Constraint 2(a) imposes on $\mathcal{M}$ the temporal constraints implied by S. Constraint 2(b) implies that any internal node in G may represent at most one transfer event. Constraint 3 determines the edges of G that are transfer edges. Constraints 4(a), 4(b), and 4(c) state the conditions under which an internal node of G may represent a speciation, duplication, and transfer respectively. Constraint 4(d) specifies which species may be designated as the recipient species for any given transfer event.

In some cases, one may wish to restrict transfer events to only occur between coexisting species. This requires that divergence time information (either absolute or relative) be available for all the internal nodes of the species tree. In such cases, the definition of a DTL-scenario remains the same, except for the additional restriction on transfer events.

DTL-scenarios correspond naturally to reconciliations and it is straightforward to infer the reconciliation of G and S implied by any DTL-scenario. Figure 1 shows two simple DTL-scenarios. Given a DTL-scenario, one can directly count the minimum number of gene losses in the corresponding reconciliation. For brevity, we refer the reader to [21] for further details on how to count losses in DTL-scenarios.

Let P_Δ, P_Θ, and P_{loss} denote the costs associated with duplication, transfer, and loss events respectively. The reconciliation cost of a DTL-scenario is defined as follows.

Definition 2 (Reconciliation cost of a DTL-scenario). *Given a DTL-scenario* $\alpha = \langle \mathcal{L}, \mathcal{M}, \Sigma, \Delta, \Theta, \Xi, \tau \rangle$ *for* G *and* S, *the reconciliation cost associated with* α *is given by* $\mathcal{R}_\alpha = P_\Delta \cdot |\Delta| + P_\Theta \cdot |\Theta| + P_{loss} \cdot Loss_\alpha$.

Given G and S, along with event costs P_Δ, P_Θ, and P_{loss}, the goal is to find a most parsimonious reconciliation of G and S. More formally,

Problem 1 (Most Parsimonious Reconciliation (MPR)). *Given* G *and* S, *the most parsimonious reconciliation (MPR) problem is to find a DTL-scenario for* G *and* S *with minimum reconciliation cost.*

We distinguish two versions of the MPR problem: (i) The *Undated MPR (U-MPR)* problem where the species tree is undated, and (ii) the *Fully-dated MPR (D-MPR)* problem where every node of the species tree has an associated divergence time estimate (or there is a known total order on the internal nodes of the species tree) and transfer events are required to occur only between coexisting species.

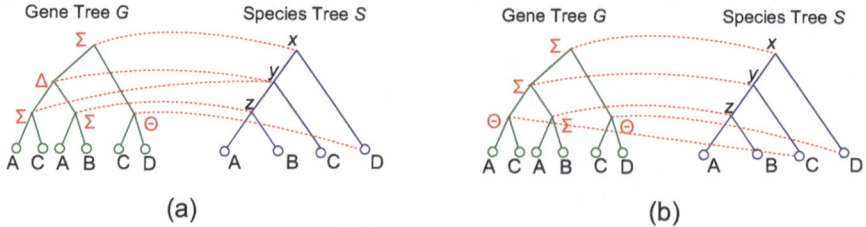

Fig. 1. Multiple optimal reconciliations. Parts (a) and (b) show two different reconciliations for the gene tree and species tree depicted in the figure. Both of the reconciliations are optimal for event costs $P_\Delta = 1$, $P_\Theta = 3$, and $P_{loss} = 1$. The reconciliation in part (a) invokes one duplication, one transfer, and two losses, while the reconciliation in part (b) invokes two transfers.

3 Multiple Optimal Solutions

In general, for any fixed values of P_Δ, P_Θ, and P_{loss}, there may be multiple equally optimal solutions to the MPR problem (both U-MPR and D-MPR). This is illustrated in Figure 1. The figure also illustrates the fundamental problem with having multiple optima: Given the different evolutionary histories implied by the different multiple optima, what is the true evolutionary history of the gene family? We address this problem in this paper. But first, in this section, we investigate the prevalence of optimal reconciliations in real datasets. For our study, we use a published biological dataset of 4735 gene trees and 100 (predominantly prokaryotic) species [8]. The gene trees in the dataset have median and average leaf-set sizes of 18 and 35.1, respectively. This dataset has been previously analyzed using DTL-reconciliation but without consideration of multiple optima. In our analysis of this dataset we used the same event costs as used in [8] (i.e., $P_\Delta = 2$, $P_\Theta = 3$, and $P_{loss} = 1$). Since the gene trees in the dataset are unrooted, we first rooted them optimally by choosing a root that minimized the reconciliation cost. In cases where there were multiple optimal rootings, we chose one of the optimal rootings at random. We computed the number of multiple optimal reconciliations for each of the rooted gene trees by augmenting the dynamic programming algorithm used to solve the MPR problem (e.g., [21]) to keep track of the number of optima for each sub-problem. Further algorithmic details appear in Section 4. Unless otherwise stated, all analyses in the manuscript were performed using the undated version of DTL-reconciliation.

Figure 2 shows the results of our analysis. As part (a) of the figure shows, only 17% of the approximately 4700 gene trees have a unique optimal reconciliation. Over half of the gene trees have over 100 optimal reconciliations and 15% have more than 10,000 optimal reconciliations. This illustrates the extent of the problem with multiple optimal reconciliations in biological datasets. As part (b) of the figure shows, the number of optimal reconciliations tends to increase exponentially with gene tree size. These results demonstrate the importance of considering multiple optima in DTL-reconciliation, and the impracticality of enumerating all optimal reconciliations for all but the smallest gene trees.

We also repeated the above analysis using the dated version of the DTL-reconciliation problem (i.e., the D-MPR problem), and observed no significant reduction in the number

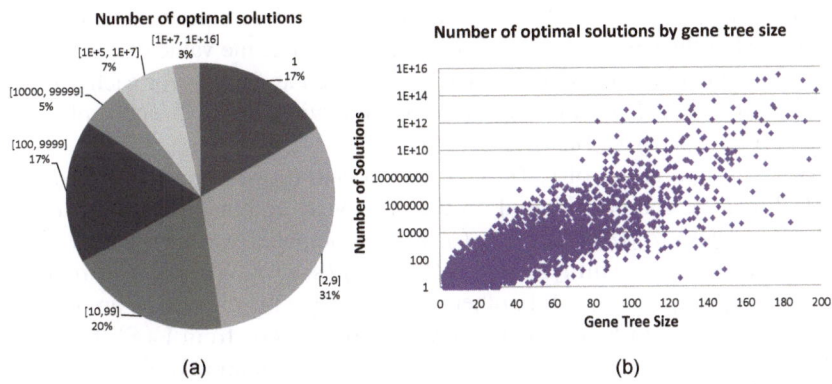

Fig. 2. Number of optimal reconciliations for the gene trees in the biological dataset. The pie chart in part (a) shows the distribution of the number of optimal reconciliations for the gene trees in the biological dataset. The dot plot in part (b) plots the size (number of internal nodes) and the number of optimal reconciliations for each gene tree. Due to arithmetic overflow concerns, results are only shown for the 4699 (out of 4735) gene trees that had fewer than 10^{16} optima.

of multiple optima. For instance, even for the dated version, 14% of the gene trees had more than 10,000 optimal reconciliations.

Recall that the gene trees in the dataset were originally unrooted. While the results above are for a fixed optimal rooting of these gene trees, we point out that about half the gene trees in the dataset have more than one optimal rooting. It may thus be necessary, in practice, to either consider all possible optimal rootings when studying multiple optimal reconciliations, or to use other information to assign a root uniquely.

4   Uniformly Random Sampling of Optimal Reconciliations

As Section 3 demonstrates, the exhaustive enumeration of all optimal reconciliations is only feasible for very small gene trees. In this section we show how to sample the space of reconciliations uniformly at random. Random sampling makes it possible to explore the space of optimal reconciliations without exhaustive enumeration, and makes it possible to understand the variability in the different reconciliations and to distinguish between the highly supported and weakly supported parts of a given optimal reconciliation. Our algorithm for random sampling is based on the dynamic programming algorithm for the MPR problem from [21]. The idea is to keep track of the number of optimal solutions for each subproblem considered in the dynamic programming algorithm. In the following, we show how to compute the number of optimal solutions at each step correctly and efficiently. First, we need a few definitions.

Given any $g \in I(G)$ and $s \in V(S)$, let $c_\Sigma(g, s)$ denote the cost of an optimal reconciliation of $G(g)$ with S such that g maps to s and $g \in \Sigma$. The terms $c_\Delta(g, s)$ and $c_\Theta(g, s)$ are defined similarly for $g \in \Delta$ and $g \in \Theta$ respectively. Given any $g \in V(G)$ and $s \in V(S)$, we define $c(g, s)$ to be the cost of an optimal reconciliation of $G(g)$ with

S such that g maps to s. The algorithm for the MPR problem performs a nested post-order traversal of the gene tree and species tree to compute the value of $c(g, s)$ for each g and s. The dynamic programming table is initialized as follows for each $g \in Le(G)$: $c(g, s) = 0$ if $s = \mathcal{M}(g)$, and $c(g, s) = \infty$ otherwise. For $g \in I(G)$, observe that $c(g, s) = \min\{c_\Sigma(g, s), c_\Delta(g, s), c_\Theta(g, s)\}$.

At each step, the values of $c_\Sigma(g, s)$, $c_\Delta(g, s)$, and $c_\Theta(g, s)$ for any $g \in I(G)$ and $s \in V(S)$, can be computed based on the previously computed values of $c(\cdot, \cdot)$. To show how $c_\Sigma(g, s)$, $c_\Delta(g, s)$, and $c_\Theta(g, s)$ are computed we need some additional notation. Let $in(g, s) = \min_{x \in V(S(s))}\{P_{loss} \cdot d_S(s, x) + c(g, x)\}$ and $out(g, s) = \min_{x \in V(S) \text{ incomparable to } s} c(g, x)$. In other words: $out(g, s)$ is the cost of an optimal reconciliation of $G(g)$ with S such that g may map to any node from $V(S)$ that is incomparable to s; and $in(g, s)$ is the cost of an optimal reconciliation of $G(g)$ with S such that g may map to any node, say x, in $V(S(s))$ but with an additional reconciliation cost of one loss event for each edge on the path from s to x. The values $c_\Sigma(g, s)$, $c_\Delta(g, s)$, and $c_\Theta(g, s)$ are computed as follows:

For any $g \in I(G)$ and $s \in I(S)$, let $\{g', g''\} = Ch_G(g)$ and $\{s', s''\} = Ch_S(s)$.

If $s \in Le(S)$ then,
$c_\Sigma(g, s) = \infty$,
$c_\Delta(g, s) = P_\Delta + c(g', s) + c(g'', s)$, and
    If $s \neq rt(S)$, then $c_\Theta(g, s) = P_\Theta + \min\{in(g', s) + out(g'', s), \ in(g'', s) + out(g', s)\}$. Else, $c_\Theta(g, s) = \infty$.

If $s \in I(S)$ then,
$c_\Sigma(g, s) = \min\{in(g', s') + in(g'', s''), \ in(g'', s') + in(g', s'')\}$.

$$c_\Delta(g, s) = P_\Delta + \min \begin{cases} c(g', s) + in(g'', s'') + P_{loss}, \ c(g', s) + in(g'', s') + P_{loss}, \\ c(g'', s) + in(g', s'') + P_{loss}, \ c(g'', s) + in(g', s') + P_{loss}, \\ c(g', s) + c(g'', s), \ in(g', s') + in(g'', s'') + 2P_{loss}, \\ in(g', s'') + in(g'', s') + 2P_{loss}, \ in(g', s') + in(g'', s') + 2P_{loss}, \\ in(g', s'') + in(g'', s'') + 2P_{loss}. \end{cases}$$

If $s \neq rt(S)$, then $c_\Theta(g, s) = P_\Theta + \min\{in(g', s) + out(g'', s), \ in(g'', s) + out(g', s)\}$. Else, $c_\Theta(g, s) = \infty$.

The optimal reconciliation cost of G and S is simply: $\min_{s \in V(S)} c(rt(G), s)$, and an optimal reconciliation with that cost can be reconstructed by backtracking in the dynamic programming table. We refer the reader to [21] for further algorithmic details.

To output optimal reconciliations uniformly at random we must keep track of the number of optimal reconciliations for each of the subproblems considered in the DP algorithm. We define the following: For any $g \in V(G)$ and $s \in V(S)$, let $N(g, s)$ denote the number of optimal solutions for reconciling $G(g)$ with S such that g maps to s. The idea is to compute $N(\cdot, \cdot)$ using the same nested post-order traversal used to compute the $c(\cdot, \cdot)$ values. The dynamic programming table for $N(\cdot, \cdot)$ is initialized as follows for each $g \in Le(G)$: $N(g, s) = 1$ if $s = \mathcal{M}(g)$, and $N(g, s) = 0$ otherwise. To compute $N(g, s)$, for $g \in I(G)$, we must consider all possible mappings of g' and g'' that yield a cost of $c(g, s)$. For the remainder of this discussion, in the interest of brevity and clarity, we will assume that $s \in I(S)$ and $s \neq rt(S)$; the cases when $s \in Le(S)$ or $s = rt(S)$ are easy to handle analogously.

Let a_1 through a_{13} denote the individual expressions in the $min\{\ \}$ blocks in the equations for $c_\Sigma(g, s)$, $c_\Delta(g, s)$, and $c_\Theta(g, s)$ above. Specifically, let a_1 denote $in(g', s') + in(g'', s'')$, a_2 denote $in(g'', s') + in(g', s'')$, a_3 through a_{11} denote the nine expressions in the $min\{\ \}$ block for $c_\Delta(g, s)$, and a_{12} and a_{13} denote the two expressions in the $min\{\ \}$ block for $c_\Theta(g, s)$. Each of these a_i's represents a certain cost, which we denote by $c(a_i)$, and a certain number of optimal reconciliations, which we denote by $N(a_i)$. Furthermore, let b_i, for $1 \leq i \leq 13$, be binary boolean variables associated with the a_i's such that $b_i = 1$ if a_i yields the minimum cost $c(g, s)$, and $b_i = 0$ otherwise. Specifically, for $i \in \{1, 2\}$, $b_i = 1$ if and only if $c(a_i) = c(g, s)$; for $i \in \{3, \ldots, 11\}$, $b_i = 1$ if and only if $c(a_i) + P_\Delta = c(g, s)$; and for $i \in \{12, 13\}$, $b_i = 1$ if and only if $c(a_i) + P_\Theta = c(g, s)$. Then, we must have:

$$N(g, s) = \sum_{i=1}^{13} b_i \times N(a_i).$$

Next, we show how to compute $N(a_i)$ for any i for which $b_i = 1$. Observe that each a_i has one term involving g' and one term involving g''. These terms take one of the three forms: $c(\cdot, \cdot)$, $in(\cdot, \cdot)$, or $out(\cdot, \cdot)$. These terms, involving g' and g'', can be viewed as representing the choice of optimal mappings for g' and g'', respectively. For instance, $c(g', s)$ implies that g' must map to s, $in(g', s)$ implies that g' may map to any node $x \in V(S(s))$ for which $(P_{loss} \cdot d_S(s, x) + c(g', x))$ is minimized (recall the definition of $in(\cdot, \cdot)$), and $out(g', s)$ implies that g' may map to any node $x \in V(S)$ that is incomparable to s, for which $c(g', x)$ is minimized. Based on this observation, for any given a_i, we can compute a set of optimal mappings for g', which we will denote by X' and a set of optimal mappings for g'', which we will denote by X''. The value of $N(a_i)$ can then be computed as follows:

$$N(a_i) = \left(\sum_{x \in X'} N(g', x) \right) \times \left(\sum_{x \in X''} N(g'', x) \right).$$

The equations for $N(g, s)$ and $N(a_i)$ above make it possible to compute the value $N(g, s)$ for each $g \in I(G)$ and $s \in V(S)$ by using the same nested post-order traversal that is used for computing the values $c(\cdot, \cdot)$. Once all the $c(\cdot, \cdot)$ and $N(\cdot, \cdot)$ have been computed, an optimal reconciliation itself can be built by backtracking through the dynamic programming table. To ensure that reconciliations are generated uniformly at random the idea is to make the choice of mapping assignments based on the number of optimal solutions contained within each choice. For instance, if a node g has already been assigned a mapping, its two children g' and g'' must be assigned mappings jointly based on their joint probability mass. In the interest of brevity, further technical and algorithmic details, as well as a formal proof of correctness, are deferred to the full version of this paper.

It is not hard to implement this algorithm for uniformly random sampling in $O(mn^2)$ time, where m and n denote the size of the gene tree and species tree respectively. This is only a factor of n slower than the fastest known algorithm for the MPR problem [21]. Our implementation of this random sampling algorithm will be made available as part of the next version of the RANGER-DTL software package [21].

5   Exploring the Space of Optimal Reconciliations

We applied our method to the biological dataset to understand the space of optimal reconciliations for the gene trees in this dataset. As before, we used event costs $P_\Delta = 2$, $P_\Theta = 3$, and $P_{loss} = 1$ for this analysis. For this study, we focused on understanding how similar the different optimal reconciliations are to each other. To that end, we used our algorithm to sample 500 optimal reconciliations for each gene tree, and wrote a program that reads in these samples and summarizes them as follows: For each internal node in the gene tree we (i) consider the fraction of times that node is mapped to the different nodes of the species tree, and (ii) consider the fraction of times that node is labeled as a speciation, duplication, and transfer event. We used this to investigate the stability of the embedding of the gene tree into the species tree (i.e., the stability of gene node mappings), and the stability of event assignments for the internal nodes of the gene tree.

We first checked to see how stable the gene node mappings were across the internal nodes in all the 4699 gene trees. Figure 3(a) shows the results of this analysis. Overall, we observed that mappings tended to be fairly well conserved across the different multiple optima. For instance, we observed that 73.15% of the internal gene tree nodes had the same mapping across all 500 samples. Recall that only 17% of the gene trees have a unique solution. We also repeated this analysis for event assignments and these results are also shown in Figure 3(a). Amazingly, we observed that 93.31% of the nodes had a consistent event assignment across all 500 samples. This suggests that event assignments tend to be highly conserved across the different multiple optima. Thus, even in those instances where there are many different optimal reconciliations it should be possible to confidently assign event types to most internal nodes of the gene tree (even though the mappings of the nodes themselves may not be consistent across the different multiple optima)). This has important implications for understanding gene family evolution, since the inference of orthologs, paralogs, and xenologs depends only on the event assignments for gene tree nodes.

In practice, users are often interested in analyzing the evolutionary history of a specific gene family. We thus asked the following question: Given a gene tree from the biological dataset, what fraction of its nodes can be expected to have (i) a consistent mapping, and (ii) a consistent event assignment, across all 500 samples. Figure 3(b) shows the results of this analysis. The results show that for most gene trees, event assignments are completely consistent across all samples for most of their internal nodes. For instance, we observed that 60.2% of the gene trees have a consistent event assignment for all of their internal nodes, and almost all gene trees had a consistent event assignment for at least half of their internal nodes. As we observed before, gene tree node mappings tend to be more variable, but still, over 91% of the gene trees had a consistent mapping for at least half of their internal nodes. We also tested to see if there was a correlation between the number of optimal reconciliations for a gene tree and fraction of its internal nodes with consistent mappings or consistent event assignments. To our surprise, we found no correlation (results not shown).

Our analyses above show that, even in the presence of multiple optimal reconciliations, most aspects of the reconciliation are highly conserved across the different multiple optima.

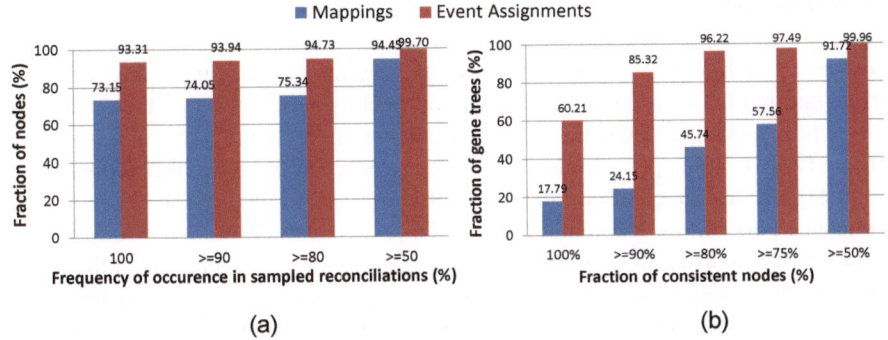

Fig. 3. Stability of mappings and event assignments. The plot in part (a) shows the fraction of internal nodes from the 4699 gene trees that have the same mapping or the same event assignment across at least a certain fraction of the 500 samples. The plot in part (b) plots the fraction of the 4699 gene trees that have at least a certain fraction of their nodes with a consistent mapping or a consistent event assignment across all 500 samples.

6 Application to Understanding Sensitivity to Event Costs

The ability to explore the space of multiple optimal reconciliations makes it possible to study the effect of using different event costs on the reconciliation. For instance, one can compare if the mapping or event assignments that are consistent across the multiple optima for a particular event cost assignment are also consistent across a different event cost assignment. Similarly, if one is unsure of which event cost assignment to use, one can try out all the different event costs, compute a set of random samples for each event cost assignment, and aggregate the samples from all event cost assignments into a single analysis to understand which aspects of the reconciliation are conserved across the different event cost assignments.

We performed a preliminary study of the effect of using different event costs on the analysis of the biological dataset. Recall that our default event costs are $P_\Delta = 2$, $P_\Theta = 3$, and $P_{loss} = 1$. For this study, we kept $P_{loss} = 1$, but considered the following combinations of the duplication and transfer costs: (i) $P_\Delta = 2$, $P_\Theta = 4$, (ii) $P_\Delta = 2$, $P_\Theta = 2$, (iii) $P_\Delta = 3$, $P_\Theta = 3$, and (iv) $P_\Delta = 1$, $P_\Theta = 1$. We computed 100 random samples for each setting of event costs. For our preliminary analysis, we asked the following question: What fraction of the gene tree nodes with consistent mappings (event assignments) under the default costs also have the same consistent mappings (resp. event assignments) under the alternative event costs? The results of this analysis for the four combinations of event costs listed aboveare as follows: For mappings, the fractions are 94%, 83.38%, 92.04%, and 63.97%, respectively. And, for event assignments, the fractions are 92.06%, 91.52%, 96.07%, and 80.37%, respectively. As the analysis indicates, consistent mappings and event assignments tend to be well conserved even when using different event costs. Even with the rather extreme event costs of $P_\Delta = P_\Theta = P_{loss} = 1$, almost 64% of the consistent mappings and over 80% of the event assignments are conserved. We defer a more detailed analysis of the differences in the space of optimal reconciliations for the different event cost assignments to the full version of the paper.

7   Conclusion

In this work, we have presented an efficient and scalable approach for the problem of multiple optimal DTL-reconciliations. Our approach is based on random sampling and we show how to sample the space of optimal reconciliations uniformly at random efficiently in $O(mn^2)$ time per sample. The sampling based approach makes it possible for users to explore the space of optimal reconciliations and to distinguish between stable and unstable parts of the reconciliation. This approach also allows users to investigate the effect of using different event costs on the reconciliation. Our analysis of the biological dataset provides the first real insight into the space of multiple optima and reveals that many, if not most, aspects of the reconciliation remain consistent across the different multiple optima and that these can be efficiently inferred. We believe that this work represents an important step towards making DTL-reconciliation a practical method for understanding gene family evolution.

Many aspects of the space of optimal reconciliations remain to be explored. For instance, it would be interesting to investigate why so many of the input instances have millions (and more) of multiple optima. In this work we did not consider the effect of alternative optimal gene tree rootings on the reconciliation space and we would like to study this further. The ability to handle multiple optima also enables the systematic evaluation of the accuracy of DTL-reconciliation at inferring evolutionary history correctly and we plan to pursue this further. Similarly, we only performed a very preliminary study of the effect of different event costs and it would be instructive to study this more thoroughly.

Funding: This work was supported by a National Science Foundation CAREER award 0644282 to MK. National Institutes of Health grant RC2 HG005639 to M.K., and National Science Foundation AToL grant 0936234 to E.J.A. and M.K.

References

1. Storm, C.E.V., Sonnhammer, E.L.L.: Automated ortholog inference from phylogenetic trees and calculation of orthology reliability. Bioinformatics 18(1), 92–99 (2002)
2. Koonin, E.V.: Orthologs, paralogs, and evolutionary genomics. Annual Review of Genetics 39(1), 309–338 (2005)
3. Wapinski, I., Pferrer, A., Friedman, N., Regev, A.: Natural history and evolutionary principles of gene duplication in fungi. Nature 449, 54–61 (2007)
4. van der Heijden, R., Snel, B., van Noort, V., Huynen, M.: Orthology prediction at scalable resolution by phylogenetic tree analysis. BMC Bioinformatics 8(1), 83 (2007)
5. Vilella, A.J., Severin, J., Ureta-Vidal, A., Heng, L., Durbin, R., Birney, E.: Ensemblcompara genetrees: Complete, duplication-aware phylogenetic trees in vertebrates. Genome Research 19(2), 327–335 (2009)
6. Sennblad, B., Lagergren, J.: Probabilistic orthology analysis. Syst. Biol. 58(4), 411–424 (2009)
7. Chen, K., Durand, D., Farach-Colton, M.: Notung: dating gene duplications using gene family trees. In: RECOMB, pp. 96–106 (2000)
8. David, L.A., Alm, E.J.: Rapid evolutionary innovation during an archaean genetic expansion. Nature 469, 93–96 (2011)

9. Rasmussen, M.D., Kellis, M.: A bayesian approach for fast and accurate gene tree reconstruction. Molecular Biology and Evolution 28(1), 273–290 (2011)
10. Burleigh, J.G., Bansal, M.S., Eulenstein, O., Hartmann, S., Wehe, A., Vision, T.J.: Genome-scale phylogenetics: Inferring the plant tree of life from 18,896 gene trees. Syst. Biol. 60(2), 117–125 (2011)
11. Goodman, M., Czelusniak, J., Moore, G.W., Romero-Herrera, A.E., Matsuda, G.: Fitting the gene lineage into its species lineage. a parsimony strategy illustrated by cladograms constructed from globin sequences. Systematic Zoology 28, 132–163 (1979)
12. Page, R.D.M.: Maps between trees and cladistic analysis of historical associations among genes, organisms, and areas. Syst. Biol. 43(1), 58–77 (1994)
13. Bonizzoni, P., Vedova, G.D., Dondi, R.: Reconciling a gene tree to a species tree under the duplication cost model. Theor. Comput. Sci. 347(1-2), 36–53 (2005)
14. Górecki, P., Tiuryn, J.: Dls-trees: A model of evolutionary scenarios. Theor. Comput. Sci. 359, 378–399 (2006)
15. Chauve, C., Doyon, J.P., El-Mabrouk, N.: Gene family evolution by duplication, speciation, and loss. J. Comput. Biol. 15(8), 1043–1062 (2008)
16. Gorbunov, K.Y., Liubetskii, V.A.: Reconstructing genes evolution along a species tree. Molekuliarnaia Biologiia 43(5), 946–958 (2009)
17. Doyon, J.-P., Scornavacca, C., Gorbunov, K.Y., Szöllősi, G.J., Ranwez, V., Berry, V.: An Efficient Algorithm for Gene/Species Trees Parsimonious Reconciliation with Losses, Duplications and Transfers. In: Tannier, E. (ed.) RECOMB-CG 2010. LNCS, vol. 6398, pp. 93–108. Springer, Heidelberg (2010)
18. Tofigh, A., Hallett, M.T., Lagergren, J.: Simultaneous identification of duplications and lateral gene transfers. IEEE/ACM Trans. Comput. Biology Bioinform. 8(2), 517–535 (2011)
19. Tofigh, A.: Using Trees to Capture Reticulate Evolution: Lateral Gene Transfers and Cancer Progression. PhD thesis, KTH Royal Institute of Technology (2009)
20. Chen, Z.Z., Deng, F., Wang, L.: Simultaneous identification of duplications, losses, and lateral gene transfers. IEEE/ACM Trans. Comput. Biology Bioinform. 9(5), 1515–1528 (2012)
21. Bansal, M.S., Alm, E.J., Kellis, M.: Efficient algorithms for the reconciliation problem with gene duplication, horizontal transfer and loss. Bioinformatics 28(12), 283–291 (2012)
22. Stolzer, M., Lai, H., Xu, M., Sathaye, D., Vernot, B., Durand, D.: Inferring duplications, losses, transfers and incomplete lineage sorting with nonbinary species trees. Bioinformatics 28(18), 409–415 (2012)
23. Charleston, M.: Jungles: A new solution to the host-parasite phylogeny reconciliation problem. Mathematical Biosciences 149, 191–223 (1998)
24. Ronquist, F.: Parsimony analysis of coevolving species associations. In: Page, R.D.M. (ed.) Tangled Trees: Phylogeny, Cospeciation and Coevolution, pp. 22–64. The University of Chicago Press (2003)
25. Merkle, D., Middendorf, M., Wieseke, N.: A parameter-adaptive dynamic programming approach for inferring cophylogenies. BMC Bioinformatics 11(suppl. 1), S60 (2010)
26. Conow, C., Fielder, D., Ovadia, Y., Libeskind-Hadas, R.: Jane: a new tool for the cophylogeny reconstruction problem. Algorithm. Mol. Biol. 5(1), 16 (2010)
27. Ovadia, Y., Fielder, D., Conow, C., Libeskind-Hadas, R.: The cophylogeny reconstruction problem is np-complete. J. Comput. Biol. 18(1), 59–65 (2011)
28. Rutschmann, F.: Molecular dating of phylogenetic trees: A brief review of current methods that estimate divergence times. Divers. Distrib. 12(1), 35–48 (2006)

SEME: A Fast Mapper of Illumina Sequencing Reads with Statistical Evaluation

Shijian Chen[1,*], Anqi Wang[1,*], and Lei M. Li[1,2,**]

[1] NCMIS, Academy of Mathematics and Systems Science,
Chinese Academy of Sciences, Beijing, 100190
[2] Molecular and Computational Biology Program, Department of Biological Sciences,
University of Southern California, Los Angeles, CA 90089
`lilei@amss.ac.cn`

Abstract. Mapping reads to a reference genome is a routine yet computationally intensive task in research based on high-throughput sequencing. In recent years, the sequencing reads of the Illumina platform get longer and their quality scores get higher. According to our calculation, this allows perfect k-mer seed match for almost all reads when a close reference genome is available subject to reasonable specificity. Our another observation is that the majority reads contain at most one short INDEL polymorphism. Based on these observations, we propose a fast mapping approach, referred to as "SEME", which has two core steps: first it scans a read sequentially in a specific order for a k-mer exact match seed; next it extends the alignment on both sides allowing at most one short-INDEL each, using a novel method "auto-match function". We decompose the evaluation of the sensitivity and specificity into two parts corresponding to the seed and extension step, and the composite result provides an approximate overall reliability estimate of each mapping. We compare SEME with some existing mapping methods on several data sets, and SEME shows better performance in terms of both running time and mapping rates.

Keywords: high-throughput sequencing, mapping, perfect match, INDEL, auto-match function.

1 Introduction

The Next Generation Sequencing (NGS) technologies are generating unprecedented large amounts of short reads in routine genome research. The high-throughput and read length of NGS make it especially suitable for re-sequencing individuals with known references and thus for detecting variations. In whole genome re-sequencing projects for mammals, NGS usually generates billions of short reads, and mapping these reads back to the reference genome is computationally intensive. Hence the design of efficient mapping algorithms is a key and challenging problem in current computational biology.

Many short-read mapping methods have been developed along the evolution of the sequencing technologies[1]. The specific read length, error rates and patterns of each

---

* These authors contribute equally.
** corresponding author.

M. Deng et al. (Eds.): RECOMB 2013, LNBI 7821, pp. 14–29, 2013.

technology at the time are the primary constraints in the design of mapping algorithms. In the early days of NGS, the short reads were only 35bp long and error rates were fairly high for the Illumina/Solexa platform. Besides, 5-6 years ago the 32-bit architecture was the main model for PCs or cluster nodes, and their memory size is limited to 4Gb. Bowtie[2] applied the Burrows-Wheeler transform and FM index to the representation of the reference, and could reduce the memory footprint to as low as 1.3Gb for the human genome. This advantage makes Bowtie very popular among high-throughput sequencing users. Although the Burrows-Wheeler transform is effective in searching perfect matches of a k-mer in a reference, we have to allow mismatches to maintain sensitivity. For instance, MAQ[3] and SeqMap[4] use spaced seeds which allow up to k mismatches. Bowtie conducts a backtracking search to allow mismatches, and mitigates excessive backtracking by "double indexing", which doubles the memory foot print. No matter what method is used for handling mismatches, complexity is substantially increased.

As chemistry and instruments of NGS are under constant improvement, the reads are getting longer with higher quality. Now the Illumina platform can generate reads longer than 100bp with fairly high quality. MiSeq[5] can even sequence reads up to 250bp. Some short read mapping programmes, like Bowtie2, have been developed for these longer reads. Bowtie2 maps multiple evenly distributed seeds of a read and uses dynamic programming to extend seed alignments into a full alignment that allows IN-DELs. We observed that INDEL errors are extremely rare compared to substitution errors for Illumina systems. Thus if an INDEL occurs in the alignment or mapping, most likely it is a result from a polymorphism between the read and the reference.

Most high-throughput sequencing applications are for conserved genomes such as human, which is the focus of this article. In [6], it is found that the size of INDEL obeys a power law distribution in Human and Rodent pseudo genes: 78 human pseudo genes have been analyzed and it shows that the average length of small INDEL is less than three; furthermore, among those INDELs with length no larger than 20bp, 95% of them are no larger than 11bp. In[7], it is found that INDELs locate throughout the genome at a frequency of one per 7.2kb on average. If we approximate the occurrences of short INDELs by a Poisson point process that matches the frequency [12], the probability of finding at most one INDEL in a 100bp window is greater than 0.9999. Most existing methods apply dynamic programming to allow general INDELs. This is unnecessary most of the time for mapping short reads when a close reference genome is available.

Partially motivated by the above considerations, in this article we propose a new short read mapping method, referred to as SEME (Sequential Exact seed-Match and Extend) hereafter, which focuses on mapping Illumina short reads generated from conserved genomes. Different from most existing Seed-and-Extend methods which map multiple seeds simultaneously, SEME scans the read according to a specific strategy and maps the seeds sequentially. Once a seed is perfectly matched to the reference we extend it on both sides to get the full alignment result or reject it. This approach avoids mapping a fixed number of seeds for each short read. The higher the sequencing quality is, the less number of seeds are needed in SEME on average. This feature is particularly favorable as sequencing technology improves. In the extension step, we introduce the AMF (Auto-Match-Function) method to detect up to two INDELs. Compared with alignment algorithms based on dynamic programming, the average complexity of the AMF method is linear. For the remaining complicated occasions, which are rare, we can incorporate the Smith-Waterman[8] algorithm for full alignments.

As important as the computational complexity of an algorithm, its mapping rate and accuracy, which is usually measured by sensitivity and specificity, needs to be statistically evaluated fairly. For example, BLAST[9] is now widely used in the search of sequence databases. Its success comes from both its efficient algorithmic implementation and the associated statistical evaluation of the alignment significance[10,11,12]. In the situation of mapping short reads, the read length, say 100bp, is so small compared to the the genome size, that the classical asymptotics of alignment cannot be applied directly. In this report we make some efforts to evaluate the accuracy of the SEME procedure. In concert with the algorithm, we start off by comparing two sequences of the same read length. If one sequence is different from the other by only substitution and small INDEL polymorphisms plus sequencing errors, then the chance of detecting matching is essentially sensitivity. On the other hand, if one sequence is sampled randomly, say according to an *i.i.d.* – independent and identically distributed – model for the sake of simplicity, then accepting a match leads to a false positive error and its chance needs to be calculated. To evaluate the overall specificity, we decompose the entire genome into many reads of the same read lengths, either overlapping or non-overlapping, and apply the above result to provide bounds to the probability of accepting at least one match by chance across the genome. We could complement the analysis based on the simple model by simulation as well. With such a probabilistic framework that takes into account of read length, read error pattern, and polymorphism rate, we can optimize the seed length by trading off sensitivity and specificity.

To enhance sensitivity, we propose a soft counting criterion for accepting or rejecting a mapping result if appropriate sequencing quality scores are available. That is, we impute "possible polymorphism" fractions from mismatches based on polymorphism rates and quality values, and use the sum of these fractions for decision.

2   Method

SEME follows a "seed-and-extend" paradigm. In the first stage, it extracts k-mers sequentially from a short read, and for each k-mer SEME searches through the reference for perfect-match locations, where the read can anchor. We will discuss the selection of k later. In the second stage, SEME extends the seed on both sides separately. If the read is indeed from a reference location, then their true alignment falls into three categories: no INDEL; one INDEL; other more complicated INDEL patterns. As we explained earlier, most short reads from a conserved genome contain no more than one INDEL and possibly some mismatches. Thus the principal task of extension can be simplified as follows: on each side of the seed, detect the possible "one-INDEL" including its type, position and length (no larger than a given upper bound). We introduce the auto-match vector and auto-match function to efficiently solve this problem.

The search of k-mer exact-match across a reference genome is a common theme in most mapping tools. Several options are available for implementation. If memory size is limited, then the Burrows-Wheeler Transform is a good choice for compressing the genome and index information. If memory is sufficiently large, then hashing can help speed up the search, [19].

2.1   Index Table of Sorted 32-Mers and Binary Search

In our scheme, we may use more than one kind of seed sizes depending on the data. Therefore we propose to use the index table of 32-mers for the human genome. That is, we encode each 32-mer subsequence of the reference genome by an integer $s_i, i = 1, \cdots, L$, where L is the genome size, and sort them by the heap sorting algorithm. Denoted the sorted 32-mer-integers by $s_{(1)} \leq s_{(2)} \leq s_{(i)} \leq s_{(L)}$, and their corresponding addresses on the genome by $a(s_{(i)}), i = 1, \cdots, L$. We keep the addresses of these sorted 32-mers in an array $u[i] = a(s_{(i)}), i = 1, \cdots, L$, referred to as "index table" hereafter. We also put the reference genome in RAM so that we can quickly find the i-th sorted k-mer by linking the i-th address in the index table with the genome, see Fig 1. Note we do not save the sorted 32-mer-integers in a vector directly because an 32-mer takes 8 bytes while the address of the 32-mer takes only 4 bytes. With such an index structure, we apply binary search, whose time complexity is $O(log_2 L)$. Take the human genome for example, as the size of the index table is about three billion, approximately 30 steps are needed to insert a k-mer into the index table.

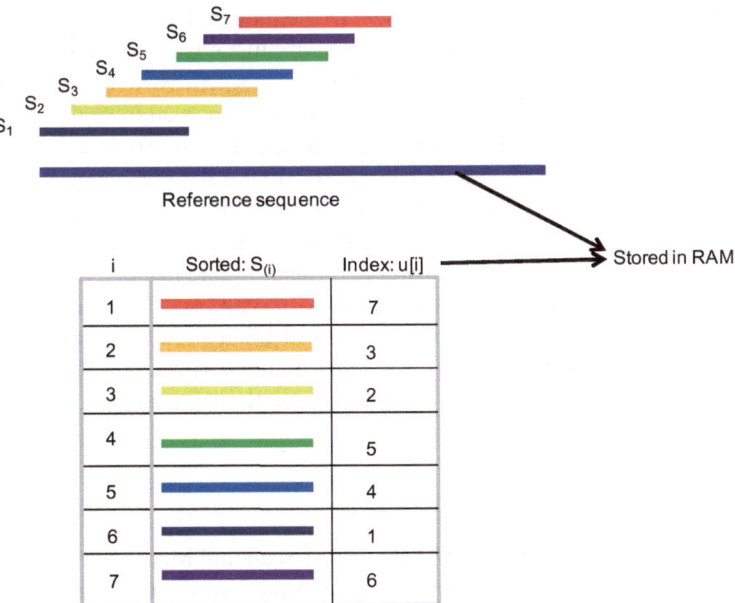

Fig. 1. Illustration of the index table. The blue bar is the reference sequence. The short bars on top of the reference represent 32-mers $s_{(i)}$ extracted from the reference. The sorted 32-mers $\{s_{(i)}\}$ and their corresponding indices $u[i]$ are listed in the table below. Only the reference genome and $\{u[i]\}$ are kept in RAM.

We make a note here. Regardless the value of k, we can carry out a search of the 32-mer starting at the same position as the k-mer. Along the binary search, the lower bound either stays or moves upwards while the upper bound either stays or moves downwards. We could have two outcomes: at some point, the 32-mer hits a match with either the lower or the upper bound; otherwise, the 32-mer matches neither of the two bounds when they meet. In the former case, the 32-mer finds a perfect match. In the latter, we check the maximum number of matching nucleotides between the target and the lower bound starting from the beginning position. Similarly we check the number for the upper bound. If this number is no smaller than k, then the k-mer has its perfect match on the genome. In comparison, search based on hashing does not have this flexibility.

Each item in the index table is a 32-bit integer which needs 4 bytes and the reference genome takes no more than 1Gb. They add up to no more than 13Gb. As the 64-bit architecture is taking over in the computer business, this memory requirement is not a serious problem. However, if we select every other 32-mer in the genome, say those at the odd addresses, then the resulted index table would be around 6Gb, and the total memory requirement is less than 7Gb. Of course, to be consistent with this configuration, we need to search two consecutive k-mers on a short read before we jump to the next seed. According to our simulation, this reduction of memory sacrifices very little in terms of performance.

To reduce the steps of binary search, we could introduce "block address" or "zip code" for each 32-mer, which is encoded into an integer in the range $[0, 4^{32} - 1]$. For a number $r < 30$, we pick up the 2^r integers $d_i = i * 2^{64-r}, i = 0, \cdots, 2^r - 1$, that divide the range uniformly, and insert each of them to the index table of the sorted 32-mers of the human genome. Denote the two indices that are just next to d_i are $(u[j_i], u[j_i + 1])$, namely, $s_{(j_i)} \leq d_i \leq s_{(j_i+1)}$ — it is possible that $s_{(j_i)} = s_{(j_i+1)}$. Now we keep the pointers $[j_i]$ in an array $q[i], i = 0, \cdots, 2^r - 1$, referred to as "block address vector" hereafter. In the practice of mapping reads, we load $q[i]$ together with $u[i]$ and the genome into computer memory. For an 32-mer-integer s, we divide it by 2^{64-r}, and the resulting integer after rounding off gives its block index denoted by i_1. Suppose $q[i_1]$ and $q[i_1 + 1]$ respectively point to $u[j_1]$ and $u[j_2]$, then the two indices $u[j_1], u[j_2 + 1]$ can serve as a more delicate starting point of the lower and upper bound respectively for the binary search of s. Since the distribution of $s_i, i = 1, \cdots, L$ can roughly be approximately by a uniform distribution, we could reduce r steps of binary search on averge using this strategy. Of course, the larger the r is, the more memory is needed. If we take $r = 15$, at the cost of 128K more memory, we could reduce the the average complexity of the binary search by half.

2.2   Seed Stage

In this stage, we use the strategy GSM (Grouped Scan and Map) to scan the short read sequentially to find a perfectly matched seed, in other words, to anchor the short read to a candidate position in the reference genome. Since we only index a single strand of the reference to save memory, we scan both short reads and their reverse complements. For the sake of simplicity, we just describe the scan scheme on one strand.

The scan function GSM puts all seeds of a short read into several groups. It scans the first seed of each group in the first round, and the second seed in the next round.

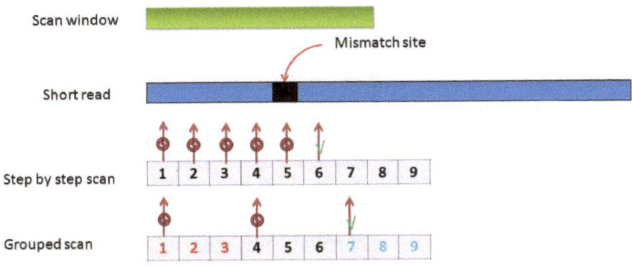

Fig. 2. Illustration of the grouped scan method. The green bar shows the scope of a scan window from which a seed is extracted. The blue bar represents a short read, on which a mismatch is marked by a black square. If we scan the read nucleotide by nucleotide, we would go though five failed mapping marked by red before the successful mapping marked by green occurs. If we scan the read with jump 3, only two failed mapping occur before a success.

This process goes on till a seed is mapped or the number of trial seeds exceeds a certain threshold. Fig. 2 is an illustration of the method. It can be seen that five seeds have to be scanned before the perfect match seed is detected by the step-by-step scan method. In contrast, we only need to map two seeds before the detection of a perfect match seed using a proper grouping strategy.

The grouped scan strategy reduces the number of trial seeds in most occasions. In our experience, if a short read is mappable(can be mapped back if all seeds are scanned) then the number of trial seeds does not exceed a certain threshold in most cases. We could experiment with a small portion of the read data to set this threshold. The principle will be discussed in section 2.4.

In addition to the scan order, seed length is another important factor we should consider. Later we will estimate the length interval which meets both sensitivity and specificity requirement on the basis of a probabilistic model. Seed lengths near the upper bound of the interval give the best specificity while seed lengths near the lower bound give the best sensitivity. If we put specificity prior to sensitivity, at each scan position we can first map a seed at the upper bound and then map a seed at the lower bound.

2.3 Extension Stage

In this stage, we detect the pattern, length and position of a possible INDEL. The core of the method are the notions of auto-match vector and auto-match function which we will define as follows.

Given two DNA segments denoted by S_1 and S_2, not necessarily of the same length, we define $V(S_1, S_2)$ to be a vector whose i-th element is 0 if the i-th elements of S_1 and S_2 are the same, and is 1 otherwise. The length of $V(S_1, S_2)$ is the shorter one of S_1 and S_2. For any string S, denote the substring of S with the first i elements removed as $S\{i\}$.

We define match vectors as: $M(0) = V(S_1, S_2)$; $M(i) = V(S_1\{i\}, S_2)$, for $i > 0$; $M(-i) = V(S_1, S_2\{i\})$, for $i > 0$, see Fig. 3. The auto-match vector $w(i)$ of S_1 and S_2 is defined as: the i-th element of $w(i)$ is the minimum of the i-th element of $M(0)$ and $M(i)$. Fig. 5 illustrates how $w(1)$ is obtained from $M(0)$ and $M(1)$. Finally, we define

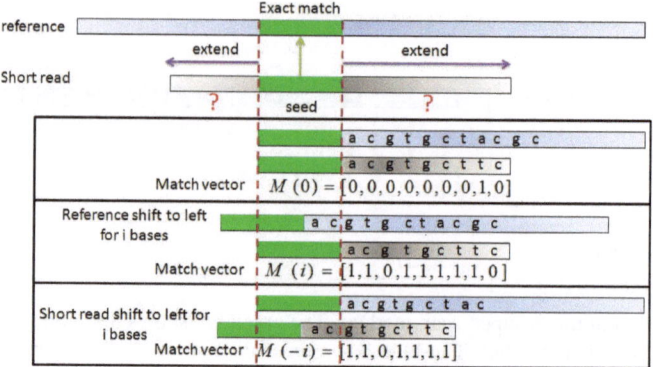

Fig. 3. Illustration of the auto-match vectors during extension

the auto-match function $AMF(i)$ to be the number of 1's in the auto-match vector $w(i)$. $AMF(0)$ is simply the number of 1's in the match vector $M(0)$.

With the help of AMF, we can detect the pattern and length of an INDEL. Fig. 4 shows a case of a two-nucleotide deletion, in which $AMF(i)$ is zero only for $i = 2$ while all other values are larger than five. We use this property to detect the pattern and length of an INDEL. Once the type and length of an INDEL is determined, we further use auto match vectors to detect its position. The idea is illustrated in Fig. 5, where we

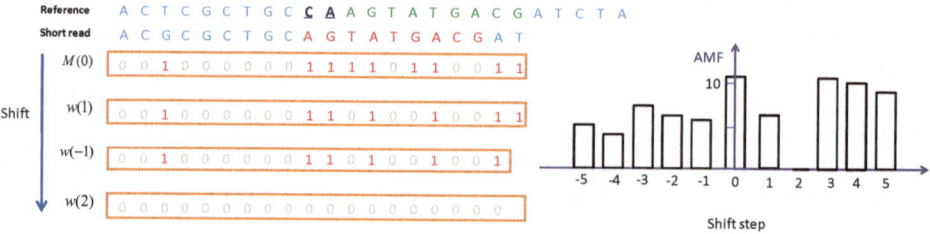

Fig. 4. The pattern of AMF corresponding to a deletion of size 2

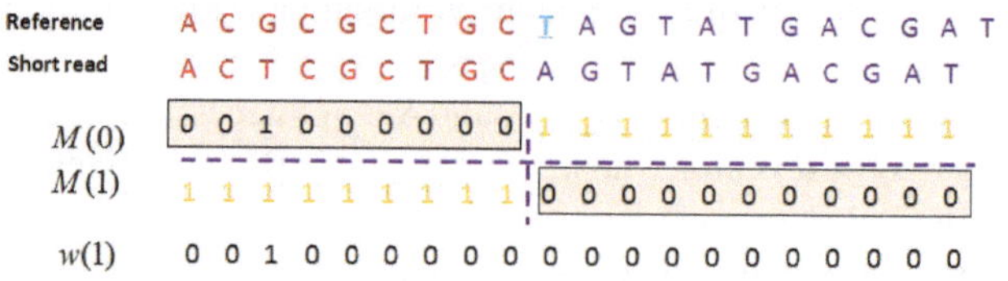

Fig. 5. Detection of the position of a DELETION. A nucleotide 'T' in green color on the reference genome is deleted. In $M(0)$ almost all elements before this nucleotide are 0 while almost all elements after it are 1. In $M(1)$ the pattern is just the opposite.

only consider $M(1)$ because the AMF calculation indicates that an 1-nucleotide-deletion exists somewhere, and the purple boundary indicates the position of the deletion.

Now we summarize the general AMF method as below.

Algorithm 1 AMF Algorithm

1. *Examine AMF(0), AMF(1), AMF(−1), ..., AMF(d), AMF(−d) sequentially (d is the maximum length of INDEL allowed) till $AMF(\mu) < \xi$ for a certain μ, where ξ is a predetermined value. A positive μ means a DELETION, and a negative μ means an INSERTION. The absolute value of μ estimates the length of the INDEL.*
2. *If such μ does not exist, we skip this extension (either a false mapping or a more complicated INDEL pattern exists); Else if $\mu = 0$, it means no INDEL; otherwise we use the pair $[M(0), M(\mu)]$ to detect the position of the INDEL in the next step.*
3. *Take the subsequence to the right of the mapped seed for example, and denote its length by l.*
   - *Initialization: let $D_0 = \sum_j w(\mu)_j$, $TMP = D_0$, $POSITION = 0$.*
   - *Recursion: For $j = 1 : l$, $D_i = D_{i-1}+[M(0)_i−M(\mu)_i]$; If $D_i < TMP$, $TMP = D_i$, and $POSITION = j$.*
   - *Output POSITION.*

For 100bp re-sequencing reads of the human genome, only a tiny fraction could be anchored by a fairly large seed, say 32bp, but could not be extended by the AMF method, and they are examined by the Smith-Waterman algorithm.

2.4 Computational Complexity

Some notations and definitions that are necessary for the complexity evaluation are listed in Table 1. We first consider those reads that can be mapped to the reference. Mapping such a short read is accomplished through: 1) finding a perfect match seed, 2) detecting the INDEL length, 3) detecting the start position of the INDEL if its length is nonzero. Next we decompose the time spent on each part in details according to the algorithm.

The time spent on exact match is $n_s T_{mp\_seed}$, where T_{mp_seed} varies depending on the algorithmic implementation and hardware. If we take the searching scheme described in

Table 1. Symbols and notations used in the complexity evaluation

Symbol	Definition
n_s	number of scanned seeds in a read, $n_s \leq 2(n − k + 1)$
n_w	number of seeds which are mapped to the reference but cannot be extended
T_{cmp_nt}	time of comparing a pair of nucleotides
T_{cmp_int}	time of comparing two integers
T_{add_int}	time of adding two integers
T_{mp_seed}	time of mapping a single seed
l	length of the read's subsequence involved in extension, $l \leq n − k + 1$
μ	length of an INDEL
Q	maximum length of an INDEL to be detected

Subsection 2.1, mapping a single seed has three steps: 1) obtaining a starting lower and upper bound in the index table for the seed using its block address; 2) binary searching for the two adjacent 32-mers between which the seed can insert; 3) finding the maximum length of perfect match up to 32 nucleotide bases. In the first step, the integer corresponding to a 32-mer seed needs to be divided by 2^{64-r}, where r is the number of binary search we would like to reduce. This can be achieved by $64 - r$ shift operations on the integer. In addition, two data access operations are required to get the two starting index bounds. The second step contains about $(30 - r)$ data access operations and $30 - r$ integer comparison. The third step can easily be implemented by shifting and comparing integers.

After finding a perfect match seed, we need to compute the values of AMF function to detect the possible "1-INDEL" length μ. First, the calculation of $M(i)$ takes l, l and $l-|i|$ comparisons of nucleotides pairs between the read and reference respectively for $i = 0$, $i > 0$ and $i < 0$. Second, calculation of $w(i)$ takes l and $l - i$ comparisons of Boolean elements in $M(0)$ and $M(i)$ respectively for $i > 0$ and $i < 0$. Third, the calculation of $AMF(i)$ takes roughly l integer additions. To detect an INDEL's start position, we can implement the third step of Algorithm 1 with the following time respectively for deletion and insertion:

$$(l-1)T_{add\_int} + l(T_{cmp\_int} + T_{add\_int}), \quad \text{and} \quad (l-\mu-1)T_{add\_int} + (l-\mu)(T_{cmp\_int} + T_{add\_int}).$$

Putting together and assuming we calculate $AMF(i)$ in the order of $i = 0, 1, -1, 2, -2, \cdots$, we have the following total time, ignoring the constant terms with respect to l.

$$T \approx n_s T_{mp\_seed} + n_w \, l[(2Q + 1)T_{cmp\_nt} + 2Q T_{cmp\_int} + (2Q + 1)T_{add\_int}] \qquad (1)$$

$$+ \begin{cases} l[T_{cmp\_nt} + T_{add\_int}] & \mu = 0 \\ l[2\mu T_{cmp\_nt} + 2\mu T_{cmp\_int} + (2\mu + 2)T_{add\_int}] & \mu > 0 \\ l[(2|\mu| + 1)T_{cmp\_nt} + (2|\mu| + 1)T_{cmp\_int} + (2|\mu| + 3)T_{add\_int}] & \mu < 0 \end{cases} \qquad (2)$$

The first term is the complexity of mapping seeds; the second term is the complexity of the unsuccessful extension of those anchored seeds. The third term is the complexity of the successful extension of the final seed. Possibly n_s includes the number of seeds that cannot be mapped anywhere, thus $n_s \geq n_w$. Later we will show that the specificity goes up as the seed length goes up. When the specificity is sufficiently large, the chance of $n_w > 0$ is small. For those reads that cannot be mapped to the reference, the third term is zero. So the time is

$$T \approx n_s T_{mp\_seed} + n_w \, l[(2Q + 1)T_{cmp\_nt} + 2Q T_{cmp\_int} + (2Q + 1)T_{add\_int}].$$

In our experience, for most of the mappable reads, the number of trial seeds is much smaller than the total number of seeds. If we set a threshold for the number of trial seeds then we avoid fruitlessly scanning. To set this threshold, we need to know the distribution of n_s for the mappable reads. Let $A(i) = \#\{n_s(\text{among mappable reads}) = i\}$, namely, the number of reads which need i trial seeds till a successful mapping, $1 \leq i \leq (n - k + 1)$, Fig. 6 shows the frequencies of $A(i)$ for an 100K-short-read data set. It is obvious that most of the mappable short reads are scanned only a few times. In fact, the 99% quantile of $A(i)$ in this example is 11, and the average number of trial seeds, for all

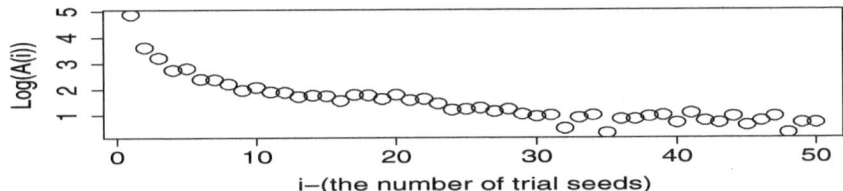

Fig. 6. Frequencies of n_s for mappable reads. The y-axis shows $log_{10}(A(i))$. The results are obtained from 100K 76bp short reads downloaded from NCBI data base, archive SRR003196. Among them 83K reads are mappable.

the short reads (including the unmapped reads) is only 2.9 for this data set. This gives an estimate of $E(n_s)$ and it explains, at least from one angle, why the sequential seeding strategy is efficient compared with that of fixed-number-seeding. The higher the quality of a read data set is, the less the average number of trial seeds are needed.

We also calculate the average length of INDELs in the example explained in the introduction section. It turns out that the average of μ is around 2.9. This means that on average, we only need to shift a short read rightwards and leftwards with respect to the reference 3 times.

3   Statistical Evaluation

In this section we evaluate the mapping accuracy of SEME based on probabilistic models. Several important statistical approaches have been developed for specific sequence alignment problems. For example, the statistic D_2[13] concentrates on the number of k-mer perfect match between two sequences of lengths m and n, and evaluate its asymptotics when m and n go to infinity. The concept of excursion in random walks and some other advanced techniques in probability were used in evaluating the significance of BLAST[14] results. In the current mapping problem, the read, say 100 bp, is much shorter than the reference genome. The asymptotics that requires both m and n go to large do not apply. In order to evaluate the sensitivity and specificity of the SEME mapping result, we propose another approach, which essentially compares the n-length read with every n-length subsequence of the reference.

Suppose that the read length is n, and we define sensitivity to be the probability that a read is mapped to where it is from, and specificity as the probability that the read does not map to any other positions – excluding repeats and possibly highly conserved homologs – on the reference. We approximate this event by any positions on a random reference of the same size. Let ν be the chance that the read is mapped to a random n-mer subsequence. According to subadditivity of probability

$$1 - specificity \leq \min\{N\nu, 1\},$$

Corresponding to the two stages of SEME, we make the following decomposition:

$$sensitivity = \tau\theta, \quad \nu = \eta\theta^*, \tag{3}$$

where τ and r are respectively the probabilities of finding a k-mer perfect match between two n-length sequences under a correct location and a random location. θ and θ^* are respectively the conditional probabilities of accepting extension under a correct location and a random situation. Assuming that the n positions are independent and the match rate is constant, we calculate τ and η precisely, and the result is accurate no matter what the read length is. θ is obtained by a soft counting method, which calculates the probability of the extension based on the distribution of the imputed "possible polymorphism" numbers that aim to adjust the effect of base-calling errors.

Lemma. For two n-length sequences, assume that bases at different positions are independent and the match rate for all positions is a constant p, then the probability that an k-mer perfect match exists is given by

$$\tau(k, p) = \sum_{m=0}^{n} [\sum_{s=1}^{K(m)} (-1)^{s+1} C_{m+1}^{s} C_{n-ks}^{m}](1 - p)^{m} p^{n-m},$$

where $K(m) = \max\{s; n - ks \geq m\} \wedge (m + 1)$.

In fact, $\tau(k, p)$ increases with p and decreases with k.

We first apply this lemma to the calculation of $\tau = \tau(k; p)$, where p is the matching rate between a read and the region where it is from and it depends on the polymorphism rate and sequencing error rate. Let X, Y, S represent the reference, individual genome and short read respectively. It can be shown that the mismatch rate at the site (X_i, Y_i, S_i) is $(1 - \beta_i)\gamma + (1 - \gamma)\beta_i + \gamma\beta_i w_i$, where $\gamma = Pr(X_i \neq Y_i)$ is the polymorphism rate, or simply the SNP rate if we skip the INDEL for the moment; and $\beta_i = Pr(Y_i \neq S_i)$, the miscall rate; $w_i = Pr(S_i \neq X_i | Y_i \neq S_i, Y_i \neq X_i)$. We note that in this context we use the jargon "polymorphism rate" γ as a measure of genomic discrepancies between the target individual and the reference, but not as a measure of population genetics. Since $\gamma\beta_i$ is small, we have the approximation to the match probability: $p_i \approx 1 - \gamma - \beta_i$. For the moment, we replace β_i by their average. We show the curves of τ under different settings in Fig 6(a). For example, the green solid line corresponds to the case of 100bp-reads with a 0.99 match rate. In this case, the sensitivity is satisfactory even when $k = 30$.

Next we apply the lemma to the random situation. In $\eta = \tau(k; p_\eta)$, we set the match rate p_η to be the sum of the squares of the base composition rates (usually around 0.25). In Fig 6(a), the red line, representing the trend of η, drops to zero quickly even when the seed is short, and to some extent it displays the specificity of SEME.

Soft Counting Criterion. The extension stage validates the anchor position by checking the the number of inconsistencies between the reference and the read. For now, we simply exclude INDEL positions. MAQ [3] evaluates a mapping result by calculating the posterior probability that the read comes from the region, and it regards the hit with the highest posterior as the correct result. The posterior can be maximized effectively by minimizing the sum of quality values of mismatched bases. We note that mismatches could be caused by miscalls as well as polymorphisms, and base-calling errors are not strong evidence of incorrect mapping. Instead of hard counting of mismatches, we propose a soft counting method that imputes the "possible polymorphism" fractions from all mismatch sites using quality values and an appropriate

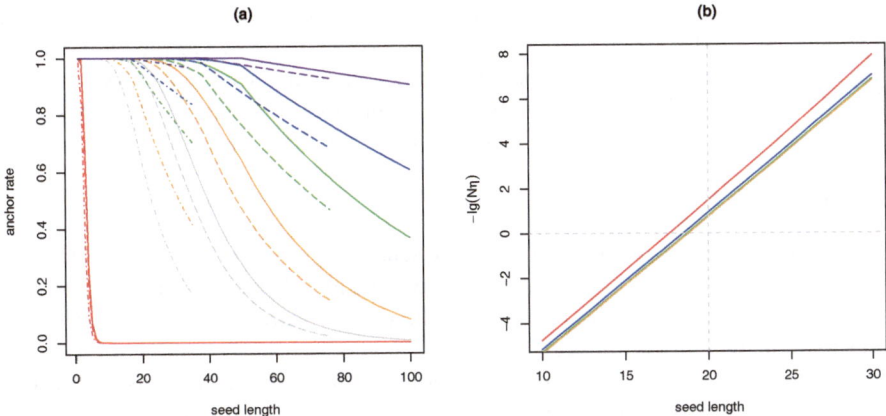

Fig. 7. **(a)** The probability of finding a perfect k-mer match between two sequences of the same length with respect to k. The solid lines, dashed lines and dotted lines represent the occasions for 100bp, 76bp, 35bp sequences; and the red, gray, orange, green, blue and purple lines respectively correspond to the common match rate 0.25, 0.95, 0.975, 0.99, 0.995, 0.999. **(b)** $-\lg(N\eta)$ vs. seed length, where η is the probability that the read is anchored to a random subsequence of the same length. The red, blue, green, and orange lines correspond to the cases for 35bp, 76bp, 100bp, and 110bp reads respectively. We use it as a measure of specificity to guide seed length selection.

polymorphism rate, aiming at reducing the effect of miscalls on mismatch sites. Consequently, we evaluate the mapping result based on the sum of the imputed "possible polymorphism" fractions. Specifically, according to the setup above, the mismatch rate is $1 - p_i = (1-\beta_i)\gamma + \beta_i[(1-\gamma) + \gamma w_i]$. We impute the "possible polymorphism" fractions at mismatch sites as

$$\frac{(1-\beta_i)\gamma}{(1-\beta_i)\gamma + \beta_i[(1-\gamma) + \gamma w_i]},$$

which can well be approximated by $\frac{\gamma}{\beta_i+\gamma}$. If quality scores are available and can be interpreted as probabilities, we have $\beta_i = 10^{-\frac{q_i}{10}}$, see [15,16]. Our statistic is defined to be

$$\sum_{i \text{ at mismatch sites}} \frac{\gamma}{\beta_i + \gamma}.$$

Under the assumption of independence, its distribution is binomial$(n - k, \gamma)$. Consequently, we can convert the statistic score of an alignment into a p-value. In this case, the larger the p-value, the stronger evidence of accepting the mapping.

Since the seed and extension part do not overlap, we can regard them as approximately independent. θ is the type one error probability of the associated test of the hypothesis: the anchor is correct. We can set the significance level α of this test to ensure a reasonable sensitivity. In fact, $\tau(1 - \alpha)$ is an upper bound for the sensitivity of SEME. The curves in Fig 6(a) show the sensitivity excluding the factor $(1 - \alpha)$ under different settings and can serve as a guidance for seed length selection.

We can similarly calculate the sum of inconsistencies of an extension alignment under the random sequence assumption, and its asymptotic distribution is normal.

The chance that we accept an incorrect anchor should be small as validated by our simulations. Essentially $1 - \theta^*$ is the power of the associated test. For the moment, we use $N\eta$ as a conservative bound of specificity when choosing seed length. As a matter of fact, $N\eta$ is also the average number of anchor places across the whole genome by chance. The curves of $-\lg(N\eta)$ with respect to seed length are shown in Fig 6(b).

3.1 Seed Length Determination

The seed length selection is a trade off between sensitivity and specificity. Shorter seeds increase sensitivity, but may lead to many incorrect anchor places; longer seeds increase specificity, but the seed may be mapped nowhere. To ensure both of them, the seed length should be in a proper range.

According to Fig 6(b), the curves corresponding to different read lengths are close to each other, especially as the size is larger than 76bp. Only when the seed length is chosen to be at least 19 or 20, the average number of anchor positions by chance would be smaller than 1. If we would ensure specificity larger than 0.999, the lower bound should be up to 24 or 25. Of course, this estimate might be conservative because $N\eta$ is a conservative bound of specificity. On the other hand, slightly larger lower bound can help avoid false positive anchors, which are expected to be rejected in the extension stage. The upper bound is chosen according to the sensitivity curve and our tolerance. For 100-bp reads with an average 0.99 match rate, [20, 32] is a proper range of seed length. In practice, the binary search algorithm simply find the maximum exact match length up to 32 nucleotide for each seed. If it is above 20, we then evaluate the significance based on this exact match length.

If the seed is 20-mer, the sensitivity for 35bp reads with a match rate 0.975 is 0.83, while it grows to $1 - 6.95 \times 10^{-4}$ for 100bp reads with a match rate 0.975. If we choose 32-mer seed and assume the match rate is 0.99, then the sensitivity is 0.75, 0.98, and 0.99587 respectively for 35bp, 76bp and 100bp reads. For shorter reads with lower quality, the seed length may even drop to less than 20 to ensure a fair sensitivity. This is the reason why in the early days of NGS, the strategy using single seed of perfect match did not work while it is feasible nowadays as the read length and sequencing quality improves.

4   Results

In our examples, the reference genome is the human genome *hg18*. We report comparisons of SEME with Bowtie2 (Version 2.0.0-beta7) and SOAP2[17,18] on three data sets from NCBI database, each of which includes 2 million reads. To make fair comparisons, we implement SEME by mimicing the parameters in the '–sensitivity' mode for Bowtie2 and '-v4 mode for SOAP2 respectively.

Mapping rates and time are two key measures for evaluating read mappers. We show the comparison of Bowtie2 and SEME on the left in Table 2. For data set 1, the running time of SEME reduces to about 1/4 whereas the mapping rate of SEME is 14.6(88.0-73.4)% higher than that of Bowtie2. Comparing with data set 1, the running time of Bowtie2 for data set 2 is a little more while that of SEME reduces further by a third. This phenomenon is due to the fact that the number of Bowtie2's trial seeds is fixed for

Table 2. Comparison of SEME, Bowtie2 and SOAP2. n is the read length of each data set. The three data sets are from NCBI database, namely, short read archives SRR003196, SRR033622 and SRR054721. They are all generated by the Illumina Platform. Each data set contain two million short reads. Left: Comparison with Bowtie2; right: Comparison with SOAP2

n	Programme	Time(s)	Map rate(%)		n	Programme	Time(s)	Map rate(%)
76	Bowtie2	508	73.4		76	SOAP2	290	39.4
	SEME	124	88.0			SEME	164	50.4
75	Bowtie2	542	96.1		75	SOAP2	207	80.5
	SEME	81	98.8			SEME	112	87.0
100	Bowtie2	787	95.5		100	SOAP2	261	74.5
	SEME	95	99.2			SEME	161	81.7

each short read while that of SEME mainly depends on the quality of each short read, that is, better quality, less trial seeds. Notice that the read lengths of data set 1 and 2 are about the same and the quality of the latter is better. From data set 2 to 3, the read length extends to 100, whereas the quality are similar. We can see that the mapping time of Bowtie2 increases, but SEME remains about the same, which verifies our analysis.

We show the comparisons of SOAP2 and SEME on the right in Table 2. Since the -v4 mode only allow 4 mismatches, mapping rates of both SOAP2 and SEME for all 3 data sets are lower than those in the comparison with BOWTIE2. Not only does the running time of SEME reduces to about one half, but also it has a 7-11% gain in mapping rates.

In sum, compared with Bowtie2, SEME runs 4.1-8.3 times faster depending on quality of data sets; Compared with SOAP2, SEME runs twice faster while the mapping rate of SEME is substantially higher.

5 Discussion

SEME has two key features. The first one is its novel mapping algorithm, which obeys the "seed-and-extend" paradigm. A common approach of the seed stage is to map multiple seeds at the same time and then make them to full alignments. The number of these multiple seeds is usually fixed from read to read. Different from this approach, SEME maps seeds sequentially. The number of seeds need to be mapped depends on the distribution of mismatch sites on the short read. The scan function of SEME efficiently minimizes the average number of trial seeds. In the extension stage, SEME can detect the pattern, position and length of small INDELs by means of auto match function and auto match vectors without enumerating all possible combinations or carrying out local alignment algorithm. Time complexity of the extension stage is linear with respect to the read length.

The second feature is that SEME has its own statistical evaluation of mapping reliability, which is critical for NGS, especially its applications to medicine. Compared to the vast amount of algorithmic development, not much associated statistics was found in the literature so far. A statistical evaluation of a mapping result is justifiable only if the model on which the analysis is based captures the data characteristics and follows the mapping algorithm closely. Our statistical analysis of the "seed-and-extend" scheme essentially boils down the evaluation of specificity and sensitivity to the matching chance of two n-length sequences, where n is the short read length. We decompose the probabilities into two parts: one corresponds to the seed stage and the other corresponds to the extension stage. Since we stick to perfect match in the seed stage, the calculation of the exact probability is relatively easy, see Lemma. In the extension stage, the sum of "possible polymorphism" fractions can approximately be described by either a binomial or a normal distribution.

The random sequence assumption is definitely far from a perfect description of any common natural genome because it ignores more complicated issues such as duplications and homologs. Appropriate simulations may complement the model-based analysis to some extent. We carried out limited simulations, and the results are quite comparable with the analytical results in terms of the values of τ and r in Equation (3).

SEME is very flexible due to its data structure and sequential scan strategy. Depending on the mapping context, the condition of the short reads and the requirement of the mapping result, we can adjust the seed length, the scan scheme and the upper bound of trial seeds. The optimization of the scan scheme depends on several factors such as the sequencing quality pattern, and we are conducting more investigations. We implement the method in C++, but the process of improving the code is ongoing.

Other than the straightforward mapping problem, we did not elaborate on SEME's variants that we are working on. For example, by encoding and decoding 'C' and 'T' with a common letter, we can use SEME to map short reads and allow methylation sites. However, with this setup, the values of sensitivity and specificity, the seed length need to be re-evaluated. For now, SEME deals with pair-end data by treating them as independent reads. How to integrate information from both ends to speed up mapping is an interesting problem to be considered. The detection of alternative splicing site inside a single seed is a more challenging task. As the Illumina read length goes beyond 160bp, the ideas of SEME described in this report may help solve the problem. Particularly, we emphasize that the statistical evaluation is important for justifying the significant of any new genomic discovery.

Acknowledgements. We thank Dr. Yong Zhang and Dr. Zhixiang Yan from BGI-Shenzhen for their help. We thank Bo Wang and Lixian Yang for their help on programming. This work was supported by the National Center for Mathematics and Interdisciplinary Sciences, CAS, the Program of "One hundred talented people", CAS, and grant 91130008 from Chinese National Science Foundation.

References

1. Li, H.: A survey of sequence alignment algorithms for next-generation sequencing. Brief Bioinformatics 11, 473–483 (2010)
2. Ben, L., et al.: Ultrafast and memory-efficient alignment of short DNA sequences to the human genome. Genome Biology 10, R25 (2009)
3. Li, H., Ruan, J., Durbin, R.: Mapping short DNA sequencing reads and calling variants using mapping quality scores. Genome Res. 18, 1851–1858 (2008)
4. Hui, J., Wing-Hung, W.: SeqMap: mapping massive amount of oligonucleotides to the genome. Bioinformatics 24, 2395–2396 (2008)

5. MiSeq Personal Sequencer - Illumina,
   `http://www.illumina.com/systems/miseq.ilmn`
6. Xun, G., Wen-Hsiung, L.: The Size Distribution of Insertions and Deletions in Human and Rodent Pseudogenes Suggests the Logarithmic Gap Penalty for Sequence Alignment. J. Mol. Evol. 40, 464–473 (1994)
7. Ryan, E.M., Christopher, T., et al.: Luttig, An initial map of insertion and deletion (INDEL) variation in the human genome. Genome Res. 16, 1182–1190 (2006)
8. Smith, T.F., Waterman, M.S.: Identification of common molecular subsequences. J. Mol. Biol. 147, 195–197 (1981)
9. Altschul, S.F., Gish, W., Miller, W., Myers, E.W., Lipman, D.J.: Basic local alignment search tool. J. Mol. Biol. 215, 403–410 (1990)
10. Karlin, S., Altschul, S.F.: Methods for assessing the statistical significance of molecular sequence features by using general scoring schemes. Proc. Natl. Acad. Sci. U S A 87, 2264–2268 (1990)
11. Waterman, M.S.: General methods of sequence comparison. Bull. Math. Biol. 46, 473–500 (1984)
12. Waterman, M.S.: Introduction to Computational Biology: Maps, Sequences and Genomes. Chapman & Hall, London (1995)
13. Ross, A.L., Haiyan, H., Waterman, M.S.: Distributional regimes for the number of k-word matches between two random sequences. PNAS 99, 13980–13989 (2002)
14. Warren, J.E., Gregory, R.G.: Statistical Methods in Bioinformatics: An introduction. Springer, New York (2001)
15. Brent, E., Phil, G.: Base-Calling of Automated Sequencer Traces Using Phred. II. Error Probabilities. Genome Res. 8, 186–194 (1998)
16. Ming, L., Magnus, N., Lei, M.L.: Adjust quality scores from alignment and improve sequencing accuracy. Nucleic Acids Research 32, 5183–5191 (2004)
17. Ruiqiang, L., Yingrui, L., Karsten, K., Jun, W.: SOAP: short oligonucleotide alignment program. Bioinformatics 24, 713–714 (2008)
18. Ruiqiang, L., Chang, Y., Yingrui, L., et al.: SOAP2: an improved ultrafast tool for short read alignment. Bioinformatics 25, 1966–1967 (2009)
19. Zaharia, M., Bolosky, W.J., Curtis, K., Fox, A., Patterson, D., Shenker, S., Stoica, I., Karp, R.M., Sittler, T.: Faster and More Accurate Sequence Alignment with SNAP. arXiv:1111.5572 [cs.DS] (2011)

Dissecting Cancer Heterogeneity with a Probabilistic Genotype-Phenotype Model

Dong-Yeon Cho and Teresa M. Przytycka[*]

National Center for Biotechnology Information, NLM, NIH
Bethesda, MD 20894 USA
przytycka@ncbi.nlm.nih.gov

Developing an approach to model heterogeneity of cancer has emerged as is an urgent need in cancer studies. To address this challenge we propose an approach for a probabilistic modeling of cancer. Starting with the assumption that each cancer case should be consider as a mixture of cancer subtypes, our model links phenotypic similarities with putative causes. Specifically, building on the idea of a topic model [1], our approach is based on two components (i) a measure of phenotypic similarity between the patients and (ii) a list of features –such as mutations, copy number variation, microRNA level etc. to be used as proposed explanations. The main idea is to define (probabilistic) disease subtypes and, for each patient, identify the mixture of the subtypes that best explain the patient similarity network. Our approach does not assume predefined subtypes nor does it assume that such subtypes have to be uniquely defined. That is, we do not assume that there exist "the" disease subtype model but rather we consider a distribution of such models providing a probabilistic context.

Our probabilistic model allows identification of genetic aberrations which are responsible for similarities and differences in patients' phenotypes, pinpointing dependencies among such aberrations, and emerging probabilistic subtypes. It provides also a probabilistic way of inferring the genotype-phenotype relationship.

We applied our approach to TCGA Glioblastoma Multiforme (GBM) data to obtain a probabilistic model of the disease, Prob_GBM. We used gene expression to describe disease phenotypes, consequently the patient network was built based on gene expression similarity. This helped us to compare results inferred from our model to the study of expression based TCGA GBM subtypes [2]. We show that while our model is largely consistent with the current knowledge about GBM, it also leads to new hypotheses, some of which we could support by the facts from the cancer literature.

To the best of our knowledge, it is the first time that a probabilistic model to explain patient similarity relation has been proposed in the context of studying of biological heterogeneity. Specifically, by building Prob_GBM we obtained for the first time an unsupervised model which explains expression similarities using mutations, copy number variations, and microRNA levels. In this work we focused on model description and demonstrating how the information represented in the model can be

---

[*] Corresponding author.

M. Deng et al. (Eds.): RECOMB 2013, LNBI 7821, pp. 30–31, 2013.
© Springer-Verlag Berlin Heidelberg 2013

leveraged to understand disease heterogeneity in the context of relatively well studied GBM. However, many interesting variations of the model are possible. For example phenotype similarity might include survival time. Features can be extended to include transcription factor biding, methylation, age, sex, or environment. Alternatively the features can be narrowed down to microRNA only, to study the impact of these molecules alone. This study opens the door to these and many other applications.

References

1. Chang, J., Blei, D.M.: Hierarchical Relational Models for Document Networks. Annals of Applied Statistics 4, 124–150 (2010)
2. Sanai, N.: Integrated Genomic Analysis Identifies Clinically Relevant Subtypes of Glioblastoma. World Neurosurgery 74, 4–5 (2010)

eALPS: Estimating Abundance Levels in Pooled Sequencing Using Available Genotyping Data

Itamar Eskin[1,*], Farhad Hormozdiari[2,*,**], Lucia Conde[3], Jacques Riby[3],
Chris Skibola[3], Eleazar Eskin[2,4], and Eran Halperin[5,6,7]

[1] Applied Mathematics Department, School of Mathematical Sciences,
Tel-Aviv University
[2] Computer Science Department, University of California, Los Angeles
[3] Division of Environmental Health Sciences, School of Public Health,
University of California, Berkeley
[4] Department of Human Genetics, University of California, Los Angeles
[5] Computer Science Department, Tel-Aviv University
[6] International Computer Science Institue, Berkeley, California
[7] Molecular Microbiology and Biotechnology Department, Tel-Aviv University
fhormoz@cs.ucla.edu

Abstract. The recent advances in high-throughput sequencing technologies bring the potential of a better characterization of the genetic variation in humans and other organisms. In many occasions, either by design or by necessity, the sequencing procedure is performed on a pool of DNA samples with different abundances, where the abundance of each sample is unknown. Such a scenario is naturally occurring in the case of metagenomics analysis where a pool of bacteria is sequenced, or in the case of population studies involving DNA pools by design. Particularly, various pooling designs were recently suggested that can identify carriers of rare alleles in large cohorts, dramatically reducing the cost of such large-scale sequencing projects.

A fundamental problem with such approaches for population studies is that the uncertainly of DNA proportions from different individuals in the pools might lead to spurious associations. Fortunately, it is often the case that the genotype data of at least some of the individuals in the pool is known. Here, we propose a method (eALPS) that uses the genotype data in conjunction with the pooled sequence data in order to accurately estimate the proportions of the samples in the pool, even in cases where not all individuals in the pool were genotyped (eALPS-LD). Using real data from a sequencing pooling study of Non-Hodgkin's Lymphoma, we demonstrate that the estimation of the proportions is crucial, since otherwise there is a risk for false discoveries. Additionally, we demonstrate that our approach is also applicable to the problem of quantification of species in metagenomics samples (eALPS-BCR), and is particularly suitable for metagenomic quantification of closely-related species.

Keywords: Relative Abundance, Pooling, Metagenomics, Expectation-Maximization.

---

* These authors contributed equally to this work.
** Corresponding author.

M. Deng et al. (Eds.): RECOMB 2013, LNBI 7821, pp. 32–44, 2013.
© Springer-Verlag Berlin Heidelberg 2013

1   Introduction

Over the past several years, genome-wide association studies (GWAS) have identified hundreds of common variants involved in dozens of common diseases [1]. These discoveries leveraged technological advances in genotyping microarrays [2, 3] which allowed for the cost effective collection of common genetic variation in large numbers of individuals. More recently, technological advances in high throughput sequencing (HTS) technologies have rapidly decreased the cost of sequencing cohorts of individuals [4]. The advantage of sequencing technologies relative to genotyping technologies is that sequencing technologies collect both rare and common variation providing the opportunity for implicating rare genetic variation, in addition to common variation, in human disease.

Unfortunately, to identify disease associations with rare variants, the cohorts that must be sequenced consist of thousands of samples. Even when considering the decrease in costs over the past decade, the cost of sequencing these cohorts is prohibitively expensive. The actual cost of sequencing a sample consists of two parts. The first part is the cost of preparing a DNA sample for sequencing which we refer to as the library preparation cost. Library preparation is also the most labor intensive part of a sequencing study. The second part is the cost of the actual sequencing which is proportional to the amount of sequence collected which we refer to as the sequencing per-base cost. Technological advances are rapidly reducing the per-base cost of sequencing while the library preparation costs are more stable. A recently proposed approach to reduce the overall sequencing cost and to avoid potential biases introduced during library preparation is to utilize sequencing pools. The basic idea behind this approach is that DNA from multiple individuals is pooled together into a single DNA mixture which is then prepared as a single library and sequenced. In this approach, the library preparation cost is reduced because one library is prepared per pool instead of one library per sample. DNA pooling has been successfully applied to GWAS data that reduce costs by one or two orders of magnitude [5–7]. However, pooling DNA from a large number of individuals can introduce a great deal of background noise in the data that may reduce the reliability of and increase the difficulty in the downstream analysis. In contrast to pooling strategies in GWAS data where a small number of pools are genotyped, each consisting of a large number of samples, in sequencing pooling studies typically a small number of individuals are sequenced in each pool, making the noise amenable to explicit modeling. Moreover, DNA pooling has been successfully applied to next generation sequencing [8], where they ran a large pooling study for the identification of rare mutations in bacterial communities.

Recent work in the area [9, 10] has focused mainly on effective designs of pooled studies that can reduce the number of pools required for the detection of causal variants. In addition, suggested association statistics for rare SNP analysis typically involve the comparison of the total number of rare mutations in the cases and controls, therefore there is no need for individual sequencing in such cases. Indeed, in this work we use as a benchmark a sequencing study of Non Hodgkin's Lymphoma, where the samples have been partitioned into sets of five samples, and each set was pooled and whole-genome sequenced. The latter study is currently ongoing, and without the tools presented in this paper, the study might result in false discoveries. Generally, such designs allow for an increased statistical power due to the increase in the sample size. However, the analysis of these studies relies on the assumption that the pools are perfectly constructed, meaning that the fraction of DNA from each sample is known; typically, each DNA mixture contains an exact amount of DNA information intended from each individual in the pool. As we show using real experimental data from non-Hodgkin's lymphoma (NHL), this assumption is wildly inaccurate and the amount of DNA in each mixture is often different from the intended amount. This might potentially lead to both false positives and reduced statistical power.

In this paper, we present a computational methodology to infer the relative abundance or the fraction of each individual in the DNA mixture of a pool directly from the sequencing data given that we have a small amount of genotyping data for the individuals. This assumption is applicable in many cases, particularly since most current sequencing studies are being performed on cohorts where a genome-wide association study has been previously performed. Our method can be applied directly to the data obtained from a pool sequencing study as the first step in the analysis. We present a formal statistical framework for the estimation of relative abundances, taking into account the presence of sequencing and genotyping errors. In practice, reliable genotypic data of all pooled samples might not be available due to separate quality control procedures for sequencing and genotyping. We therefore propose an extension that handles missing genotypic data by leveraging the linkage disequilibrium structure of the genome. We demonstrate using real data that a naive analysis without applying our method would lead to false positive associations.

The computational problem of estimating the relative abundances in DNA pools is closely related to the computational problem of estimating the abundance of species in metagenomic samples. Bacteria are vital for humans, affecting a wide range of food and health industries. Known to reside in the human body in numbers higher than the number of human cells [11], the set of bacteria and their interactions are an indication to the physical condition of a person, and were shown to be correlated with various diseases [12–14]. In 2008, the National Institute of Health (NIH) launched the Human Microbiome Project (HMP) to examine all existing microorganisms in the human body. Following the HMP, another project named Metagenomics of the Human Intestinal Tract project (MetaHIT) was launched with the goal of studying gut bacteria. Both projects aim to increase our knowledge of bacterial community effects on our body. The first step, however, is to understand which bacteria are available in each sample and the fraction of each bacterium. The latter problem is mathematically very similar to the estimation of DNA fractions in a pooled sample, and we therefore apply our methods to metagenomic instances (eALPS-BCR).

Bacterial fraction estimation in the context of metagenomics has already been addressed by Amir and Zuk [15], who used an approach based on Sanger-sequencing of a highly-preserved genomic region (16S) found in all bacteria.

They obtain a sequence-based profile of the bacterial community in that region, and use a Compressed Sensing (CS) framework to compute the fraction of each bacterium in the sample. Due to its decreasing cost and increasing throughput, High-throughput sequencing (HTS) has also been widely applied to metagenomic samples to infer species abundance [16–19]. The kind of data generated by a single Sanger sequencing reaction is very different from HTS data, and the compressed sensing approach is not specifically designed for such data. Therefore, methods such as GAAS [18] and GRAMMY [19] are based on similarity scores of high-throughput sequencing reads which are mapped to a database of known bacterial genomes. Examination of whole-genome reads as opposed to a single highly-preserved region is more suitable to the analysis of homogeneous bacterial communities, considering that very few mutations might be present in the 16S region of closely related species. In general, the problem of estimating relative abundances becomes increasingly difficult in lower taxonomic levels, and is particularly hard when considering strains of the same species. We show that with minor adjustment, our method (eALPS-BSR) can be directly applied to HTS data, and argue that modeling of linkage-disequilibrium patterns of bacteria greatly improves estimation accuracy in such scenarios. Experimental results on various simulated arrangements of bacterial communities will be available in the full version of this paper.

2   Methods

Description of the Data Generating Process. We first set the stage by describing a mathematical model for the generation of sequencing data in a pooling scheme. As always, the model might be an oversimplified abstraction of reality, however in the results section we show that our estimates are highly accurate on real data, and we therefore argue that the model approximates to an adequate degree the realistic mechanism of sequencing data generation.

Consider a scenario in which the DNA of N individuals is pooled and then sequenced. In addition, assume that these N individuals have genotype information in M positions, described by a matrix $\mathbf{H}_{N \times M}$, where $h_{ij} \in \{0, 0.5, 1\}$ is the minor allele count of the i-th individual in the j-th position. Such a scenario may appear in pooled sequencing studies, such as the one we describe in the Results (for non-Hodgkin's lymphoma), or in scenarios where a set of DNA pools is used to detect rare variants (such as in [20–22]). In addition, as we discuss below, this scenario also occurs in metagnomic analysis where a set of bacteria are sequenced together.

Ideally, at least in the case of human studies, one would aim at specific relative abundances for each of the samples, which are typically equal amounts of DNA for each sample, but in some cases there are other designs (e.g., [9]). However, in practice the actual relative abundances may be quite different from the desired levels. Particularly, we demonstrate in the Results section that for some pools with presumably equal amounts of DNA from each individual, the actual fractions of the samples often deviate considerably, and this has to be taken into account in any subsequent analysis.

We denote the unknown relative abundances $\alpha = (\alpha_1, \ldots, \alpha_N)$ where α_i is the relative abundance of the i-th individual. The pooled sample undergoes high-throughput sequencing, resulting in the collection $\mathbf{X} = \{\mathbf{x_j}\}$, where every $\mathbf{x_j}$ is a vector of length t_j (the coverage at position j), and the elements x_{jr} represent the minor/major allele status of the r-th read in the j-th position:

$$x_{jr} = \begin{cases} 1 & \text{read } r \text{ in position } j \text{ shows a minor allele} \\ 0 & \text{o/w} \end{cases}$$

We assume that for each position j, the number of reads t_j in that position is generated from a Poisson distribution with some parameter C, the mean coverage over the entire genome. The reads for every position are then distributed according to a mixture of N Bernoulli distributions with parameters $h_{1j}, \ldots, h_{Nj}$, the mixture weights being the relative abundances $(\alpha_1, \ldots, \alpha_N)$. Formally, our model assumes that a read x_{jr} is generated by randomly picking an individual i according to the proportions $(\alpha_1, \ldots, \alpha_N)$, and assigning the allele status 0/1 according to the minor allele probability h_{ij}. To specify the identity of the mixture components, we introduce the (unknown) latent variables $\mathbf{Z} = \{z_{ijr}\}$, where z_{ijr} are indicator functions that determine the individual every read originated from, i.e.:

$$z_{ijr} = \begin{cases} 1 & \text{read } r \text{ in position } j \text{ originates from individual } i \\ 0 & \text{o/w} \end{cases}$$

We model the sequencing technology as an error-prone process, with a probability ε for a sequencing error that switches the read from minor to major or vice-versa. Thus, in our model the unknown parameters of the model are α, ε and $\mathbf{H}$, and the observed data is $\mathbf{X}$. We are mostly interested in α in this paper, although we also show how to estimate ϵ. Under this model, the likelihood of the data is given by:

$$p(\mathbf{X}|\mathbf{H}; \alpha, \varepsilon) = \prod_{j=1}^{M} p(\mathbf{x_j}|\mathbf{h_j}; \alpha, \varepsilon) = \prod_{j=1}^{M} \prod_{r=1}^{t_j} \sum_{i=1}^{N} \alpha_i p_i(x_{jr}|\varepsilon) \tag{1}$$

where $p_i(x_{jr}|\varepsilon)$ is the probability to observe read x_{jr} given that it originated from individual i and with sequencing error ε, thus: $p_i(x_{jr}|\varepsilon) = p(x_{jr}|\mathbf{h_j}, z_{ijr} = 1; \varepsilon) = (1 - \varepsilon)h_{ij}^{x_{jr}}(1 - h_{ij})^{1-x_{jr}} + \varepsilon(1 - h_{ij})^{x_{jr}}h_{ij}^{1-x_{jr}}$.

It is important to notice that the likelihood formulation in (1) relies on the assumption that reads do not span more than a single variant. In reality this is of course not the case, but occurrences of closely positioned SNPs is infrequent enough as to allow us to overlook this possibility, without substantially undermining the correctness of our model. Given the genotypes, the reads x_{jr} are therefore generated independently across the different positions in the genome, as they only depend on the value of $\mathbf{h_j}$. In the case where some of the genotypes are unknown (as discussed below) this is not true and should be addressed properly.

Relative Abundance Estimation. We now present the algorithm for estimation of relative abundances in the full genotypic data scenario (eALPS), where genotypes of all N sequenced individuals are given. Our objective is to find a maximum-likelihood estimate of the model parameters, i.e. the relative abundances α and the sequencing error ε. Since $\mathbf{Z}$ is unknown, we use an Expectation-Maximization (EM) approach, which instead of trying to maximize the likelihood given in equation (1), considers the marginal likelihood of the observed data:

$$p(\mathbf{X}, \mathbf{Z}|\mathbf{H}; \alpha, \varepsilon) = \prod_{j=1}^{M} \prod_{r=1}^{t_j} \prod_{i=1}^{N} \left(\alpha_i p_i(x_{jr}|\varepsilon)\right)^{z_{ijr}} \tag{2}$$

The EM algorithm is an iterative algorithm, where in each iteration the algorithm searches for parameters that maximize the expected value of the marginal log-likelihood function given a current estimate of the parameters. This procedure is repeated until a convergence of either the log-likelihood or the parameters is achieved. Following the standard notation for EM, we call this quantity the Q-function (i.e. the marginal log-likelihood function), and write it as:

$$Q(\alpha, \varepsilon|\alpha^{(t)}, \varepsilon^{(t)}) = \mathbb{E}_{\mathbf{Z}|\mathbf{X},\mathbf{H},\alpha^{(t)},\varepsilon^{(t)}}\left[\log L(\alpha, \varepsilon; \mathbf{X}, \mathbf{Z}, \mathbf{H})\right]$$

$$= \sum_{j=1}^{M} \sum_{r=1}^{t_j} \sum_{i=1}^{N} \beta_{ijr} \log \alpha_i + \sum_{j=1}^{M} \sum_{r=1}^{t_j} \sum_{i=1}^{N} \beta_{ijr} \log \left(p_i(x_{jr}|\varepsilon)\right) \tag{3}$$

Where $\beta_{ijr} = \mathbb{E}[z_{ijr}|x_{jr}, \alpha^{(t)}, \varepsilon^{(t)}]$. The maximization over α involves only the first term in (3), which is clearly a concave function of α and can be solved easily using Gibbs' inequality, while enforcing the constraint that $\sum_{i=1}^{N} \alpha_i = 1$. Finding a closed form expression for $\varepsilon^{(t+1)}$ is not possible, however simple numerical methods such as gradient descent can be applied to produce the next estimate for the sequencing error. The update rules are then:

$$\alpha_i^{(t+1)} = \frac{\sum_{j=1}^{M} \sum_{r=1}^{t_j} \beta_{ijr}}{\sum_{i'=1}^{N} \sum_{j=1}^{M} \sum_{r=1}^{t_j} \beta_{i'jr}}; \quad \varepsilon^{(t+1)} = \underset{\varepsilon}{\operatorname{argmax}} \sum_{j=1}^{M} \sum_{r=1}^{t_j} \sum_{i=1}^{N} \beta_{ijr} \log \left(p_i(x_{jr}|\varepsilon)\right)$$

Missing Genotypes. In practice, it is often the case that genotype information is only available to a subset of the data, specifically to the samples that were previously genotyped for a genome-wide association study in the pre-high-throughput sequencing era. Moreover, even in the case where all individuals are genotyped, some of the SNPs are not called for some of the individuals, and in such cases our approach is not applicable. We therefore developed an improved method that can handle missing genotype data without compromising the accuracy of estimated parameters. Formally, suppose that for a pool of N individuals, we have only $N' < N$ genotyped individuals, and we wish to estimate the relative abundances $\alpha = (\alpha_1, \ldots, \alpha_N)$ given the observed genotypes $\mathbf{G}_{N' \times M}$, $g_{ij} \in \{0, 0.5, 1\}$, and the observed read counts $\mathbf{X}$ as in the previous section. Regarding the true genotypes $\mathbf{H}$ as a set of latent variables in addition to $\mathbf{Z}$, we can follow a similar derivation of the EM algorithm to maximize the new likelihood function:

$$p(\mathbf{X}, \mathbf{G}, \mathbf{Z}, \mathbf{H} | \alpha, \varepsilon) \propto p(\mathbf{G}|\mathbf{H}) \cdot p(\mathbf{X}, \mathbf{Z}|\mathbf{H}; \alpha, \varepsilon) \propto \prod_{j=1}^{M} \prod_{i=1}^{N} \prod_{r=1}^{t_j} (\alpha_i \cdot p_i(x_{jr}|h_{ij}, \varepsilon))^{z_{ijr}}$$

Maximization of this likelihood function can be achieved in a similar fashion to the previous case where all genotypes are known, with the expectation step involving an extra iteration on all possible realizations of the missing genotype. This approach, however, fails to take into account the presence of Linkage Disequilibrium (LD) between adjacent loci, which renders invalid the assumption of independence between the $\mathbf{h_j}$'s, producing suboptimal estimates of the model parameters. Particularly, we show in the Results section that this method (eALPS-MIS) systematically underestimates the relative abundances of the missing individuals.

Fortunately, leveraging the information of LD available in population samples, as well as the known genotypes themselves, allows for very accurate estimations of the conditional probability of the latent variable H, given the observed data and the current estimate of the parameters. In fact, when LD information is utilized, most possible values of h_{ij} have negligible probabilities, and can be omitted from the expectation step. We continue to show that even a hard assignment of $\mathbf{h_j}$ to the most likely value in every iteration of the EM algorithm conserves its desirable convergence properties.

The algorithm we propose (eALPS-LD) therefore uses the following scheme: Given a current estimate of the parameters α and ε, find a maximum likelihood estimate for the missing genotypes h_{ij}, $N' < i \leq N$, using the LD model that will be described shortly. Using this estimate of h_{ij}, continue the EM iteration as in the previous EM derivation for known genotypes, i.e. calculate the expectation over the latent variables ($\mathbf{Z}$), and maximize the log-likelihood function. This approach can be justified from a statistical point of view using the same arguments presented in [23]. The hard assignment of $\mathbf{h_j}$ is also computationally advantageous, as it eliminates the need for an exhaustive enumeration of all realizations of possible genotypes.

To find the most likely missing genotype, we need to model population haplotype frequencies, and we do so using a Markov model with a similar structure to those recently used by [24–26]. The basic structure of this LD model is that of a left-to-right directed graph, with M disjoint sets of nodes corresponding to the M loci. Edges in the directed graph correspond to the transition probabilities, and only connect nodes in consecutive sets. Every node in the graph corresponds to one of the two possible alleles, with potentially multiple nodes representing each allele in a specific locus, allowing for multiple haplotypes (more accurately, haplotype clusters) with the same allele in that position to be represented. The edges carry the population frequency of transition from a haplotype in one position to a haplotype in the next position, meaning that every haplotype in the population corresponds to a path in the graph. Training of the model according to population samples can be done either with the Baum-Welch algorithm for HMMs, like in [25], or in the constructive approach described in [26]. In our implementation, we used the BEAGLE genetic analysis software package (version 3.3.2) to build the LD model.

We now turn to define the full model used to infer the missing genotypes, with the above LD model as a basic building block. In the interest of simplicity, we consider the case where only one genotype is unobserved, though a straightforward extension to handle multiple missing genotypes is applicable. The overall model is a hidden Markov model composed of two copies of the LD model, i.e. every state is represented by a pair (q_1, q_2) with q_1 expressing the first haplotype and q_2 the second haplotype of the missing genotype. Assuming Hardy-Weinberg equilibrium, the two haplotypes of the missing genotype are independent, therefore the transition probabilities are simply the product of the frequencies carried by each of the corresponding edges in the LD model. Each node in the HMM emits the minor allele read count of that position, c_j, with probability $\bar{h}_j^{c_j}(1 - \bar{h}_j)^{t_j - c_j}$ where $\bar{h}_j = \sum_{i=1}^{N} \alpha_i^{(t)}\big((1 - \varepsilon)h_{ij} + \varepsilon(1 - h_{ij})\big)$. The posterior probability of every possible haplotype can be computed using the standard forward-backward algorithms in $O(MS^2E^2)$ time, where S is an upper bound on the number of states for each position in the basic LD model, and E is an upper bound on the indegree of nodes in the graph (i.e. number of incoming edges). Recall that edges in the graph connect only those nodes lying on a path that represents a haplotype in the reference population, therefore E is expected to be a small number. We refer the reader to further discussion of algorithm complexity in the full version of this paper.

Bacterial Community Reconstruction. The estimation of relative abundance levels in DNA pools is naturally applicable to metagenomic analysis, particularly to the reconstruction of bacterial communities. Given a mixture of known bacteria, the goal of Bacterial Community Reconstruction (BCR) is to detect which bacterial species are present in the sample and to estimate their fractions. A number of methods accomplish this task by exploiting the 16S region, which is a highly conserved 1.5kb segment found in all known bacteria. Shown to be effective in reconstruction of various bacterial communities [15], this approach is naturally limited to bacteria that exhibit sufficient dissimilarity in the 16S region, as the ability to distinguish between different species diminishes with increasing inter-relatedness. If one is interested in reconstructing a community of closely related organisms, e.g. same-species strands in a microbial gut sample, considerably longer genomic segments need to be analyzed.

Recently, a novel method (GRAMMy) based on high throughput sequencing of the entire genome was introduced in [19] and tested on various standard datasets. Somewhat similar in character to the method in this paper, the authors consider the metagenomic reads as arising from a finite mixture model, where the mixing parameters are the relative abundances and the component distributions of reads are approximated using k-mer frequencies in the reference genomes. Expectation-maximization is then applied to estimate the mixing parameters. We hereby propose an efficient method based on common SNPs in orthologous genes, that eliminates the necessity to handle whole-genome read data, and focuses only on the highly informative SNPs that reside in homologous genes. A major benefit of this approach is that it allows, in the same manner as with human genomes, to take advantage of existing LD structure in closely-related bacteria to account for possibly unknown species of bacteria in the sample.

Suppose we have N sequences of known bacteria that we wish to use as references, and a metagenomic sample that is sequenced to produce short reads. To be able to use the same formulation as before for this setting, a preprocessing step is performed on the bacterial reference genomes. First, genes that are homologous in all N genomes are extracted and aligned against an arbitrarily selected reference genome. Subsequently, SNP calling is performed on the aligned regions resulting in a set of M SNPs. The total number of SNPs we will acquire in this procedure depends on the similarity of the reference genomes - high relatedness of the samples means more orthologs, albeit fewer variants in every single gene. BCR thus reduces to the previously presented problem of relative abundance estimation: we regard the available database of orthologous regions as the collection of true genotypes present in the sample.

3   Results

Non-Hodgkin's Lymphoma Dataset. Our method was applied to a real population study of non-Hodgkin lymphoma (NHL) for which genome-wide association data were available. In this dataset, a whole-genome sequencing on a group of lymphoma cases was conducted, with the aim of identifying additional common and rare lymphoma associated variants undetected by previous genome-wide association studies (GWAS). The studied samples consisted of a subset of follicular lymphoma samples that were part of a recent GWAS conducted in the San Francisco Bay Area. Full details of the GWAS, including the process and criteria for subject selection, genotyping, quality control and statistical analysis have been described elsewhere [27]. A total of 312,768 markers genotyped in 1,431 individuals passed the quality control criteria and were used for genome-wide association analysis. Among the follicular lymphoma cases for which GWAS data was available, 155 were used in this study. To construct each pool, equal amounts of DNA (1,320ng) were combined from 5 individuals of the same sex and age in a total volume of 110 uL. Importantly, we demonstrate below that in reality the amounts of DNA were not equal even though the pooling protocol aimed at exact amounts of 1,320ng of DNA from each pool. Sequencing was outsourced to Illumina FastTrack Services (San Diego, CA). gDNA samples were used to generate short-insert (target 300 bp) paired-end libraries and a HiSeq2000 instrument was used to generate paired 100 base reads according to the manufacturer instructions. The software ELAND was used for sequence alignment, and the coverage was 35 per base for the pool, thus 7 per base for each sample.

Simulated Data. We used the Non-Hodgkin's Lymphoma genotype data as a starting point in order to simulate data according to the following model. We assume that the genotype values g_{ij} are given by the Non-Hodgkin's Lymphoma genotype data. Then, for every position we draw a random sample t_j (the total number of reads covering the j-th SNP) from a Poisson distribution $Pois(T)$,

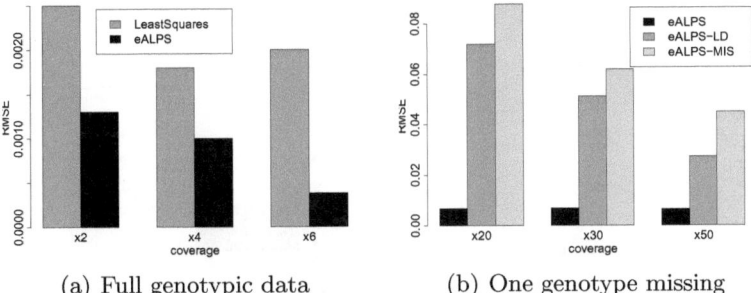

(a) Full genotypic data   (b) One genotype missing

Fig. 1. Relative abundances estimations in simulated pools with 5 individuals, using 230,000 SNPs. (a) Assuming all 5 genotypes are known (b) Assuming one of the individuals' genotypes is missing. Additional results for 1,000 SNPs are found in the full version of this paper. The methods compared are: eALPS - full genotypic data; eALPS-LD - missing genotype & utilizing LD; eALPS-MIS - missing genotype & full (soft) EM.

where the mean is equal to the desired coverage T. The minor allele counts, c_j are then drawn from a Binomial distribution $B(t_j, \sum_i \alpha_i((1 - \varepsilon)g_{ij} + \varepsilon(1 - g_{ij})))$, and the major allele counts are just $t_j - c_j$. We calculate the Root Mean Squared Error (RMSE) of the predicted α, and compare to a simple least square estimation of the relative abundances. Results for this comparison are shown in Figure 1. The least squares method is based on the assumption of normally distributed noise, which is clearly violated for low coverage sequencing. Indeed, we observe that least squares tends to perform poorly as the coverage goes down, while our method (eALPS) achieves significantly better performance in coverages lower than 4X.

We note that the least squares estimation is only applicable when all individuals have genotype information. We also explored the scenario in which at least one of the individual's genotype is unknown. Particularly, we randomly picked one of the pools that has full genotype data, generated major and minor allele counts as mentioned in the previous experiment, and compared the performance of the full genotypic data method (eALPS) to the methods discussed in Section 2 (eALPS-MIS and eALPS-LD) when one of the individuals' genotype data is omitted. We examined the effect of different coverages and the number of sampled SNPS on the RMSE measured, summarized in Figure 1. Evidently, utilizing the linkage disequilibrium information considerably improves the accuracy as observed by comparing the performance of eALPS-MIS and eALPS-LD.

Results on Real Data

Complete genotype information. In the NHL data, we have 31 pools where the genotype information for all individuals is available. We use eALPS to estimate the relative abundances of each individual in each pool. Figure 2 illustrates how some pools contain individuals with relative abundances that are significantly higher (or lower) compared to other individuals in that pool. Performing

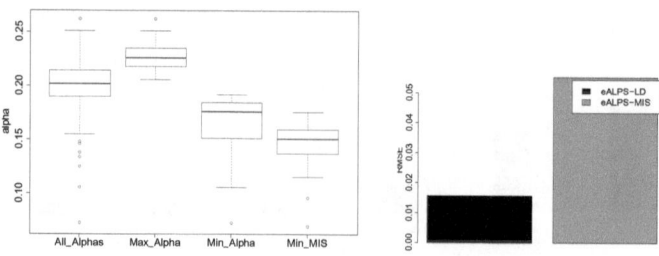

(a) Empirical distribution of esti- (b) Missing genotypes estimated abundances   mation error

Fig. 2. Relative abundances in individuals from the NHL study, estimated using eALPS, eALPS-LD and eALPS-MIS. All pools contain 5 individuals, and where intended to have uniform relative abundances. (a) summarizes the distribution of alphas estimated using eALPS on the NHL data, demonstrating that in practice relative abundances vary. The blue boxplots are (from left to right): all relative abundances, the maximal and minimal abundances for every pool, estimated using eALPS. The red boxplot is the minimal relative abundance estimated by eALPS-MIS, showing that the method systematically underestimates the relative abundance of the missing genotype (minimal values were always achieved for the missing individual). Examples of concrete values in NHL pools can be found in the full version of this paper. (b) compares the error on the NHL data with one masked genotype. eALPS-LD was given four of the genotypes in every position, while eALPS was given the full genotype information.

any analysis (i.e. association study) on these pools requires careful consideration. More rigorous validations of the proposed model, i.e. model selection and Goodness-of-fit tests, were performed and are fully described in the complete version of this paper. These tests suggest that applying simple statistics to the NHL pools under the wrong assumption of equal DNA quantities will definitely lead to a large number of false positives.

Missing Genotypes. To assess the accuracy of the missing genotype methods on real data, we masked one genotype of each individual from each of a set of 14 pools, and we ran both eALPS-MIS and eALPS-LD. Figures 2(b) and 2(a) presents the results for these experiments, where eALPS is used as a baseline for the calculation of RMSE. As can be clearly observed from Figure 2(b), eALPS-LD outperforms eALPS-MIS. Moreover, eALPS-MIS tends to systematically underestimate the relative abundances of the missing individual, which can be explained by the unrealistic uniform prior on possible genotypes. In a sense, incorporating LD is equivalent to applying a very informative position-specific prior on the possible genotypes of the missing individual. The results strongly demonstrate that this approach is highly beneficial.

Acknowledgments. I.E, F.H, L.C, J.R, C.S, and E.H were partially supported by the National Institue of Health (NIH) grant number 1R01CA154643-01A1. The genotype data was also generated as part of (NIH) grant number

1R01CA154643-01A1. I.E was also supported by the German-Israeli Foundation (GIF) grant number 1094-33.2/2010. FE and EE were also supported by National Science Foundation (NSF) grants 0513612, 0731455, 0729049, 0916676 and 1065276, and NIH grants HL080079, DA024417. E.H. is a faculty fellow of the Edmond J. Safra Center for Bioinformatics at Tel-Aviv university and was supported in part by the Israeli Science Foundation (grant 04514831).

References

[1] Manolio, T.A., et al.: A HapMap harvest of insights into the genetics of common disease. The Journal of Clinical Investigation 118(5), 1590–1605 (2008)

[2] Matsuzaki, H., et al.: Genotyping over 100,000 SNPs on a pair of oligonucleotide arrays. Nature Methods 1(2), 109–111 (2004)

[3] Gunderson, K.L., et al.: A genome-wide scalable SNP genotyping assay using microarray technology. Nature Genetics 37(5), 549–554 (2005)

[4] Wheeler, D.A., et al.: The complete genome of an individual by massively parallel DNA sequencing. Nature 452(7189), 872–876 (2008)

[5] Skibola, C.F., et al.: Genetic variants at 6p21.33 are associated with susceptibility to follicular lymphoma. Nature Genetics 41(8), 873–875 (2010)

[6] Brown, K.M., et al.: Common sequence variants on 20q11.22 confer melanoma susceptibility. Nature Genetics 40(7), 838–840 (2008)

[7] Hanson, R.L., et al.: Identification of PVT1 as a candidate gene for end-stage renal disease in type 2 diabetes using a pooling-based genome-wide single nucleotide polymorphism association study. Diabetes 56(4), 975–983 (2007)

[8] Erlich, Y., et al.: DNA Sudoku–harnessing high-throughput sequencing for multi-plexed specimen analysis. Genome Research 19(7), 1243–1253 (2009)

[9] Golan, D., et al.: Weighted pooling–practical and cost-effective techniques for pooled high-throughput sequencing. Bioinformatics 28(12), i197–i206 (2012)

[10] Prabhu, S., Pe'er, I.: Overlapping pools for high-throughput targeted resequencing. Genome Research 19(1), 1254–1261 (2009)

[11] Savage, D.C., et al.: The Gastrointestinal Epithelium and its Autochthonous Bacterial Flora. The Journal of Experimental Medicine 127(1), 67–76 (1968)

[12] Guarner, F., Malagelada, J.R.: Gut flora in health and disease. Lancet 361(9356), 512–519 (2003)

[13] Heselmans, M., et al.: Gut Flora in Health and Disease: Potential Role of Probiotics. Current Issues in Intestinal Microbiology 6(1), 0–8 (2005)

[14] Mahida, Y.R.: Epithelial cell responses. Best Practice & Research Clinical Gastroenterology 18(2), 241–253 (2004)

[15] Amir, A., Zuk, O.: Bacterial community reconstruction using compressed sensing. Journal of Computational Biology 18(11), 1723–1741 (2011)

[16] Hamady, M., et al.: Error-correcting barcoded primers allow hundreds of samples to be pyrosequenced in multiplex. Nature Methods 5(3), 235–237 (2008)

[17] Dethlefsen, L., et al.: The Pervasive Effects of an Antibiotic on the Human Gut Microbiota, as Revealed by Deep 16S rRNA Sequencing. PLoS Biology 6(11), e280 (2008)

[18] Angly, F.E., et al.: The GAAS metagenomic tool and its estimations of viral and microbial average genome size in four major biomes. PLoS Computational Biology 5(12), e1000593 (2009)

[19] Xia, L.C., et al.: Accurate genome relative abundance estimation based on shotgun metagenomic reads. PloS One 6(12), e27992 (2011)

[20] Lin, W.Y., et al.: Evaluation of pooled association tests for rare variant identification. BMC Proceedings 5(suppl. 9), S118 (2011)

[21] Price, A.L., et al.: Pooled association tests for rare variants in exon-resequencing studies. American Journal of Human Genetics 86(6), 832–838 (2010)

[22] Lee, J.S., et al.: On Optimal Pooling Designs to Identify Rare Variants Through Massive Resequencing. Genetic Epidemiology 35(3), 139–147 (2011)

[23] Neal, R.M., Hinton, G.E.: A view of the EM algorithm that justifies incremental, sparse, and other variants. In: Learning in Graphical Models, 1977, pp. 355–368. Kluwer Academic Publishers (1998)

[24] Kimmel, G., Shamir, R.: A block-free hidden Markov model for genotypes and its application to disease association. Journal of Computational Biology 12(10), 1243–1260 (2005)

[25] Kennedy, J., et al.: Genotype error detection using Hidden Markov Models of haplotype diversity. Journal of Computational Biology 15(9), 1155–1171 (2008)

[26] Browning, S.R.: Multilocus association mapping using variable-length Markov chains. American Journal of Human Genetics 78(6), 903–913 (2006)

[27] Conde, L., et al.: Genome-wide association study of follicular lymphoma identifies a risk locus at 6p21.32. Nature Genetics 42(8), 661–664 (2010)

Analysis of Metabolic Evolution in Bacteria Using Whole-Genome Metabolic Models

Ali A. Faruqi, William A. Bryant, and John W. Pinney*

Imperial College, South Kensington Campus
London, SW7 2AZ
j.pinney@imperial.ac.uk
http://www.theosysbio.bio.ic.ac.uk/bacterial-metabolism/

Abstract. Recent advances in the automation of metabolic model reconstruction have led to the availability of draft-quality metabolic models (predicted reaction complements) for multiple bacterial species. These reaction complements can be considered as trait representations and can be used for ancestral state reconstruction, to infer the most likely metabolic complements of common ancestors of all bacteria with generated metabolic models. We present here an ancestral state reconstruction for 141 extant bacteria and analyse the reaction gains and losses for these bacteria with respect to their lifestyles and pathogenic nature. A simulated annealing approach is used to look at coordinated metabolic gains and losses in two bacteria. The main losses of *Onion yellows phytoplasma* OY-M, an obligate intracellular pathogen, are shown (as expected) to be in cell wall biosynthesis. The metabolic gains made by *Clostridium difficile* CD196 in adapting to its current habitat in the human colon is also analysed. Our analysis shows that the capability to utilize N-Acetyl-neuraminic acid as a carbon source has been gained, rather than having been present in the *Clostridium* ancestor, as has the capability to synthesise phthiocerol dimycocerosate which could potentially aid the evasion of the host immune response. We have shown that the availability of large numbers of metabolic models, along with conventional approaches, has enabled a systematic method to analyse metabolic evolution in the bacterial domain.

Keywords: Metabolic Evolution, Ancestral State Reconstruction, Metabolic Models, Hierarchical Clustering, Simulated Annealing, Pathogenicity.

1 Introduction

One of the aims of systems biology has been to integrate information regarding metabolism in order to construct metabolic models and thus to analyse the effects of genetic perturbations on metabolism at the system level. In recent years, a number of attempts have been made to study the evolution of metabolic networks

---

* Corresponding author.

M. Deng et al. (Eds.): RECOMB 2013, LNBI 7821, pp. 45–57, 2013.

and these have provided insights into the mechanisms of evolution of various extant bacteria [1–3]. Understanding the evolution of bacterial metabolism is of great importance for a number of reasons. In particular, it has the potential to provide insights into the evolution of pathogenicity and its relationship with metabolism.

Bacteria not only evolve through vertical inheritance, but also through horizontal gene transfer (HGT). Often HGT can provide metabolic genes [4,5], and potentially antibiotic resistance and toxin encoding genes [6] to bacteria. On the other hand evolution through gene loss can occur in some environments [7]. These processes directly involve gene losses and gains, but it is not the genes themselves that are of most interest, but their function and how they interrelate with the functions of all other genes in the system.

Evolution is often studied through Ancestral State Reconstruction (ASR) for various biological traits [8,9]. ASR relies on biological trait information from extant organisms to infer trait occurrence in the common ancestors of those organisms. This information can be provided in the form of a character matrix for the characteristics under investigation. Depending on the context, a parsimony or maximum likelihood approach can be used on a phylogenetic tree to obtain the probabilities of different ancestral nodes possessing the considered traits. This approach has been taken in looking at gene families in the metabolic context [10], and metabolic reaction occurrences have been compared according to inferred metabolic models for a small set of 16 *E. coli* strains to investigate the evolution of these strains [11].

Previously genomic comparisons have been done using information from the WIT database, examining differences between the metabolic pathway complements of various extant organisms [12]. Additionally phylogenetic profiles have been inferred based on enzyme evolutionary predictions [13] to establish the ancestral relationships between a large number of prokaryotes and eukaryotes.

With the advent of automatic methods for bacterial metabolic model reconstruction – such as the Model SEED pipeline [14] – it is possible for the first time to establish direct reaction complements for any bacterium for which there is a complete genome sequence. Data from these draft-quality automatically generated metabolic models can be used as the input to ASR, since these models make direct assertions about which reactions are present and absent in each bacterium. Consequently, it is possible to infer ancestral metabolic complements directly and to investigate the precise metabolic changes accumulated by bacteria in the evolution towards their current lifestyles and ecological niches at the system level. This improves on previous approaches by being reaction-specific, rather than at a pathway level. Also, information about specific reactions can be made based on enhanced inferences (achieved through the Model SEED pipeline) about reaction presence and absence, not just based on direct observation of annotated enzymes.

Here we present an ancestral state reconstruction of the metabolic reactions inferred to be present in 141 bacteria by the Model SEED server. A hierarchical clustering was used to establish the metabolic similarity of these 141 bacteria, and this was compared to the 23S rRNA phylogenetic tree inferred for these same bacteria. Further we related this clustering to the lifestyles of the bacteria according to three categorisations: habitat, respiratory mode and pathogenic mode and showed that each of these categorisations encapsulates information about how the evolution of these bacteria has proceeded.

Results for the gain and loss of reactions for each of the extant bacteria have been produced using the metabolic networks inferred for the common ancestors of these bacteria and for two cases these gains and losses have been investigated at the system level to look for coordinated sets of reactions (those reactions adjacent in the metabolic networks inferred from their respective metabolic models) that have been lost (in the case of an obligate intracellular pathogen) and gained (in the case of a free-living pathogen). This has been achieved by using an approach based on active modules [15] called ambient which finds connected subnetworks in the bipartite network of reactions and metabolites associated with strong evidence for reaction gain or loss for both these bacteria [Bryant et al. - in submission]. ambient has picked out several reaction pathways in *C. difficile* CD196 that would not be seen by gene-based analysis (since several of the reactions have no gene association) but are clearly found by taking advantage of the generated metabolic model used here.

2   Methods

2.1   23S rRNA Phylogeny Construction

23S rRNA sequences for all 141 organisms in the current analysis along with an out-group organism (*Thermoplasma acidophilum*) were obtained from the NCBI Nucleotide Database. Multiple sequence alignment of the 23S rRNA sequences of these organisms was obtained using MAFFT [16]. A threshold score of $E = 8.4e-11$ was used (the default threshold value used by MAFFT).

Based on the results of multiple sequence alignment, a phylogeny was constructed using PhyML 2.4.4 [17]. Bootstrapping was performed 100 times on the tree to obtain the most likely phylogeny. After rooting the tree, the out-group was removed. For visualisation of the phylogeny obtained and for the creation of phylogeny images Dendroscope was used [18]. The phylogeny can be seen in Supplementary Fig. 1 available at our website[1].

2.2   Comparison of Reaction Numbers and Lifestyle

Three lifestyle classifications were used to assess how they related to the evolutionary histories of the bacteria in this study. The classifications are named i) habitat, indicating the usual environment the bacteria experience, ii) respiratory mode, indicating their ability to tolerate oxygen and iii) pathogenic mode, each bacterium falling into one of four categories: free-living, facultative host-associated, obligate intracellular mutualists and obligate intracellular pathogens.

---

[1] http://www.theosysbio.bio.ic.ac.uk/bacterial-metabolism/

These classifications were taken from work by Zientz et al. [19] and Merhej et al. [10]. It should be noted that although the last classification is termed 'pathogenic mode' this is just an alternative classification of habitat, based on the types of environment experienced by bacteria in their eukaryotic hosts.

A Mann Whitney U test was conducted between each category for each classification to establish correlations between reaction numbers and lifestyles. The Benjamini-Hochberg multiple testing correction was used to control for false positives and the corrected p-values were used to establish significance.

2.3   Ancestral State Reconstruction

For Ancestral State Reconstruction (ASR), `Mesquite` was used [20]. The Ancestral State Reconstruction algorithm, as implemented in Mesquite, looks for ancestral states which maximize the probability of the observed characteristics in extant organisms.

Maximum likelihood reconstruction methods look for ancestor states that maximise the probability of producing the current state, having evolved under a defined model of evolution [8,21]. It is equivalent to the marginal reconstruction method as implemented in PAUP [22]. Every reaction was classified as present or absent according to the Model SEED metabolic model creation server [14]. The Asymmetrical Markov k-state 2 parameter model (AsymmMK model) in `Mesquite` was chosen as it allows different rates for reaction gains and losses. In the ASR, the out-group organism was removed from the phylogeny and the reaction traits for the out-group were not specified in the character matrix.

A boolean character matrix was created for all the 2526 metabolic reactions that were present in at least one of the bacteria under investigation. Maximum Likelihood ASR was performed for this categorical, discrete dataset of reactions. Values for the probability (P_q^j) of the presence of a particular reaction (j) in a particular ancestral organism (q) were calculated by Mesquite based on the AsymmMK model.

2.4   Correlation between Dendrograms and Lifestyle Classifications

Two dendrograms were obtained from the 23S rRNA alignment and the metabolic traits comparison. The `cutree` package in R was used to examine every possible clustering of each dendrogram and the maximum Adjusted Rand Index [23] from all possible clusterings was obtained for each dendrogram against the three classifications in this analysis: habitat, respiratory mode and pathogenic mode. Adjusted Rand Index measures the similarity of different partitions of a set; in this case the partitions are the three classifications and the set is all bacteria under consideration.

2.5   Inference of Gains and Losses in Extant Bacteria

Each branch in the phylogeny connects two nodes. One node is the parent (ancestor) node and other node is the child (descendant) node. In order to assess the gain and loss of reactions from an ancestor to its descendant, δP^j values were calculated according to the following formula:

$$\delta P^j = P^j_{child} - P^j_{parent} \tag{1}$$

where P^j_{parent} and P^j_{child} are the probabilities of the presence of reaction j in the parent (ancestral organism)and child (descendant organism) nodes of a particular branch respectively. Therefore, a δP^j value close to 1 indicates a high likelihood of gain of reaction j in a branch and a δP^j value close to -1 indicates a loss of reaction.

There are a total of 140 internal (parent) nodes in the phylogeny. Each node gives rise to 2 branches giving a total of 280 branches. δP^j values were calculated for all the reactions on all of the branches. Thus, there are a total of 707,280 δP^j values (280 x 2526) for the entire phylogeny.

A δP cutoff of ± 0.9 was used to define those reactions gained or lost. Using this threshold δP value, ancestral state reconstruction predicted a total of 10,396 gain and loss events. $\delta P \leq$ -0.9 (loss) had 5001 events and $\delta P \geq$ +0.9 (gain) had 5395 events.

2.6 Metabolic Traits Hierarchical Clustering

The construction of a metabolic trait-based hierarchical clustering was done using the `Pars` programme in the `PHYLIP` package [24]. Each reaction present in at least one, but less than 141 of the bacteria under investigation, was used as a metabolic trait, as for the ASR. The `Pars` programme produced a total of 12 trees, from which a consensus tree was obtained using the `CONSENSE` program in the `PHYLIP` package.

2.7 Analysis of Coordinated Metabolic Changes

`ambient` [Bryant et al. - in submission] was used to run simulated annealing on the bipartite network of reactions and metabolites to find the 100 most significant coordinated metabolic changes in two bacteria representing the obligate intracellular (*Onion yellows phytoplasma* OY-M) and free-living (*Clostridium difficile* CD196) pathogenic lifestyles adopted by many of the bacteria investigated here.

The metabolic network used for both bacteria was the complete 'meta-' metabolic network consisting of the union of all 141 networks used in this paper. This allowed both gains and losses to be seen for each bacterium. `ambient` uses scores for each reaction and metabolite in its attempt to find connected network components encompassing many highly changed reactions. In this case the scores for reactions were taken from δP values for the relevant bacteria. Metabolites were scored in the using the default `ambient` scoring method - with a penalty equal to their connectivity in the metabolic network, to select against currency metabolites.

ambient was run to look for coordinated areas of loss of reactions in *Onion yellows phytoplasma* and gain of reactions in *C. difficile*. ambient was run with the following non-default parameters: maximum number of steps (-N) was set to 2,500,000, temperature gradient (-U) to 0.95, initial temperature factor (-T) to 3, reaction score offset (-Y) to -0.15 and number of steps between equilibrium tests (-i) to 6000.

3   Results and Discussion

3.1   Distribution of Organism Lifestyles and Reactions

Information about the number and types of organisms and reactions [14] was integrated with data about the lifestyles of those organisms [10,19]. Fig. 1 shows the distribution of the number of reactions in each organism with respect to their lifestyles: habitats, respiratory modes and pathogenic mode. The median number of reactions in the organisms is 1014. The reactions common to all 141 organisms account for about 1% of the total number of reactions.

As can be seen from Fig. 1, most of the organisms that have fewer than 700 reactions are host-associated; indeed from the distribution of pathogenic modes these bacteria represent the vast majority of obligate intracellular symbionts and pathogens. A Mann Whitney U test was conducted to establish whether there was any statistically significant relationship between lifestyle and number of reactions present in each bacterium. Results for each individual test and their p-values corrected for multiple testing can be seen in Supplementary Table 1.

The results show that differing habitats do not necessarily have a large impact on numbers of reactions that the bacteria maintain, except when comparing the free-living bacteria with those which are host-associated. There is also some impact of respiratory mode on number of reactions, but this could be due to a dependence of respiratory mode on bacterial habitat.

The most significant results come from the comparison of the different pathogenic lifestyles of these organisms, as classified by Merhej et al. [10]. Supplementary Table 1C clearly shows, as expected from observations of symbiotic and parasitic bacteria, that the number of reactions available for each bacterium is strongly dependent on their relationship with their eukaryotic host. This is not just true for obligate intracellular bacteria, but also to an extent for host-associated pathogenic bacteria. Unsurprisingly, obligate intracellular mutualists and parasites do not differ significantly in the size of their metabolic network, since their lifestyles, restricted to within a eukaryotic host, mean they experience the same nutrient availability and limitations.

3.2   Ancestral State Reconstruction

Ancestral state reconstruction for each reaction (trait) was performed on the phylogenetic tree inferred from the 23S rRNA alignment. A total of 30 metabolic reactions were present in all the 141 bacteria and these were excluded from the analysis so 2526 reactions were considered.

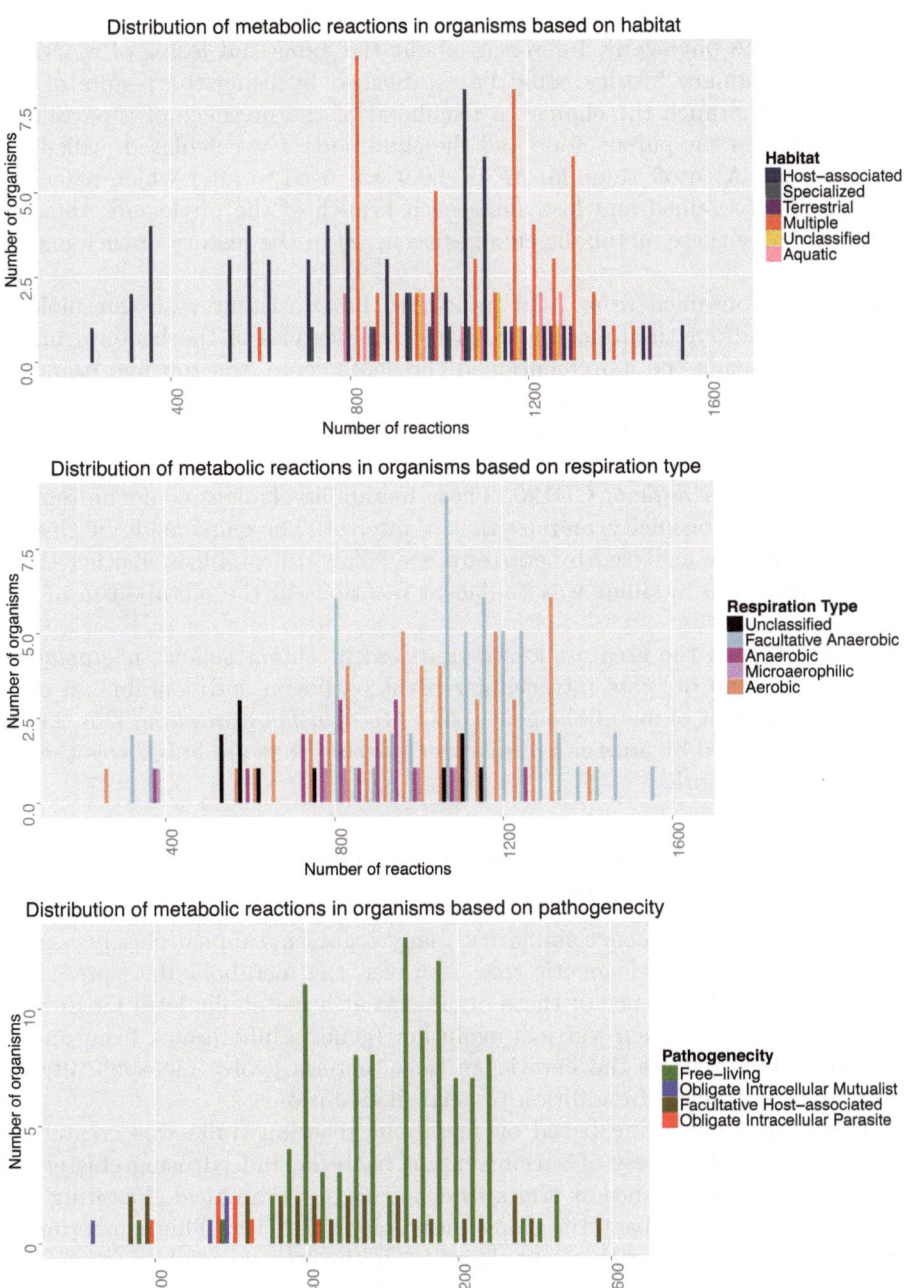

Fig. 1. Histograms showing the relationship between reaction numbers and bacterial lifestyles. Each diagram shows the distribution of total number of reactions in organisms based on habitat type (top), respiratory type (middle) and pathogenicity (bottom) according to Model SEED reconstructions of 141 bacteria.

ASR predicted the presence and absence of every reaction at every ancestral node on the phylogeny. Inferences about the gains and losses of reactions through evolutionary history could be established by using the results of the ASR. For each branch the change in likelihood of the presence of a particular reaction between the parent node and the child node was calculated, called δP (see Methods). A cutoff value for δP of ± 0.9 was used to infer which reactions were most likely gained and lost along each branch of the phylogeny, thus establishing where these metabolic changes occurred in the history of each extant bacterium.

The results obtained from ASR appear to be consistent with our biological knowledge about the different habitats and lifestyles of the bacteria under investigation. Using the aforementioned threshold score, the top five branches that showed the greatest number of gain and loss events were terminal branches leading to various extant bacteria.

The greatest gain was observed in the terminal branch leading to the bacterium *Clostridium difficile* CD196. These metabolic changes could be related to *C. difficile* pathogenicity, and are thus of interest. The gains made by this *C. difficile* strain were analysed by **ambient**, see below, to establish whether these gains occurred in a modular way (adjacent reactions in the metabolism of the bacterium).

In terms of loss, the greatest loss is observed in *Onion yellows phytoplasma* OY-M which is an obligate intracellular plant pathogen and contains an even smaller complement of metabolic genes than *Mycoplasma genitalium* [25]. These losses were analysed by **ambient** to discover whether they are linked together in the metabolic network.

3.3   Metabolic Traits Hierarchical Clustering (MHC) Compared with 23S rRNA Phylogeny

Based on 23S rRNA sequence similarity, many organisms appear closely related to each other on the phylogenetic tree. However, the metabolic data presented here indicate the divergence of these organisms at a metabolic level far greater than that implied by their vertical evolution (genetic inheritance from ancestors) alone. This reflects the knowledge that bacteria evolve metabolically by horizontal gene transfer in addition to vertical evolution.

A hierarchical clustering based on metabolic reaction traits was created to show metabolic relatedness of various extant bacteria. Indeed using clusters of orthologous groups of genes as traits to construct a hierarchical clustering has been shown to cluster bacteria along metabolic lines [10]. This clustering is based on metabolic similarity, so should reflect both vertical evolution (where the bulk of metabolic capabilities are inherited from) and horizontal gene transfer, depending on the importance of each of these mechanisms in the evolution of each organism.

The clustering was constructed using the character matrix of metabolic traits, to gain a better understanding of the evolutionary relationships as revealed through the ASR results presented above. Supplementary Fig. 2 shows the consensus tree obtained based on the metabolic traits of the organisms. The results obtained here clearly show that even though two organisms may be distantly related based on 23S rRNA sequence similarity, they can be very closely related in terms of their metabolic capabilities, i.e. that they have been subject to convergent evolution. A clear example here is between *Mycoplasma pulmonis* UAB CTIP and *Onion yellows phytoplasma* OY-M. They appear evolutionarily distant on the 23S rRNA phylogeny but are very closely related according to their metabolic trait profiles. Supplementary Figures 1 and 2 show the phylogeny and the metabolic hierarchical clustering respectively.

Dendrograms were produced from the trait-based tree and the RNA-based phylogeny and these were analysed to find whether the clusterings in the dendrograms corresponded to the three lifestyle classifications considered here. Maximum Adjusted Rand Indices (ARIs) were produced for each dendrogram / classification pair to quantify their relatedness. For the RNA-based tree none of the maximum ARIs were greater than 0.1, indicating little or no correlation between vertical evolution and current lifestyle. However, when the metabolic traits (i.e. metabolic reaction complement) and the lifestyles were compared a value of 0.15 was obtained for respiratory mode and a value of 0.37 for pathogenic lifestyle (free-living, host-associated, obligate intracellular mutualists and obligate intracellular pathogens). This indicates that the pathogenic mode adopted by a bacterium has a clear influence on its metabolic network.

3.4   Active Module Analysis

While overall gains and losses of reactions in bacteria are informative in establishing some of the principles of metabolic evolution, the specific changes and how coordinated these changes are might shed more light on the dependence of metabolic evolution on bacterial lifestyles and pathogenicity. Most metabolic processes rely on multiple distinct reactions, therefore on multiple genes encoding those enzymatic functions, so gains and losses of adjacent metabolic functions (pathways) might be expected to occur simultaneously. Here we used `ambient` [Bryant et al. - in submission] to look for reaction gains and losses that form connected components of the metabolic networks of the bacteria under consideration. Two bacteria were analysed, representing two different lifestyles: the obligate intracellular (*Onion yellows phytoplasma* OY-M) and the free-living (*Clostridium difficile* CD196).

The analysis of *C. difficile* produced 14 metabolic modules significant at the $q = 0.001$ level, which can be seen in Supplementary Fig. 3. Table 1 shows a summary of the functions of the modules found. Several modules are involved in monosaccharide utilisation and some in cell wall biosynthesis. Of particular interest is the apparent gain of phthiocerol dimycocerosate biosynthesis capabilities; this lipid has been shown to protect *Mycobacterium tuberculosis* when growing in a mammalian host [26], so could potentially perform the same function for *C. difficile*.

It has been established previously that *C. difficile* CD196 utilises as carbon sources N-Acetyl-glucosamine and N-Acetyl-neuraminic acid, which are both

Table 1. A summary of the metabolic functions gained by *C. difficile* since branching from the rest of the bacteria of the genus *Clostridium* represented in this analysis. Each line is an individual module (connected metabolic component) that has significantly higher scores for gains than would be expected in the whole metabolic network (at the corrected $p = 0.001$ level). The 'Metabolic Function' column represents a summary of the enzymatic functions present in the module.

AMBIENT Module ID	Number of Reactions	Metabolic Function	Corrected p-value
1	12	Methylamine metabolism	$< 1e-5$
2	15	Polyamine metabolism	$< 1e-5$
3	12	Phthiocerol dimycocerosate biosynthesis	$< 1e-5$
4	15	Salicin metabolism	$< 1e-5$
5	8	Niacin, Cob(I)alamin metabolism	$< 1e-5$
6	6	Fatty acid biosynthesis	$< 1e-5$
7	6	4-Hydroxybuanoate metabolism	$< 1e-5$
8	5	Monosaccharide metabolism	$< 1e-5$
9	5	Lipid metabolism	$3.6e-4$
10	5	Amino acid metabolism	$3.6e-4$
11	5	Monosaccharide utilisation	$2.2e-4$
12	5	D-Lactate metabolism	$3.6e-4$
13	4	D-Proline metabolism	$2.2e-4$
14	6	N-Acetyl-D-neuraminic acid utilisation	$8.6e-4$

represented in the metabolic network used here. It appears that the reactions around N-Acetyl-glucosamine are shared with the other *Clostridium* strains in this study. One of the significant modules found by **ambient** shown in Fig. 2, shows that *C. difficile* gained the ability to utilise N-Acetyl-neuraminic acid since its divergence from the other *Clostridia* in the study. The assimilation of N-Acetyl-neuraminic acid proceeds by conversion through several intermediates to Fructose-6-Phosphate, which is part of central carbon metabolism.

The reactions responsible for this interconversion, allowing *C. difficile* to utilise this carbon source, have been inferred by Model SEED to be present in this *C. difficile* strain. Some of the reactions in the model were predicted to be present without having a gene associated with them. In the case of this module two genes, nanA and CD196_2092, were associated with two of the reactions, ATPN-acyl-D-mannosamine 6-phosphotransferase and N-Acetylneuraminate pyruvate-lyase, in the module. These genes are transcribed in the same direction and have just three closely spaced same-sense genes between them, each of unknown function. This establishes the tantalising possibility that these three intervening genes could encode proteins with other functions within this coordinately gained metabolic module.

As expected from an obligate intracellular pathogen, **ambient** finds extensive coordinated losses in the *Onion yellows phytoplasma* OY-M metabolic network, with over 350 reactions lost in connected metabolic modules (as shown in Supplementary Fig. 4). The closest relatives of *Onion yellows* in this study share only the same Phylum (Firmicutes), so this represents a long period of evolutionary

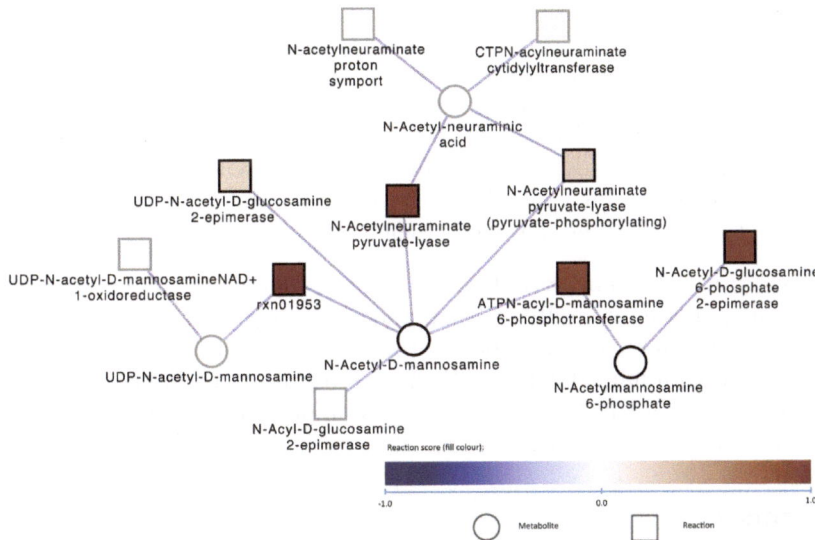

Fig. 2. Metabolic module 14 gained in *C. difficile* CD196 and its metabolic neighbour-hood, according to **ambient** analysis of the reaction gains and losses from its closest ancestor on the 13S rRNA phylogenetic tree. Members of module 14 are outlined in black and those not in the module are outlined in grey. The fill colours of the reactions correspond to δP values.

history. Nonetheless *Onion yellows* has only gained (and retained) 91 reactions in the same period, indicating a very strong bias towards metabolic function loss, as expected from the bacterium's lifestyle. By far the largest module shows the complete loss of lipid biosynthesis, as expected since *Phytoplasmas* lack a cell wall.

4   Conclusion

The ancestral state reconstruction results and metabolic traits phylogeny have been able to unpick and clarify the significant gains and losses of metabolic capabilities in various organisms during their evolutionary history. The findings have correlated well with previous biological knowledge of the lifestyles of these organisms. The hierarchical clustering of these bacteria using metabolic traits has shown that as expected metabolic evolution is far more intimately linked with current lifestyle than is bacterial ancestry.

The adaptation of bacteria to different conditions has led to a consider-able gain and/or loss of reactions over time. Considerable gain has been ob-served in *Clostridium difficile*, which is consistent with the expectations for a non-intracellular opportunistic pathogen. Considerable losses, including those of lipid biosynthesis, have been observed in *Onion yellows phytoplasma*, which is a known obligate intracellular plant pathogen which does not produce a cell wall.

The metabolic traits based hierarchical clustering has provided insight into examples of convergent evolution with respect to bacterial metabolism.

The **ambient** analysis presented here has clearly picked out some relevant and biologically meaningful metabolic modules that have been gained or lost in a coordinated fashion. This approach, combined with the multiple metabolic models produced by Model SEED, which can infer reaction presence even in the absence of known enzymes, is a powerful tool that goes beyond previous approaches to investigating metabolic evolution.

Acknowledgments. Thanks to David Hughes, Lesley Hoyles, Pakorn Aiewsakun and Ghazal A Milani for kindly collating the reaction presence/absence tables for the 141 bacteria analysed in this paper from the Model SEED website. AAF was supported by the Mohamedali Habib Welfare Trust, Karachi. WAB was supported by the BBSRC, grant BB/G020434/1. JWP is supported by a University Research Fellowship from the Royal Society.

References

1. Mithani, A., Preston, G.M., Hein, J.: A Bayesian Approach to the Evolution of Metabolic Networks on a Phylogeny. PLoS Computational Biology 6(8), e1000868 (2010)
2. Mazurie, A., Bonchev, D., Schwikowski, B.: Evolution of metabolic network organization. BMC Systems Biology 4(59) (2010)
3. Pfeiffer, T., Soyer, O.S., Bonhoeffer, S.: The evolution of connectivity in metabolic networks. PLoS Biology 3(7) (2005)
4. Pál, C., Papp, B., Lercher, M.J.: Adaptive evolution of bacterial metabolic networks by horizontal gene transfer. Nature Genetics 37(12), 1372–1375 (2005)
5. Yagi, J.M., Sims, D., Brettin, T., Bruce, D., Madsen, E.L.: The genome of Polaromonas naphthalenivorans strain CJ2, isolated from coal tar-contaminated sediment, reveals physiological and metabolic versatility and evolution through extensive horizontal gene transfer. Environmental Microbiology 11(9), 2253–2270 (2009)
6. Petridis, M., Bagdasarian, M., Waldor, M.K., Walker, E.: Horizontal transfer of Shiga toxin and antibiotic resistance genes among Escherichia coli strains in house fly (Diptera: Muscidae) gut. Journal of Medical Entomology 43(2), 288–295 (2006)
7. Zomorodipour, A., Andersson, S.G.E.: Obligate intracellular parasites: *Rickettsia prowazekii* and *Chlamydia trachomatis*. FEBS Letters 452(1), 11–15 (1999)
8. Schluter, D., Price, T., Mooers, A.Ø., Ludwig, D.: Likelihood of ancestor states in adaptive radiation. Evolution 51, 1699–1711 (1997)
9. Latysheva, N., Junker, V.L., Palmer, W.J., Codd, G.A., Barker, D.: The evolution of nitrogen fixation in cyanobacteria. Bioinformatics 28(5), 603–606 (2012)
10. Merhej, V., Royer-Carenzi, M., Pontarotti, P., Raoult, D.: Massive comparative genomic analysis reveals convergent evolution of specialized bacteria. Biology Direct 4(13) (2009)
11. Baumler, D.J., Peplinski, R.G., Reed, J.L., Glasner, J.D., Perna, N.T.: The evolution of metabolic networks of E. coli. BMC Systems Biology 5(1), 182 (2011)
12. Liao, L., Kim, S., Francois Tomb, J.: Genome comparisons based on profiles of metabolic pathways. In: Proceedings of the 6th International Conference on Knowledge-Based Intelligent Information and Engineering Systems, KES 2002, pp. 469–476 (2002)

13. Whitaker, J.W., Letunic, I., McConkey, G.A., Westhead, D.R.: metaTIGER: a metabolic evolution resource. Nucleic Acids Research 37(Database issue), D531–D538 (2009)
14. Henry, C.S., DeJongh, M., Best, A.A., Frybarger, P.M., Linsay, B., Stevens, R.L.: High-throughput generation, optimization and analysis of genome-scale metabolic models. Nature Biotechnology 28(9), 969–974 (2010)
15. Ideker, T., Ozier, O., Schwikowski, B., Siegel, A.F.: Discovering regulatory and signalling circuits in molecular interaction networks. Bioinformatics 18(suppl. 1), S233–S240 (2002)
16. Katoh, K., Asimenos, G., Toh, H.: Multiple alignment of DNA sequences with MAFFT. Methods in Molecular Biology 537, 39–64 (2009)
17. Guindon, S., Dufayard, J.F., Lefort, V., Anisimova, M., Hordijk, W., Gascuel, O.: New algorithms and methods to estimate maximum-likelihood phylogenies: assessing the performance of PhyML 3.0. Systematic Biology 59(3), 307–321 (2010)
18. Huson, D., Richter, D., Rausch, C., Dezulian, T.: Dendroscope: An interactive viewer for large phylogenetic trees. BMC Bioinformatics 8(460) (2007)
19. Zientz, E., Dandekar, T., Gross, R.: Metabolic interdependence of obligate intracellular bacteria and their insect hosts. Microbiology and Molecular Biology Reviews 68(4), 745–770 (2004)
20. Maddison, W.P., Maddison, D.R.: Mesquite: a modular system for evolutionary analysis. Version 2.75 (2011), http://mesquiteproject.org
21. Pagel, M.: The maximum likelihood approach to reconstructing ancestral character states of discrete characters on phylogenies. Systematic Biology 48(3) (1999)
22. Swofford, D.L.: PAUP*: Phylogenetic Analysis Using Parsimony (*and Other Methods). Version 4. Sinauer Associates, Sunderland, Massachusetts (2003)
23. Hubert, L., Arabie, P.: Comparing partitions. Journal of Classification 2(1), 193–218 (1985)
24. Felsenstein, J.: Phylip, http://evolution.genetics.washington.edu/phylip.html
25. Oshima, K., Kakizawa, S., Nishigawa, H., Jung, H.Y., Wei, W.: Reductive evolution suggested from the complete genome sequence of a plant-pathogenic phytoplasma. Nature Genetics 36(1), 27–29 (2003)
26. Cox, J.S., Chen, B., McNeil, M., Jacobs, W.R.: Complex lipid determines tissue-specific replication of Mycobacterium tuberculosis in mice. Nature 402(6757), 79–83 (1999)

Detecting Protein Conformational Changes in Interactions via Scaling Known Structures

Fei Guo, Shuai Cheng Li*, Wenji Ma, and Lusheng Wang*

Department of Computer Science, City University of Hong Kong,
83 Tat Chee Avenue, Kowloon, Hong Kong
{shuaicli,cswangl}@cityu.edu.hk

Abstract. Conformational changes frequently occur when proteins interact with other proteins. How to detect such changes in silico is a major problem. Existing methods for docking with conformational changes remain time-consuming, and they solve the problem only for a small portion of protein-protein complexes accurately. This work presents a more accurate method (FlexDoBi) for docking with conformational changes. FlexDoBi generates the possible conformational changes of the interface residues that transform the proteins from their unbound states to bound states. Based on the generated conformational changes, multidimensional scaling is performed to construct candidates for the bound structure. We develop the new energy items for determining the orientation of proteins and selecting of plausible conformational changes. Experimental results illustrate that FlexDoBi achieves better results than other methods for the same purpose. On 20 complexes, we obtained an average iRMSD of 1.55Å, which compares favorably with the average iRMSD of 1.94Å in the predictions from FiberDock. Compared with ZDOCK, our results are of 0.35Å less in average iRMSD on the medium difficulty group, and 0.81Å less on the difficulty group.

Keywords: Flexible Docking, Backbone Flexibility, Database Method, Weighted Multi-Dimensional Scaling, Energy Function.

1 Introduction

Many proteins realize their biological functions through interacting with other proteins to form complexes. In forming a complex, the protein structures involved frequently undergo conformational changes. Modeling and detecting these conformational changes in docking problems is a challenging task, and is a topic under active research, since a solution to the problem will help to remove bottlenecks in various biological studies.

Protein docking is the task of calculating the three dimensional structure of a complex starting from the individual structures of proteins. There are many techniques for predicting protein-protein docking configurations. Broadly, they can be grouped into two categories. The first we call *rigid molecule docking*

---

* Corresponding authors.

M. Deng et al. (Eds.): RECOMB 2013, LNBI 7821, pp. 58–74, 2013.

methods. They work by sampling the effective positions and orientations of a
rigid-body protein around another one. Among these, methods based on fast
Fourier transformation [4,9], geometric surface matching [21], as well as inter-
molecular energy [7,5,1] have been proposed. In addition, other existing methods
to identify the interface residues are based on analyzing the differences between
interface residues and non-interface residues in known complexes, often through
the use of statistical techniques [20,2] and 3D structural algorithms [23,13].

The second category of docking techniques is the *flexible molecule docking*
methods. These methods work by changing the backbone and/or side-chain conformations to refine flexible structures of complexes. The flexible docking methods can be divided into three groups according to their treatment of structural
flexibility. The first group, including FiberDock and RosettaDock, searches for
energetically favored conformations in a wide conformational search space. Fiber-
Dock [19] combines a novel normal mode analysis (NMA)-based backbone refinement with side-chain optimization and rigid-body minimization. It minimizes
the backbone conformation along a few degrees of freedom, which are carefully
picked by NMA. The side-chain flexibility of interface residues is modeled by
a rotamer library. After refining all docking solutions, the predicted structures
are ranked according to an energy function. RosettaDock [18] is a Monte Carlo-
based docking method. It optimizes both rigid-body orientation and side-chain
conformation via rotamer packing. RosettaDock refines the flexible backbone by
minimizing the energy functions via varying the backbone torsional angles. The
second group deals with hinge bending motions in the docked molecules, such
as FlexDock [22]. It first detects hinge regions, rigid parts and motion directions
in the flexible structure. Then, each rigid part of the flexible molecule is docked
with the rigid molecule, and the directions generate more conformations of the
flexible molecule. Finally, all the partial docking solutions are assembled with
good shape complementarity, and the top scoring ones are selected. The last one,
HADDOCK [5], is an experimental data-driven method by using the biochemical and biophysical interaction data, such as chemical shift perturbation data
resulting from NMR titration experiments, mutagenesis data or bioinformatics
predictions. This information is introduced as Ambiguous Interaction Restraints
(AIRs) to drive the docking process. An AIR is defined as an ambiguous distance
between all residues shown to be involved in the interaction. The method uses
simulated annealing in torsion angle space to refine the structure, allowing for
both backbone and side-chain flexibility on the interface. The final structures
are clustered and ranked according to their average interaction energies.

In this paper, we present a more accurate method, FlexDoBi, for docking with
conformational changes. We develop an approach to detect the conformational
changes from unbound states to bound states. Our approach examines a set of
scaled structures as candidates for the bound structure (possibly with conformational changes), and uses a new energy function to select the best solutions.

To obtain the set of scaled structures, we maintain a database of structures,
from which raw candidates for the conformationally changed residues can be
rapidly selected. These candidates are then refined through an efficient method

based on multi-dimensional scaling. This allows accurate near-native structures to be constructed with very minimal number of sampling steps. One advantage in this approach is that, whereas the large search space of existing methods requires intensive computational power and produces a large portion of conformations different from the native complex, in our method the geometrical constraints —imposed by the distance between two residues respectively at both ends of an interface fragment— eliminate a substantial number of unlikely candidate structures. One caveat is that for our method to work, the regions far from the interface should be almost unchanged in the protein complex.

The energy function used in FlexDoBi for structural evaluation is carefully constructed, since the effectiveness of the function is a crucial factor in determining the resultant structure. In this work we developed a new statistical energy item, which is combined linearly with four other energy items to rank the poses from the first step, and to direct the search of the plausible conformations in the second step.

Experimental results show that FlexDoBi achieves better results than other methods for the same purpose. On 20 complexes, we obtained an average iRMSD of 1.55Å, which compares favorably with the average iRMSD of 1.94Å in the predictions from FiberDock. Compared with ZDOCK, our results are of 0.35Å less in average iRMSD on the medium difficulty group, and 0.81Å less on the difficulty group.

2   Method Overview

Our method for the flexible docking problem contains two steps. In the first step, we find the relative orientation and position between two subunits. That is, we determine roughly where the two subunits bind. Each relative orientation and position combination is referred to as a *configuration* or *pose*. Once a pose is given, we can determine the interface region between two subunits, and fix the orientation as well as position of the regions far from the interface. In the second step, we use an efficient way to compute the (possibly changed) conformation of the interface. Here our method examines only thousands of structure candidates for the bound conformation of the interface, which is significantly less than existing methods.

To perform the first step, we modify P-Binder [8], a tool we have developed recently. P-Binder utilizes an enumeration method to identify the docking configurations of two subunits. It first performs a large number of rigid transformations to enumerate the poses. For each configuration, the side-chain conformation on the interface is built for energy evaluation. The problem of modeling side-chain is a well-studied one [25,3,14], and we use SCWRL4 [14] for this purpose. Side-chain conformations are packed on the structures at this stage and are modified in the second step. The poses are evaluated through a linear combination of five energy items, one of which is newly developed in this paper. The top ranking poses are selected for the second step processing.

A   B

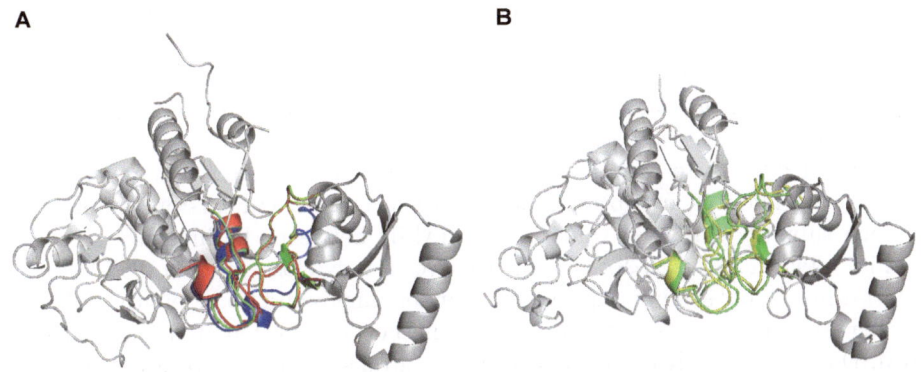

Fig. 1. The refinement of the case 2z0e(A:B). (A) The unbound structure of interface is colored in green, and some fragment candidates selected by FlexDoBi are colored in blue and red. (B) The unbound structure of interface is colored in green, and the refined structure created by FlexDoBi is colored in yellow.

In the second step, we assume that only the interface region in a given configuration of the unbound structures will experience conformational changes. Hence to obtain a near-native structure of complex, one only needs to modify the residues in the interface region. Our strategy is to replace each fragment formed by the consecutive residues in the interface region with some similar fragments. A residue is to be replaced if any of its atom is within 10Å to any atoms in the partner subunits. In each subunit, four or more consecutive residues to be replaced form a *replaceable fragment*. A database of known structure fragments is maintained to search for suitable replacement candidate structures. Referring to the pair of residues respectively at both ends of a fragment as *stems*, we use the following two measures in our selection of candidate structures: (1) the Root Mean Square Deviation (RMSD) of the heavy backbone atoms in the stems, and (2) the sequence similarity between the replaceable fragment and the candidate.

Some processing is required in replacing the fragments, since selected fragment candidates may result in unreasonable bond lengths, bond angles and even collisions in the protein structure. Hence, in our structural modification, we scale all fragment candidates to reduce these inconsistencies. This is formulated as a Weighted Multi-Dimensional Scaling (WMDS) [15] problem, and solved by using a heuristic method, which aims to reduce the unreasonable bond length on the interface as well as remove most of the clashes between pairs of subunits in complex.

Each docking orientation and position is to be evaluated by a new energy function. This energy function is a combination of the following energy items: side-chain energy [14], dDFIRE energy function [26]), Atomic Contact Energy [28,27], Secondary Structure Energy (our newly developed energy item), and the Gromacs force field [16]. We use a trained SVM model to rank the docking solutions and report best ones with the lowest energy values.

Throughout this paper, a complex may contain several subunits and multiple binding interfaces. Each binding interface in a complex occurs in a pair of subunits. Two residues in a pair of subunits are called *interface residues* if any two atoms, one from each residue, interact. By interact, we mean the distance between two atoms is less than 6Å.

Figure 1 depicts an example of our result. In (A) we give a case where many fragment candidates are obtained for the replaceable fragments on the interface of each subunit. The best value of $C\alpha$ RMSD between the interface structure and its replacement is 2.78Å; FlexDoBi gives a very close value of 2.57Å. In (B) multi-dimensional scaling improves the torsion angles and bond lengths, allowing for better candidate ranking. We select the best prediction structure with the iRMSD of 2.21Å(the RMSD between the C_α atoms of interface residues) between the best refined structure and the bound complex.

3   Results

To evaluate our method, we have done three kinds of experiments. Recall that our method replaces the fragment formed by the consecutive residues in each interface region with similar fragments in a database. To test the feasibility of our method, we show that for each native fragment on the interface of the bound subunit, there are some similar fragments in the database. The second kind of experiments is designed to test the accuracy of the second step of our method, i.e., the ability of identifying the conformational changes from unbound state to bound state. The idea is to use native bound complex to fix the pose of the regions far from the interface and use our approach to compute the conformation change of interface (See Section 3.2). In Section 3.3, we compare our method (FlexDoBi) with FiberDock [19], which also assumes that the pose of subunits is given. Finally, to test the ability of finding the pose and the conformational change, we compare our method with ZDOCK [4].

3.1   Similarity between Native Interface and Selected Candidates

Observations of protein complexes show that for many complexes, the major structural changes between the bound and unbound states occur on the interface regions. Our sample data set is extracted from the medium difficulty group in protein-protein docking Benchmark 4.0 [11], which contains 29 complexes. We calculate the C_α RMSD values on the whole structures and on the interface residues within distance 10Å. The average C_α RMSD value between the complex structures and the unbound proteins in native binding orientation is 1.32Å. However, the average RMSD between the interface residues of these two states is 2.54Å. These details are shown in Figure 2. Clearly, the interfaces are more flexible than the rest of the structures. This justifies our method for transforming an unbound structure into its bound state by substituting only the fragments in interface.

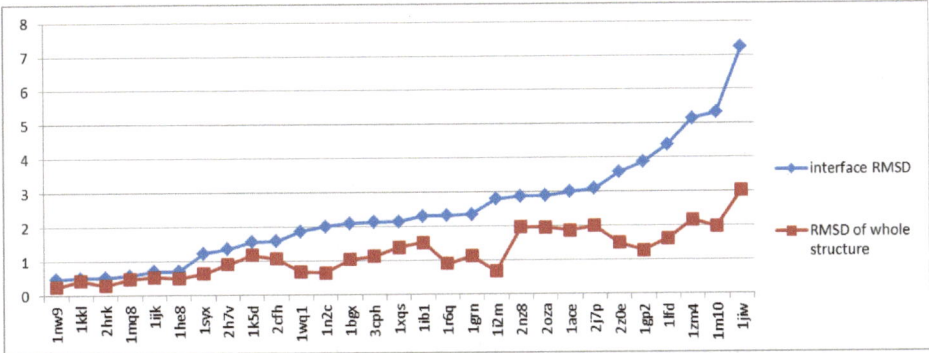

Fig. 2. The C_α RMSD between the complex structures and the unbound proteins in native binding orientation: interface RMSD (blue) and RMSD for the whole structure (red)

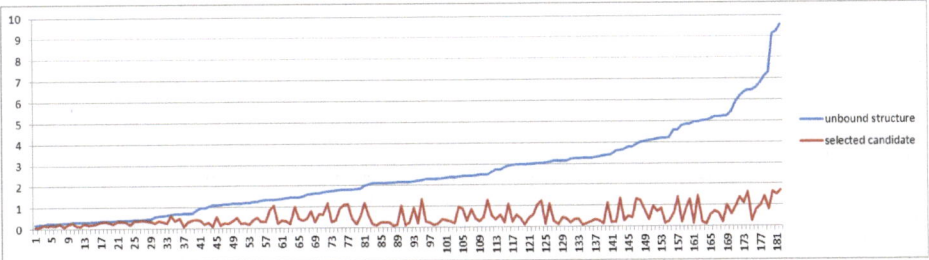

Fig. 3. The C_α RMSD between the interface fragments on bound conformations and unbound structures (blue) or best candidates selected by FlexDoBi (red)

Suitable replacement fragment candidates are selected from a database. We use a database comprising roughly 13255 protein chains, selected by using PISCES [24] with cutoff values being 90 percent identity, 2.0 Angstrom resolution, and 0.25 R-value. Fragment candidates are selected from this database without the homologous proteins. We find that the homologous candidates appear in the fragment candidates for 21 complexes among 29 complexes, and filter out those fragments to make a fair assessment of our method. Among the medium difficulty complexes, 184 replaceable fragments are extracted from the interfaces of bound states. We search the candidates for the bound state of the replaceable fragment. As shown in Figure 3, for all the fragments, the best candidates are found within 1.87Å.

3.2 Conformational Changes of Native Poses

In this experiment, we verify that suitable fragment candidates can be identified from the database, and reshaped properly for interface fragments. We assume that the native pose is given, and two subunits are unbound. Now the task is to

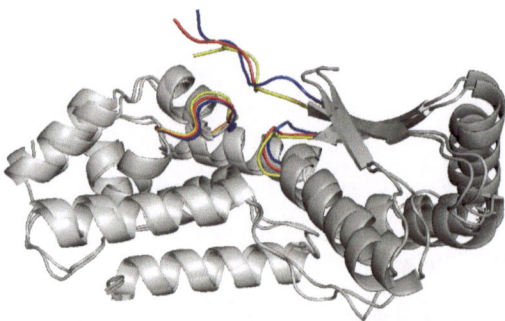

Fig. 4. The refinement of the case 1r6q(A:C). The unbound structure of interface is colored in yellow and the bound structure is in blue. The refined structure, created by FlexDoBi, is in red.

Table 1. Refinement of the unbound conformations in their native binding orientations

Complex ID	Unbound Receptor[a]	Unbound Ligand[a]	FlexDoBi		Unbound[b]	
			iRMSD[c]	Energy Value[d]	iRMSD	Energy Value
1bgx(HL:T)	1ay1HL	1cmwA	1.97	-287.02	2.10	-233.53
1acb(E:I)	2cgaB	1egl_	2.63	-282.94	2.79	-229.68
1ijk(A:BC)	1auq_	1fvuAB	0.68	-120.15	0.70	-116.70
1jiw(P:I)	1aklA	2rn4A	6.82	-369.11	7.23	-165.38
1kkl(ABC:H)	1jb1AB	2hpr_	0.48	-112.72	0.51	-117.16
1m10(A:B)	1auq_	1m0zB	4.56	-239.47	5.32	-107.04
1nw9(B:A)	1jxqA	2opyA	0.40	-359.64	0.47	-300.06
1gp2(A:BG)	1gia_	1tbgDH	3.76	-236.95	3.86	-160.61
1grn(A:B)	1a4rA	1rgp_	2.44	-366.41	2.35	-228.89
1he8(B:A)	821p_	1e8zA	0.52	-242.02	0.70	-185.29
1i2m(A:B)	1qg4A	1a12A	2.59	-410.51	2.80	-367.41
1ibl(AB:E)	1qjbAB	1kuyA	1.80	-298.37	2.30	-271.32
1k5d(AB:C)	1rrpAB	1yrgB	1.49	-378.72	1.57	-214.85
1lfd(A:B)	5p21A	1lxdA	4.21	-203.56	4.38	-144.88
1mq8(A:B)	1iamA	1mq9A	0.55	-127.40	0.58	-85.25
1n2c(ABCD:EF)	3minABCD	2nipAB	1.68	-234.86	2.01	-169.06
1r6q(A:C)	1r6cX	2w9rA	1.70	-256.47	2.32	-186.50
1syx(A:B)	1qgvA	1l2zA	1.10	-203.26	1.24	-76.83
1wq1(R:G)	6q21D	1wer_	1.61	-379.54	1.87	-328.65
1xqs(A:C)	1xqrA	1s3xA	2.13	-363.10	2.15	-278.70
1zm4(A:B)	1n0vC	1xk9A	5.03	-278.57	5.15	-180.03
2cfh(A:C)	1sz7A	2bjnA	1.49	-298.42	1.59	-248.20
2h7v(A:C)	1mh1_	2h7oA	1.12	-263.02	1.36	-208.97
2hrk(A:B)	2hraA	2hqtA	0.76	-241.75	0.52	-237.32
2j7p(A:D)	1ng1A	2iylD	3.15	-491.82	3.09	-393.70
2nz8(A:B)	1mh1_	1ntyA	2.67	-383.72	2.88	-312.00
2oza(B:A)	3hecA	3fykX	2.69	-549.15	2.89	-221.08
2z0e(A:B)	2d1iA	1v49A	2.21	-343.12	3.57	-229.21
3cph(A:G)	3cpiG	1g16A	2.07	-303.29	2.13	-309.73

[a] unbound structure of receptor or ligand in the complex.
[b] unbound structure is superimposed on the bound conformation by the orientation of lowest $C\alpha$ RMSD for the whole structure.
[c] C_α RMSD between the interface in the predicted structure and in the native complex.
[d] energy value for the prediction complex.

transform the unbound subunits into bound states. To obtain the native pose, the unbound structure is superimposed on the native bound complex by the orientation of lowest C_α RMSD for the whole structure. The value of $iRMSD$ is to denote the RMSD between the C_α atoms of interface in the predicted structure and in the native complex after superimposing the interfaces.

The medium difficulty group in Benchmark 4.0 is used for this study. Details are in Table 1. Among the 29 instances, we identify better conformations for 22; that is, FlexDoBi discovers better conformations of the interfaces than simply putting two unbound subunits together. The iRMSD value becomes worse for three instances, and are similar in four instances; by similar, we mean the difference between the iRMSD of the prediction structures and that of the unbound ones is less than 0.05Å. The average C_α iRMSD value between the interface predicted by FlexDoBi and the corresponding portion of the native complex is 2.29Å. Yet, the average iRMSD value between the interface of unbound structures and that of the native complex is 2.51Å.

The best instances, predicted by FlexDoBi, are 1m10(A:B), 1r6q(A:C) and 2z0e(A:B), where the values of $C\alpha$ iRMSD are reduced by 0.7Å, 0.6Å and 1.3Å, respectively. Figure 4 displays the conformation discovered by FlexDoBi for 1r6q(A:C). FlexDoBi predicts the interface conformation with 1.70Å iRMSD, however, the value of iRMSD for the unbound structures on the native orientation is 2.32Å. The energy of the conformation predicted by FlexDoBi, -256.47, is lower than the initial energy of the unbound structure, -186.50. We should notice that lower energy does not always imply better conformation in terms of iRMSD.

3.3 Comparison with FiberDock

In this subsection, we compare the results of FlexDoBi with FiberDock [19]. FiberDock is a novel NMA-based backbone flexibility treatment, which refines

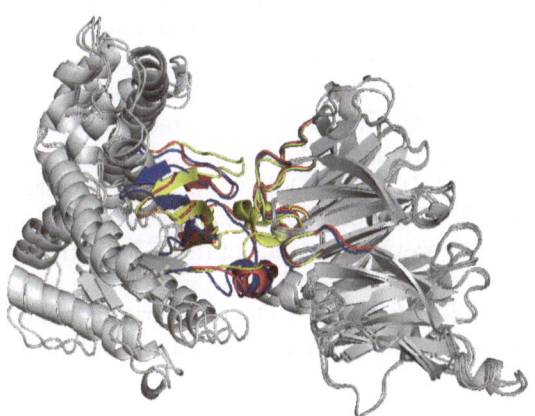

Fig. 5. The refinement of the case 1got(A:B). The unbound structure of interface is colored in yellow and the bound structure is in blue. The refined structure, created by FlexDoBi, is in red.

Table 2. Local docking results of FlexDoBi and FiberDock

Complex ID	Unbound Receptor[a]	Unbound Ligand[a]	FlexDoBi		FiberDock		Unbound[b]
			iRMSD[c]	rec-iRMSD[d]	iRMSD	rec-iRMSD	
1a0o(A:B)	1chn_	1fwpA	3.19	3.68	2.44	2.12	3.27
1acb(E:I)	2cgaB	1egl_	2.63	2.85	2.58	2.54	2.79
1ay7(A:B)	1rghB	1a19B	0.47	0.40	1.30	0.59	0.43
1bth(H:P)	2hnt_	6ptiA	1.34	1.67	1.16	1.31	1.49
1cgi(E:I)	2cgaB	1hpt_	2.09	2.28	2.08	2.26	2.53
1dfj(E:I)	9rsa_	2bnh_	0.56	0.53	1.12	1.11	0.56
1e6e(A:B)	1e1nA	1cjeD	0.64	0.84	1.21	0.62	0.73
1fin(A:B)	1hcl_	1vin_	5.47	7.47	6.06	6.16	5.17
1ggi(L:H)	1ggcL	1cgiH	0.66	1.08	1.95	1.26	0.71
1got(A:B)	1tag_	1tbgA	0.92	1.35	4.68	3.78	3.62
1ibr(A:B)	1qg4A	1f59A	2.37	1.27	2.63	2.56	2.53
1oaz(H:L)	1oaqH	1oazL	0.75	0.50	1.00	1.07	0.70
1pxv(A:C)	1x9yA	1nycA	2.90	3.79	3.42	3.31	3.85
1t6g(C:A)	1ukr_	1t6e_	0.48	0.37	0.88	0.66	1.10
1tgs(Z:I)	2ptn_	1hpt_	0.64	0.56	1.57	1.54	1.38
1wq1(R:G)	6q21D	1wer_	1.61	1.79	1.50	0.93	1.87
1zhi(A:B)	1m4zA	1z1aA	0.75	1.10	1.24	0.74	0.94
2buo(A:T)	1a43_	2buoT	1.24	0.54	4.05	4.30	1.96
2kai(A:I)	2pka_	6pti_	0.38	0.34	0.74	0.72	0.31
3hhr(A:B)	1hgu_	3hhrB	2.93	3.56	1.98	2.56	2.94

[a] unbound structure of receptor or ligand in the complex.
[b] unbound structure is superimposed on the bound conformation by the orientation of lowest C_α RMSD for the whole structure.
[c] C_α RMSD between the interface in the predicted structure and in the native complex.
[d] C_α RMSD between the interface in the predicted structure of receptor and in its bound conformation.

the structure of complex from a given docking configuration. We evaluate the performance of two methods by using the unbound native pose. The data set is extracted from FiberDock's paper. We obtain much better result than that of FiberDock. The comparison result is detailed in Table 2. Among 20 instances, FlexDoBi produces better results for 14 cases. By better, we mean that the iRMSD value is at least 0.05Å smaller than the iRMSD of FiberDock method. Only for four instances, FiberDock produces better results. The average values of C_α iRMSD between the predicted structures and the native complexes are 1.55Å (FlexDoBi) and 1.94Å (FiberDock), respectively.

Rec-iRMSD is to denote the iRMSD value of receptor, which is the subunit of more residues. The average values of Rec-iRMSD between the predictions and the bound conformations are 1.71Å (FlexDoBi) and 2.01Å (FiberDock), respectively. In case of 1got(A:B), FlexDoBi predicts new interface conformation in complex with 0.92Å C_α iRMSD, however, the value of iRMSD for the unbound structures on the native orientation is 3.62Å. Figure 5 displays the docking configuration discovered by FlexDoBi for 1got(A:B). The comparisons indicate that FlexDoBi produces better interface conformations while changing the unbound states into bound states.

3.4   Evaluation on Benchmark v4.0

In this study, we assume that the native pose is unknown. We perform a search which finds both pose as well as identifies the structural changes. For each complex,

Table 3. Docking results of FlexDoBi, ZDOCK and FiberDock on medium difficulty group

Complex	FlexDoBi[a]	ZDOCK[a]	FiberDock	Complex	FlexDoBi	ZDOCK	FiberDock	Complex	FlexDoBi	ZDOCK	FiberDock
1bgx	9.83	11.90	9.82	1i2m	4.24	2.21	2.96	1zm4	5.96	2.44	3.67
1ace	5.46	2.61	3.16	1ib1	7.19	5.89	7.60	2cfh	4.18	1.53	1.69
1ijk	4.17	1.86	1.22	1k5d	2.16	2.51	4.94	2h7v	4.02	2.64	2.36
1jiw	7.49	8.22	5.93	1lfd	6.21	4.94	4.04	2hrk	2.35	2.06	1.56
1kkl	3.18	27.92	23.55	1mq8	2.99	6.72	8.19	2j7p	4.86	6.89	8.65
1m10	5.87	9.42	6.26	1n2c	6.75	3.21	3.51	2nz8	5.17	2.87	1.81
1nw9	3.49	3.19	5.06	1r6q	3.93	5.20	4.38	2oza	4.89	8.49	8.69
1gp2	4.18	3.39	1.70	1syx	6.97	4.81	2.04	2z0e	3.64	4.24	6.87
1grn	3.49	1.81	2.31	1wq1	2.06	1.82	2.64	3cph	3.27	3.91	4.16
1he8	4.76	2.38	2.34	1xqs	2.76	2.67	2.25				

[a] C_α iRMSD between the predicted configuration by each method and the native complex.

Table 4. Docking results of FlexDoBi, ZDOCK and FiberDock on difficulty group

Complex	FlexDoBi[a]	ZDOCK[a]	FiberDock	Complex	FlexDoBi	ZDOCK	FiberDock	Complex	FlexDoBi	ZDOCK	FiberDock
1e4k	9.42	15.20	8.71	1bkd	7.04	7.33	6.38	1jmo	11.01	15.99	10.52
2hmi	6.14	16.99	13.46	1de4	6.76	1.77	1.49	1jzd	7.92	16.70	11.59
1f6m	5.76	12.24	12.33	1eer	7.49	7.90	5.40	1r8s	6.23	6.48	6.86
1fq1	5.54	8.05	7.61	1fak	6.73	7.73	7.44	1y64	6.42	14.37	15.31
1pxv	5.17	3.81	3.82	1h1v	16.13	16.72	14.53	2c0l	5.14	4.36	5.05
1zli	6.97	12.25	9.86	1ibr	8.23	9.83	8.86	2i9b	4.18	5.58	4.75
2o3b	9.15	14.16	9.37	1ira	20.13	16.42	12.48	2ido	5.48	5.09	3.42
1atn	4.70	4.74	4.27	1jk9	5.69	2.16	2.77	2ot3	9.11	4.40	3.25

[a] C_α iRMSD between the predicted configuration by each method and the native complex.

we adopt a similar procedure as in P-Binder to predict the poses. The top 100 poses according to our new energy function are chosen and are fed into our method for modeling conformational changes. The top ten results from the method according to energy value are reported. These are finally compared with the docking results from ZDOCK [4] and the flexible docking solutions from FiberDock.

We calculate the medium difficulty group and the difficulty group in Benchmark v4.0. The values of C_α iRMSD between the unbound structures in the native poses and the native complexes range from 1.48Å to 16.76Å. Several proteins in difficulty group undergo significant conformational changes upon binding. The results are presented in Table 3 and Table 4. For 29 complexes in medium difficulty group, the average C_α iRMSD values between the predictions and the native complexes are 4.61Å (FlexDoBi), 4.96Å (ZDOCK) and 4.88Å (FiberDock), respectively. For 24 complexes in difficulty group, the average C_α iRMSD values between the predictions and the native complexes are 7.78Å (FlexDoBi), 8.59Å (ZDOCK) and 7.84Å (FiberDock), respectively.

In several unbound subunits, the coordinates of some backbone atoms are missing. We add the coordinates of the missing residues by using MODELLER [6]. MODELLER is a tool for homology or comparative modeling of protein three-dimensional structures. In two groups, the missing residues appear in the unbound structures of four complexes: residues 36-43 in 1fq1B, residues 206-215 in 1grnB, residues 72-94 in 1jmoA, and residues 46-58 in 3cphA. After the gaps are filled in, the accuracy of the predictions is improved. The docking configuration discovered by FlexDoBi for 3cph(A:G), after the gap is filled, is displayed in Figure 6. The complex predicted by FlexDoBi has an iRMSD of 3.27Å C_α, which is better than the iRMSD of 3.91Å from ZDOCK.

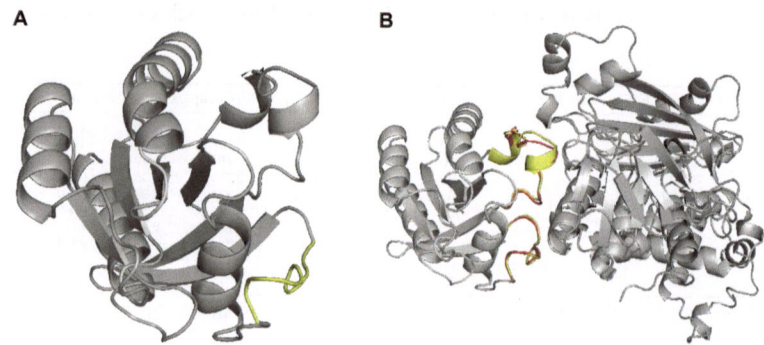

Fig. 6. The refinement of the case 3cph(A:G). (A) The missing residues in unbound structure of 3cphA are filled by MODELLER (yellow). (B) The unbound structure of interface is colored in yellow. The refined structure, created by FlexDoBi, is in red.

Table 5. Docking results of FlexDoBi and ZDOCK on the rigid-body group

Complex	FlexDoBi[a]	ZDOCK	Complex	FlexDoBi	ZDOCK	Complex	FlexDoBi	ZDOCK	Complex	FlexDoBi	ZDOCK
1ahw	2.51	8.91	1ewy	1.18	2.54	7cei	1.64	0.95	1pvh	2.37	6.59
1bvk	1.58	3.73	1ezu	2.75	3.28	1a2k	18.60	3.81	1qa9	2.23	12.15
1dqj	3.62	5.60	1f34	5.41	10.59	1ak4	16.90	5.89	1rlb	14.50	1.68
1e6j	1.45	1.71	1fle	1.78	2.67	1akj	5.89	14.89	1rv6	10.55	1.60
1jps	16.36	7.88	1gl1	2.13	1.46	1azs	1.06	1.18	1s1q	16.10	7.10
1mlc	8.48	1.54	1gxd	6.97	10.01	1b6c	1.97	2.30	1sbb	3.54	8.77
1vfb	2.61	4.10	1hia	1.26	4.25	1buh	15.92	1.53	1t6b	3.90	10.27
1wej	1.32	1.16	1jtg	2.30	1.33	1e96	2.47	3.20	1us7	1.03	3.58
2fd6	0.82	2.04	1mah	0.69	1.02	1efn	3.95	5.94	1wdw	2.18	1.54
2i25	1.49	1.74	1n8o	1.12	1.27	1f51	2.98	1.13	1xd3	1.68	1.90
2vis	17.31	7.71	1oc0	1.48	3.20	1fc2	3.10	11.44	1xu1	18.83	2.92
1bj1	1.20	1.07	1oph	2.33	4.16	1fcc	2.56	10.97	1z0k	2.62	1.94
1fsk	0.67	1.11	1oyv[c]	2.23	1.30	1ffw	1.81	3.50	1z5y	1.99	1.78
1i9r	1.26	2.28	1oyv[c]	2.89	1.68	1fqj	4.03	9.75	1zhh	15.70	14.96
1iqd	0.78	0.79	1ppe	1.54	0.77	1gcq	12.12	8.03	1zhi	1.83	1.72
1k4c	2.50	4.90	1r0r	2.77	6.29	1ghq	6.46	12.40	2a5t	3.19	7.68
1kxq	0.90	1.16	1tmq	1.15	1.78	1gla	5.63	4.11	2a9k	1.02	8.72
1nca	1.38	1.04	1udi	1.09	1.46	1gpw	2.40	1.41	2ajf	4.29	3.26
1nsn	3.37	5.41	1yvb	2.60	1.07	1h9d	1.62	4.05	2ayo	4.89	1.89
1qfw[b]	2.76	14.24	2abz	3.18	5.94	1hcf	17.80	2.42	2b4j	4.13	5.86
1qfw[b]	10.30	10.12	2b42	1.05	1.07	1he1	1.21	2.30	2btf	1.25	6.62
2jel	1.19	1.53	2j0t	1.46	3.26	1i4d	1.98	1.96	2fju	2.47	5.81
1avx	0.65	1.48	2mta	2.26	2.48	1j2j	1.76	2.18	2g77	2.38	2.44
1ay7	1.75	4.17	2o8v	1.65	3.66	1jwh	17.52	1.90	2hle	2.02	2.58
1bvn	1.16	1.39	2oul	0.81	1.24	1k74	0.75	2.30	2hqs	1.18	8.59
1cgi	3.19	2.27	2pcc	14.13	3.45	1kac	2.10	6.82	2oob	4.98	7.94
1clv	1.11	1.38	2sic	1.53	0.64	1klu	2.97	6.77	2oor	3.68	6.90
1d6r	2.13	5.42	2sni	0.61	1.91	1ktz	3.51	7.06	2vdb	3.64	5.68
1dfj	1.11	1.37	2uuy	2.86	3.74	1kxp	1.60	1.92	3bp8	4.50	8.84
1e6e	3.12	1.42	3sgq	1.38	2.60	1ml0	0.88	1.23	3d5s	1.70	1.73
1eaw	1.60	1.49	4cpa	2.03	2.39	1ofu	7.29	1.89			

[a] C_α iRMSD between the predicted configuration by each method and the native complex.
[b] The first complex is 1qfw(HL:AB), and the second complex is 1qfw(IM:AB).
[c] The first complex is 1oyv(B:I), and the second complex is 1oyv(A:I).

We also compare our method with ZDOCK on the rigid-body group in Benchmark v4.0. The values of C_α iRMSD between the unbound structures in the native poses and the native complexes range from 0.24Å to 2.02Å. The results are presented in Table 5. For 123 complexes in rigid-body group, the average C_α iRMSD values between the predictions and the native complexes are 3.96Å (FlexDoBi) and 4.15Å (ZDOCK), respectively.

4 Method Details

4.1 Selecting Candidates from Database

We examine the known protein structures, and identify suitable candidates to replace each fragment on the interface. We use a database comprising roughly 13255 protein chains, selected by using PISCES [24] with cutoff values 90 percent identity, 2.0 Angstrom resolution, and 0.25 R-value (Sept. of 2012). Fragment candidates are selected from this database without the homologous proteins. We look for the fragment candidates whose stems are similar to the stems of replaceable fragment; by similar, we mean the value of RMSD between the stems on the fragment candidate and the replaceable fragment is less than 3Å. Once fragment candidates are obtained, we take the top 50 fragments according to sequence similarity as the matching candidates. We apply the BLOSUM matrix on each pair of the replaceable fragment and the candidate with the same number of residues in database, and select the most similar fragment candidates close to the replaceable fragments.

4.2 Fitting Candidates on Replaceable Fragment

We cannot replace the fragment by the candidates directly, as it will result in unrealistic atomic distances and clashes. We scale the candidates to resolve those issues.

We formulate this structure problem as an instance of weighted multi-dimensional scaling (WMDS). For a given d dimension and n points of data, we have a distance matrix D and a weighted matrix W, both symmetric $n \times n$ matrixes, and wish to find $X = x_1, x_2, ..., x_n$ where x_i is a coordinate in d dimension, such that we minimize the stress, defined as $\delta(X) = \sum_{0 < i < j \leq n} W_{i,j}(||x_i - x_j|| - D_{i,j})^2$. WMDS can be used to turn high dimensional data into 2 or 3 dimensional data suitable for graphing. It has also been used in LoopWeaver [10] for modeling loop structures, and MUFOLD [29] for assembling protein fragments.

For our problem instances, $d = 3$ and n is the total number of backbone atoms in all replaceable fragments and the stems on the interface of two subunits A and B. We define the distance matrix D as

$$d_{i,j} = \begin{cases} ||t_i - t_j|| & i,j \in stem \\ ||c_i - c_j|| & otherwise \end{cases}$$

where $T = t_1, t_2, ..., t_n$ is the set of atomic coordinates in the protein, and $C = c_1, c_2, ..., c_n$ is the set of atomic coordinates in the candidate structure, using the same numbering system. In the candidate structure, we choose one of the matching candidates instead of each replaceable fragment.

The weighted matrix is defined as

$$
w_{i,j} = \begin{cases}
1000 & i,j \in stem \\
T(i \mod 4, j - i) & i,j \in f_s, j - i \leq 4 \\
(\min\{d_{i,j}, r - \Phi d_{i,j}\})^{-2} & i,j \in f_s, j - i > 4 \\
d_{i,j}^{-2} & i,j \in f_d \\
0 & otherwise
\end{cases}
$$

where $0 \leq i < j \leq n$, T is a 4×4 lookup table as defined in LoopWeaver [10], r is the largest pairwise distance between any two atoms in the corresponding matching candidate, and Φ is the golden ratio conjugate.

The restriction of $i, j \in f_s$ means that two atoms i and j belong to the same fragment. The restriction of $i, j \in f_d$ means that two atoms i and j belong to two different fragments, and they must satisfy one of the following requirements: (1) two fragments, one from each subunit, interact with each other; (2) two fragments, both from one subunit, interact with the same fragment of another subunit. The weight between atoms of the same fragment is the same as that defined in LoopWeaver. For two different fragments, the interacting residues and the surrounding regions must move relatively to each other, while having a minor effect on the contribution to the stress function. We set the weight to $d_{i,j}^{-2}$ when atoms i and j belong to different fragments, because pairs of closer atoms are more meaningful than pairs of relatively farther atoms when refining the conformation structures.

We use the SMACOF algorithm [15] for solving the WMDS problem. This algorithm works by minimizing the stress function, yielding a fast, deterministic heuristic. By performing the iterative generation, the quality of interface refinement often gets better, and the unrealistic atomic distances are eliminated in the candidate structure.

Searching Best Conformations. Given a pair of subunits, we extract several pairs of replaceable fragments. For each fragment, at most 50 candidates are chosen. Then we replace the fragments by the corresponding candidates randomly. If a better conformation according to energy function is found, we keep it. Otherwise we try to replace a fragment by other candidates. This process is repeated until there are no improvements. We repeat this to generate multiple structure candidates. We use SCWRL4 to build the side-chain conformation of these structure candidates, and evaluate them by the dDFIRE energy function.

4.3   Energy Items

Our method will generate a large number of structure candidates. Here we develop a new energy function to select the best structures. Our energy function contains the following energy items:

(1) The side-chain atoms of interface residues are packed by SCWRL4 [14] and the corresponding energy item is extracted.

(2) The dDFIRE energy is an all-atom statistical function [26], based on the atom distance and three orientation angles involved in dipole-dipole interactions.

(3) The item of Atomic Contact Energy is produced by an atomic energy measure in [28,27]. The free energy for a pair of interacting atoms has been calculated on atom-pairing frequencies in known complexes.

(4) We use DSSP [12] to determine the type of secondary structure for each residue, and construct the item of *Secondary Structure Energy* by using the statistical method in [8]. The improvement is that we consider three types of secondary structure and 20 types of amino acid, and one solvent contacting the residues in protein surfaces. The Secondary Structure Energy item takes 60×60 possible residue pairs, obtained from the statistical analysis of residue-pairing frequencies in a complex database. We select roughly 6323 complexes from PDB database, and these complexes are made up of two or more protein subunits. Their structures are determined by X-ray with cutoff values being resolution 2.2, 30% identity (Sept. of 2012). We calculate the free energy for all pairs of interacting residues in candidate structures.

(5) The Gromacs force field is built up from two distinct subunits to describe the interaction between their atoms [16]. Gromacs calculates electrostatic interactions in the standard coulomb potential as

$$F(r_{ij}) = f \frac{q_i q_j}{\varepsilon_r r_{ij}^2} \hat{r}_{ij}$$

where $\hat{r}_{ij}$ is the unit vector, parallel with the line from charge q_i to charge q_j, and $r_{ij} = r_j - r_i$; $f = \frac{1}{4\pi\varepsilon_0} = 138.9$, and ε_r is the relative dielectric constant in Gromacs.

The energy items are used at three places. First, we use a linear combination of these energy items to rank the poses from step one. Second, we use the same linear function to direct the search for finding the plausible conformations. The coefficient of each item is optimized by using the linear combination method in [8]. Finally, we use a trained SVM model to rank the docking solutions and report best ones with the lowest energy values. To obtain the parameters, we use 36 unbound-unbound complexes from Dockground [17] as the training set, which are not including in the testing set.

5   Conclusion and Discussion

In this article, we present a new method for flexible refinement of docking solutions. We formulate the backbone flexibility problem on the interface as an instance of the Weighted Multi-Dimensional Scaling problem, which is able to model the local conformational changes. The results show that FlexDoBi models the backbone motions on the protein-protein interface. The backbone refinement procedure improves the accuracy of near-native docking solution candidates.

Our method can eliminate a larger number of inaccurate candidate structures, due to the geometrical constraints imposed by the distance between two residues

72   F. Guo et al.

respectively at both ends of each interface fragment. However, we only deal with the case where the regions far from the interface should be almost unchanged in complex.

We notice that large conformation changes can occur and result in a whole structure of the interacting proteins. On the difficulty group, the large changes appear in the unbound structures of three complexes: 1y64(A:B), 1f6m(A:C) and 1ira(Y:X). In the case of 1y64B, the conformational change occurs in loop region (residues 1396-1416). First, we replace this loop region with all loop candidates in above protein database, regardless of the stem RMSD. Then, we also refine the interface conformation of complex by using our flexible docking method (FlexDoBi) in this paper. The best discovered configuration is displayed in Figure 7. We predict a new configuration of complex with 6.42Å$C\alpha$ iRMSD, whereas the value of iRMSD for the predicted complex without the replaced loop is 11.77Å. Those issues will be our further investigations in recent future.

A   **B**

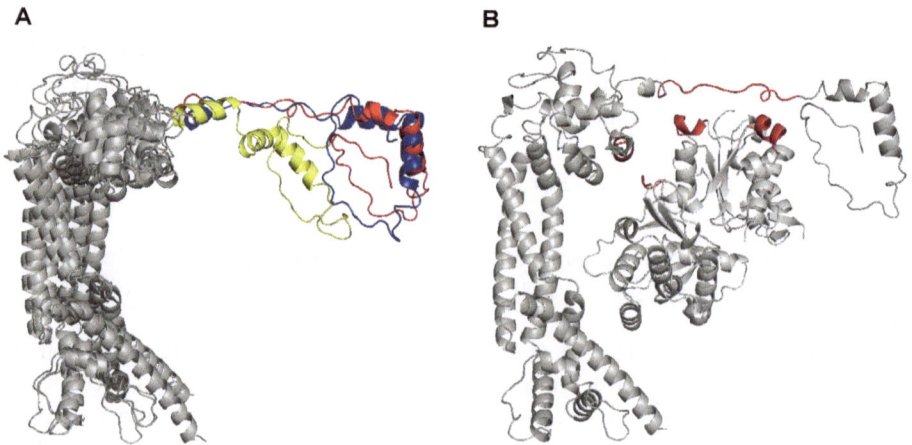

Fig. 7. The refinement of the case 1y64(A:B). (A) The unbound structure is colored in yellow and the bound structure is in blue. The replaced loop is in red. (B) The refined interface structure is in red.

Availability. The test set of complexes and the predictions are available for download from http://www.cs.cityu.edu.hk/~fguo22/FlexDoBi.html.

Acknowledgments. This work is supported by the grants from the Research Grants Council of the Hong Kong Special Administrative Region, China [Project No. CityU 121608, 124512] and the startup fund [Project No. CityU 7200276].

References

1. Alcaro, S., Gasparrini, F., Incani, O., Caglioti, L., Pierini, M., Villani, C.: "quasi flexible" automatic docking processing for studying stereoselective recognition mechanisms, part 2: Prediction of deltadeltag of complexation and 1h-nmr noe correlation. Journal of Computational Chemistry 28(6), 1119–1128 (2007)

2. Bradford, J.R., Westhead, D.R.: Improved prediction of protein-protein binding sites using a support vector machines approach. Bioinformatics 21(8), 1487–1494 (2005)

3. Brown, J.B., Bahadur, D., Tomita, E., Akutsu, T.: Multiple methods for protein side chain packing using maximum weight cliques. Genome Informatics 3(12), 191–200 (2006)

4. Chen, R., Li, L., Weng, Z.: Zdock: an initial-stage protein-docking algorithm. Proteins 52, 80–87 (2003)

5. Dominguez, C., Boelens, R., Bonvin, A.M.J.J.: Haddock: a protein-protein docking approach based on biochemical or biophysical information. Journal of the American Chemical Society 125, 1731–1737 (2003)

6. Eswar, N., Marti-Renom, M.A., Webb, B., Madhusudhan, M.S., Eramian, D., Shen, M., Pieper, U., Sali, A.: Comparative protein structure modeling with modeller. Current Protocols in Bioinformatics Supp. 15 (2006)

7. Fernández-Recio, J., Totrov, M., Abagyan, R.: Identification of protein-protein interaction sites from docking energy landscapes. Journal of Molecular Biology 335(3), 843–865 (2004)

8. Guo, F., Li, S.C., Wang, L.: P-Binder: A System for the Protein-Protein Binding Sites Identification. In: Bleris, L., Măndoiu, I., Schwartz, R., Wang, J. (eds.) ISBRA 2012. LNCS, vol. 7292, pp. 127–138. Springer, Heidelberg (2012)

9. Heifetz, A., Katchalski-Katzir, E., Eisenstein, M.: Electrostatics in protein-protein docking. Protein Science 11(3), 571–587 (2002)

10. Holtby, D., Li, S.C., Li, M.: LoopWeaver – Loop Modeling by the Weighted Scaling of Verified Proteins. In: Chor, B. (ed.) RECOMB 2012. LNCS, vol. 7262, pp. 113–126. Springer, Heidelberg (2012)

11. Hwang, H., Vreven, T., Janin, J., Weng, Z.: Protein-protein docking benchmark version 4.0. Proteins 78, 3111–3114 (2010)

12. Kabsch, W., Sander, C.: Dictionary of protein secondary structure: pattern recognition of hydrogen-bonded and geometrical features. Biopolymers 22, 2577–2637 (1983)

13. Konc, J., Janežič, D.: Probis algorithm for detection of structurally similar protein binding sites by local structural alignment. Bioinformatics 26(9), 1160–1168 (2010)

14. Krivov, G.G., Shapovalov, M.V., Dunbrack, R.L.: Improved prediction of protein side-chain conformations with scwrl4. Proteins 77(4), 778–795 (2009)

15. de Leeuw, J.: Applications of convex analysis to multidimensional scaling. In: Recent Developments in Statistics, pp. 133–146. North Holland Publishing Company (1977)

16. Lindahl, E., Hess, B., Spoel, D.: Gromacs 3.0: a package for molecular simulation and trajectory analysis. Journal of Molecular Modeling 7(8), 306–317 (2001)

17. Liu, S., Gao, Y., Vakser, I.: Dockground protein-protein docking decoy set. Bioinformatics 24, 2634–2635 (2008)

18. Lyskov, S., Gray, J.: The rosettadock server for local protein-protein docking. Nucleic Acids Research 36, W233–W238 (2008)

19. Mashiach, E., Nussinov, R., Wolfson, H.J.: Fiberdock: Flexible induced-fit backbone refinement in molecular docking. Proteins 78(6), 1503–1519 (2009)

20. Neuvirth, H., Raz, R., Schreiber, G.: Promate: a structure based prediction program to identify the location of protein-protein binding sites. Journal of Molecular Biology 338, 181–199 (2004)

21. Schneidman-Duhovny, D., Inbar, Y., Nussinov, R., Wolfson, H.J.: Geometry-based flexible and symmetric protein docking. Proteins 60(2), 224–231 (2005)

22. Schneidman-Duhovny, D., Nussinov, R., Wolfson, H.J.: Automatic prediction of protein interactions with large scale motion. Proteins 69, 764–773 (2007)
23. Shulman-Peleg, A., Nussinov, R., Wolfson, H.J.: Siteengines: recognition and comparison of binding sites and protein-protein interfaces. Nucleic Acids Research 1(33), W337–W341 (2005)
24. Wang, G., Dunbrack, R.L.: Pisces: a protein sequence culling server. Bioinformatics 19(2), 1589–1591 (2003)
25. Xu, J., Berger, B.: Fast and accurate algorithms for protein side-chain packing. Journal of the ACM 53, 533–557 (2006)
26. Yang, Y., Zhou, Y.: Specific interactions for ab initio folding of protein terminal regions with secondary structures. Proteins 72, 793–803 (2008)
27. Zhang, C.: Extracting contact energies from protein structures: A study using a simplified model. Proteins 31(3), 299–308 (1998)
28. Zhang, C., Vasmatzis, G., Cornette, J.L., DeLisi, C.: Determination of atomic desolvation energies from the structures of crystallized protein. Journal of Molecular Biology 267(3), 707–726 (1997)
29. Zhang, J., Wang, Q., Barz, B., He, Z., Kosztin, I., Shang, Y., Xu, D.: Mufold: A new solution for protein 3d structure prediction. Proteins 78, 1137–1152 (2010)

IPED: Inheritance Path Based Pedigree Reconstruction Algorithm Using Genotype Data

Dan He[1,*], Zhanyong Wang[2], Buhm Han[3,4],
Laxmi Parida[1], and Eleazar Eskin[2]

[1] IBM T.J. Watson Research, Yorktown Heights, NY 10598, USA
[2] Department of Computer Science, University of California Los Angeles,
Los Angeles, CA 90095, USA
[3] Division of Genetics, Brigham and Women's Hospital, Harvard Medical School,
Boston, MA, USA
[4] Program in Medical and Population Genetics, Broad Institute of Harvard and MIT,
Cambridge, MA, USA
dhe@us.ibm.com

Abstract. The problem of inference of family trees, or pedigree reconstruction, for a group of individuals is a fundamental problem in genetics. Various methods have been proposed to automate the process of pedigree reconstruction given the genotypes or haplotypes of a set of individuals. Current methods, unfortunately, are very time consuming and inaccurate for complicated pedigrees such as pedigrees with inbreeding. In this work, we propose an efficient algorithm which is able to reconstruct large pedigrees with reasonable accuracy. Our algorithm reconstructs the pedigrees generation by generation backwards in time from the extant generation. We predict the relationships between individuals in the same generation using an inheritance path based approach implemented using an efficient dynamic programming algorithm. Experiments show that our algorithm runs in linear time with respect to the number of reconstructed generations and therefore it can reconstruct pedigrees which have a large number of generations. Indeed it is the first practical method for reconstruction of large pedigrees from genotype data.

1   Introduction

Inferring genetic relationships from genotype data is a fundamental problem in genetics and has a long history [5,9,1,6,10,12]. Pedigree reconstruction is a hard problem and even constructing sibling relationships is known to be NP-hard [7]. In this work, we focus on reconstruction methods using genotype data. Various methods have been proposed for automatically reconstructing pedigrees using genotype data, which can be categorized into two categories. The first category is methods which reconstruct the haplotypes of the unknown ancestors in the pedigree. Thompson [14] proposed a machine learning approach to find the pedigree that maximizes the probability of observing the data. As the method

---

* Corresponding author.

M. Deng et al. (Eds.): RECOMB 2013, LNBI 7821, pp. 75–87, 2013.
© Springer-Verlag Berlin Heidelberg 2013

reconstructs both the pedigree graph and the ancestor haplotypes at the same time, it is very time-consuming and can be only applied to small families of size 4-8 people. The second category is methods which reconstruct the pedigree directly without reconstructing ancestor haplotypes. Thatte and Steel [13] proposed a HMM based model to reconstruct arbitrary pedigree graphs. However, their model, in which every individual passes on a trace of their haplotypes to all of their descendants is unrealistic. Kirkpatrick et al. [7] proposed an algorithm to reconstruct pedigrees based on pairwise IBD (identity-by-descent) information without reconstructing the ancestral haplotypes. A generation-by-generation approach is employed and the pedigree is reconstructed backwards in time, one generation at a time. The input of the algorithm is the set of extant individuals with haplotype and IBD information available. At each generation, a compatibility graph is constructed, where the nodes are individuals and the edges indicate the pair of individuals which could be siblings. The edges are defined via a statistical test such that an edge is constructed only when the test score between the pair of individuals is less than a pre-defined threshold. Sibling sets are identified in the compatibility graph using a Max-clique algorithm iteratively to partition the graph into disjoint sets of vertices. The vertices in the same set have edges connecting to all the other vertices of the same set. Both categories of methods encounter difficulties depending on the structure of pedigree. When the individuals are not related through inbreeding, these methods are fast and accurate. However, when inbreeding is present, the reconstruction becomes much more complicated and these methods perform poorly.

In this work, we propose an efficient algorithm, IPED (Inheritance **P**ath based **Ped**igree Reconstruction), which enables the reconstruction of very large pedigrees, with and without the presence of inbreeding. Our algorithm follows the approach of [7] and starts from extant individuals and reconstructs the pedigree generation by generation backwards in time. For each generation, we predict the pairwise relationships between the individuals at the current generation and create parents for them according to their relationships. When we evaluate the pairwise relationships for a pair of individuals, we consider the pairwise IBD length for their extant descendants, namely the leaf individuals in the pedigree. We then apply a statistical test on the two individuals to determine if they are siblings or not siblings.

One of the challenges in our approach is to compute the expected IBD length between a pair of extant individuals efficiently, in the presence of inbreeding. The CIP and COP methods of [7] are efficient for outbreed pedigrees but very inefficient for inbred pedigrees. This is because for the inbreeding case the alleles from an extant individual can be inherited in an exponential number of ways from his or her ancestors with respect to the number of nodes in the pedigree graph. The CIP algorithm applies a random walk from the ancestor to sample these exponential number of ways to estimate the expected IBD length between a pair of extant individuals. In addition, the pedigree needs to be explored multiple times when constructing each generation. Therefore the algorithm is inefficient even for relatively small number of generations. In our experiments, CIP can not finish for a family of size around 50 individuals with 4 generations.

In order to address this problem, we consider the inheritance paths between the ancestor and the extant individuals, where each inheritance path corresponds to one path in the pedigree from the ancestor to the extant individual. If we know all the inheritance paths from the ancestor to the extant individuals, we can estimate the probability that an allele of the extant individual is inherited from the ancestor. The probability can be further utilized to compute the expected average IBD length between a pair of extant individuals. However, the number of inheritance paths can be exponential. We observed although the number of inheritance paths can be exponential, their lengths are bounded by the height of the pedigree. Therefore we use a hash data structure to hash all the inheritance paths of the same length into a bucket and the number of buckets is bounded by the height of the pedigree and thus is usually small. We save the hash tables for each individual and we develop a dynamic programming algorithm to populate the hash table of the individuals generation by generation. By doing this, we avoid redundant computation of the inheritance paths where the entire pedigree needs to be explored repeatedly and thus the dynamic programming algorithm is very efficient. Also because we avoid the time-consuming sampling step by using the inheritance path, our algorithm IPED is extremely efficient and it does not need to specify whether or not inbreeding is present, which is a big advantage over COP and CIP. Our experiments show that our algorithm is able to reconstruct the pedigree with inbreeding for a family of size 340 individuals with 10 generations in just 14 seconds. To our knowledge, this is the first algorithm that is able to reconstruct such large pedigrees with inbreeding using genotype data.

2   Methods

2.1   Pedigrees

A pedigree graph consists of nodes and edges where nodes are diploid individuals and edges are between parents and children. Circle nodes are females and boxes are males. An example of pedigree graph is shown in Figure 1. Parent nodes are also called *founders*. In the example, individual 13,14,15 are *extant individuals* and their founders are individuals 9, 10 and 11, 12, respectively. *Outbreeding* means an individual mates with another individual from different family. In the example, 3,4 and 6, 7 are both outbreeding cases. *Inbreeding* means an individual mates with another individual from the same family. In the example, 9, 10 is inbreeding case. We can see inbreeding case is usually more complicated as an individual can inherit from his ancestors in multiple ways. For example, 13, 14 can inherit from 1, 2 in two ways but 15 can inherit from 1,2 in only one way.

As we only have extant individuals and we reconstruct their ancestors, the pedigree is reconstructed backwards in time. We use the same notion of generations in [7], namely generations are numbered backwards in time, with larger numbers being older generations. Every individual in the graph is associated with a generation g. All the extant individuals are associated with $g=1$ and their direct parents are associated with generation $g=2$. The *height* of a pedigree is the biggest g. We define an *inheritance path* between a child and his ancestor the same as it is defined in [10], namely as a path between the two corresponding nodes in the pedigree graph. For example, the inheritance path between 1 and 15 consists of nodes 1-6-11-15. There are two inheritance paths between 1 and 13: 1-4-9-13 and 1-6-10-13. Also we assume the inheritance paths are not directed. In this work, we do not consider pedigrees with *half-siblings*, namely we assume an individual only mates with another individual in the same generation.

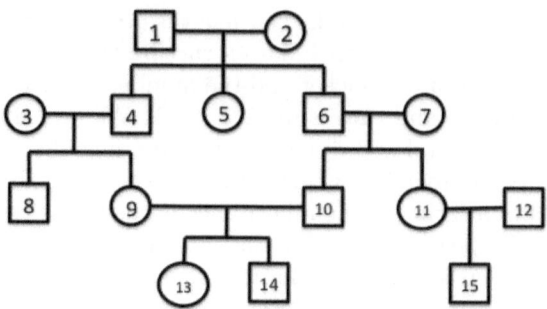

Fig. 1. An example of pedigree graph

2.2   Metrics to Evaluate the Relationship of a Pair of Individuals

As our algorithm reconstructs the pedigree generation by generation, we need to determine the relationship of any pair of individuals at a generation. We consider two different metrics for extant individuals and ancestral individuals, respectively.

To determine the relationship of a pair of extant individuals, we consider the IBD (identity-by-descent) length of the two individuals. In order to be distinguished from IBS (identity-by-state), the IBD region needs to be long enough, for example, of size 1Mb. If we are given the genotypes of the extant individuals, we can compute the IBD regions between a pair of individuals using existing tools such as Beagle [3]. In this work, in our simulation, we assume we are given haplotypes of the extant individuals and we consider identical regions of length greater than 1Mb between the two individuals as their IBD regions. We consider the averaged IBD length instead of total length of IBD to handle the cases where IBD regions are unevenly distributed. For simplicity, we use "IBD length" to denote "averaged IBD length".

Then for a pair of extant individuals i, j, we conduct a statistical test and compute a score $v_{i,j}$ as the following:

$$v_{i,j} = \frac{\left(estimate(IBD_{i,j}) - E(IBD_{i,j})\right)^2}{var(IBD_{i,j})} \qquad (1)$$

where $estimate(IBD_{i,j})$ is the estimated IBD length between individuals i and j, $E(IBD_{i,j})$ is the expected IBD length between i and j, $var(IBD_{i,j})$ is the variance of the IBD length between i and j. $estimate(IBD_{i,j})$ can be computed easily given genotypes or haplotypes of individual i and j. As recombination occurs in meioses, it is shown [4] that the length of IBD between i and j follows an exponential distribution $exp(Mr)$, where M is the number of meioses between i and j, r is the recombination rate which is set as 10^{-8}, namely the probability for recombination occurs at any loci is 10^{-8}. Therefore, $E(IBD_{i,j})$ and $var(IBD_{i,j})$ are computed as the following:

$$E(IBD_{i,j}) = \frac{1}{M \times r} \tag{2}$$

$$var(IBD_{i,j}) = \frac{1}{(M \times r)^2} \tag{3}$$

For outbreeding case, $M = 2(g - 1)$ where g is the generation. So for extant individuals, as we are constructing the second generation, $g = 2$. For inbreeding case, a random walk algorithm whose complexity is exponential is applied. More details will be given in the next section.

As we need to consider both paternal and maternal alleles, our IBD estimation is chromosome-wise instead of individual-wise. As i, j both have a pair of chromosomes noted as i_1, i_2, j_1, j_2, there are two possible ways to compare them for IBD, namely $[(i_1, j_1), (i_2, j_2)]$ or $[(i_1, j_2), (i_2, j_1)]$. We select the way that maximizes the sum of the averaged IBD length for both chromosomes. Without losing generality, assuming we select $[(i_1, j_1), (i_2, j_2)]$. Then we compute $v_{i,j} = \frac{v_{i_1,j_1} + v_{i_2,j_2}}{2}$, where v_{i_1,j_1} is computed according to Formula 1 by considering the estimated IBD between i_1, j_1. Notice $E(IBD_{i,j})$ and $var(IBD_{i,j})$ don't depend on the chromosomes of i and j.

In the method of Kirkpatrick et al. [7], if the test score $v_{i,j}$ is less than a pre-defined threshold value S, i, j are considered as siblings. However, it is not clear how to determine the value S and the threshold usually varies for individuals of different relatedness. In [7], the threshold is determined empirically by simulating many pedigrees. As we show in our experiments, the performance of the algorithms varies with the threshold.

In our work, we try to avoid using a threshold. As the pair of nodes are either siblings or non-siblings, we can compute the number of meioses between them for each case. For the case that they are siblings, the number of meioses is 2 and we can compute the length of the expected IBD using Formula 3. For non- sibling cases, we don't know exactly how many meioses there are between the pair of nodes. However, we can compute a lower bound for such number: namely the two nodes are first-cousin and the number of meioses is 4, which is the minimum number for a pair of non-sibling nodes. Then we can compute the length of the expected IBD for non-sibling again using Formula 3. We compare the two test scores and determine the pair of nodes are siblings if the test score for sibling case is lower.

To determine the relationship of a pair of ancestral individuals, we use a similar strategy as the one in [7]. Assuming individuals k and l are at generation $g > 1$. The sets of all extant descendants of k and l are K and L, respectively. We compute a score $v_{k,l}$ between k and l as

$$
\begin{aligned}
v_{k,l} &= \frac{1}{|K||L|} \sum_{i \in K} \sum_{j \in L} v_{i,j} \\
&= \frac{1}{|K||L|} \sum_{i \in K} \sum_{j \in L} \frac{\left(estimate(IBD_{i,j}) - E(IBD_{i,j})\right)^2}{var(IBD_{i,j})}
\end{aligned}
\tag{4}
$$

where $|K|$ is the size of K, the number of extant descendants of k, $i \in K$ is an extant individual in K, $v_{i,j}$ is computed via Formula 1. Again, we compute $v_{k,l}$ for both sibling case and first-cousin case and determine k, l are siblings if the score for sibling case is lower. More details will be given in the next section on how to compute $E(IBD_{i,j})$ and $var(IBD_{i,j})$.

2.3   IPED: Inheritance Path Based Pedigree Reconstruction Algorithm

The computation of $E(IBD_{i,j})$ and $var(IBD_{i,j})$ is complicated in that the number of possible meioses between i and j can be exponential with respect to the nodes in the pedigree graph. To estimate the expected length of IBD between a pair of extant individuals, we need to consider all possible options for a pair of alleles to inherit from the shared ancestor, which is also exponential to the number of nodes in the pedigree. A random walk algorithm *CIP* from the founders with sampling is applied in [7]. However, the sampling is still time consuming in an exponential search space. What's more, as the reconstruction is generation-by-generation, from generation 2 to higher generation, the sampling strategy needs to be conducted every time when we move from one generation to the next generation backwards, which obviously involves redundant computation. Therefore, CIP is not efficient for inbreeding case. In our experiments, CIP can not finish for a family of size around 50 individuals with 4 generations.

To address the aforementioned two problems, we proposed a very efficient algorithm IPED (Inheritance Path based Pedigree Reconstruction Algorithm), which is based on the idea that the probability that a pair of alleles from two individuals are inherited from shared ancestor depends on the number of possible *inheritance paths* and their corresponding lengths from the shared ancestor. An example of inheritance path is shown in Figure 1. We can see the length of inheritance path determines the number of meioses between the two individuals and thus determines the probability of a pair of alleles from the two extant individuals inherited from the same ancestor. For example, the number of meioses between 8, 9 is 2 as they are siblings and the distance between them in the pedigree is 2. The number of meioses between 13 and 15 can be either 6 or 4, as there are multiple paths in the pedigree graph between them. In our algorithm, if there are multiple possible numbers of meioses, we used the averaged value to approximate the IBD length. So for 13 and 15 the average number of meioses is 5.

Therefore, to determine the number of possible distances, or possible meioses between the extant individuals, for any founder in the current generation, we save the number of inheritance paths and the length of these inheritance paths from the founder to all the extant descendant individuals. Notice for inbreeding, there maybe an exponential number of inheritance paths with respect to the number of nodes in the pedigree. However, the length of the inheritance paths is finite, which is bounded by the height of the pedigree. Therefore, what we need to save is just a hash table with (length, number) pairs where the length of the inheritance path is the key and the number of inheritance paths with such length is the value. For example, there are 2 length-2 paths, 5 length-3 paths, 6 length-4 paths, then we just need to save three pairs (2,2), (3,5), (4,6), instead of saving all 9 paths separately. Therefore, we don't need to save exponential number of paths. Instead, we save only a small number of pairs, which is bounded by the height of the pedigree. Notice we need to save such pairs $[i, ((l_{i_1}, n_{i_2}), \ldots, (l_{i_k}, n_{i_k})]$ between the founder and every extant descendant of it, where i is the i-th extant descendant, (l_{i_k}, n_{i_k}) is the k-th (length, number) pair between the founder and the descendant. We call such pairs *Inheritance Path Pair (IPP)*. Given the number of extant individuals is fixed and is usually not a big number, the complexity is bounded by a constant.

The inheritance path pairs can be used to compute the possible distances, or the average number of meioses of a pair of extant individuals. Assuming a pair of founders G and K with inheritance path pairs $[i, ((l_{g_1}, n_{g_1}), \ldots, (l_{g_h}, n_{g_h})]$ and $[j, ((l_{k_1}, n_{k_1}), \ldots, (l_{k_f}, n_{k_f}))]$. The average number of meioses between individual i, j can be computed with Algorithm 1, where t is a test option. For sibling case, $t = 1$ and for first-cousin case, $t = 2$. Once the number of meioses is computed, it can be applied to Formula 3 directly to compute the statistic test score.

Algorithm 1. Calculate the average number of meioses between i, j

---
Input: t (test option), $[i, ((l_{g_1}, n_{g_1}), \ldots, (l_{g_h}, n_{g_h})]$ and $[j, ((l_{k_1}, n_{k_1}), \ldots, (l_{k_f}, n_{k_f}))]$
Output: The average number of meioses between i, j
  $Length \leftarrow 0$
  $Num \leftarrow 0$
  for $a = 1$ to h **do**
    for $b = 1$ to f **do**
      $Num \leftarrow Num + n_{g_a} \times n_{k_b}$
      $Length \leftarrow Length + (l_{g_a} + t + l_{k_b} + t) \times (n_{g_a} \times n_{k_b})$
    end for
  end for
  $number\ of\ meioses \leftarrow \frac{Length}{Num}$

Notice some of the inheritance paths may be shared by two extant individuals for inbreeding case. For example, in Figure 1, the inheritance paths between 1 and 15 1-6-11-15 and between 1 and 13 1-6-10-13 share one edge 1-11. Thus the number of meioses is 4 instead of 6. Using the above algorithm, we will have 6 as the number of meioses. However, as we want to avoid saving the exponential paths explicitly, we just assume the paths do not overlap. Therefore, IPED is not optimal. Instead, it is an approximation algorithm. Another approximation our method is employing is that we approximate the mean and variance of the IBD length by using the average number of meioses (Algorithm 1). We also assume that if there are multiple paths between two individuals, it is not possible for the individuals to be IBD through one path at a locus and IBD through another path at the next locus. Such case should be rare in practice because multiple recombination events should simultaneously occur in the pedigree at one locus. Despite of these approximations, our experiments show that IPED achieves good reconstruction accuracy.

Once we save such pairs for each founder at one generation, when we reconstruct the next generation (the parents of the current generation) backwards, we need to compute such pairs between all the possible founders in the next generation and all the extant individuals. A naive algorithm is to compute the IPPs between every founder and every extant individual on each generation. However, this requires significant redundant computations since all the nodes of lower generation will be explored multiple times when computing the inheritance paths. We developed a dynamic programming algorithm where the IPPs of the current generation can be used to compute the IPPs of the next generation.

The dynamic programming algorithm starts the reconstruction from generation 2 as generation 1 consists of all the known extant individuals. Then at generation 2, assuming we have a founder G_2^i (without losing generality, assuming he is father) and his k children in generation 1 as $G_1^{i_1}, G_1^{i_2}, \ldots, G_1^{i_k}$. Then for every paternal allele of each child, obviously we have 1 possible length 1 inheritance path from the founder. Therefore, we save $[G_1^{i_j}, (1, 1)]$ for G_2^i for $1 \leq j \leq k$. Now let's assume we are at generation T, and we are reconstructing generation $T + 1$. Again, assuming we have a founder G_{T+1}^i as father and his k children in generation T as $G_T^{i_1}, G_T^{i_2}, \ldots, G_T^{i_k}$. We then obtain the IPPs for G_{T+1}^i by merging the IPPs for $G_T^{i_1}, G_T^{i_2}, \ldots, G_T^{i_k}$. The recursion is shown as below:

$$IPP(G_{T+1}^i) = \sum_{j=1}^{k} IPP(G_T^{i_j}) + 1$$

where $IPP(G_{T+1}^i)$ is the set of IPPs for node G_{T+1}^i. Assuming for $G_T^{i_j}$, we have IPPs
$[G_1^t, ((L_{j_1}, N_{j_1}), \ldots, (L_{j_m}, N_{j_m}))]$, $IPP(G_T^{i_j}) + 1$ is to update these pairs as
$[G_1^t, (L_{j_1}+1, N_{j_1}), \ldots, (L_{j_m}+1, N_{j_m})]$. $IPP(G_T^a) + IPP(G_T^b)$ is to merge two sets of IPPs. When we merge two pairs (L_a, N_a) and (L_b, N_b), if $L_a = L_b$, we obtain a merged pair $(L_a, N_a + N_b)$. Otherwise we keep the two pairs. Therefore, after the merge, we obtain $[G_1^t, ((L_1, N_1), \ldots, (L_m, N_m))]$ for each extant individual G_1^t who is the descendant of G_{T+1}^i, where $L_1, \ldots, L_m$ are all unique and $m \leq T+1$. The summation ($\sum$) is similarly defined as the repeated merging operation over multiple sets of IPPs.

An example of the dynamic programming algorithm is shown in Figure 2. As we can see in the example, when we merge the IPPs, we increase the length of the paths by 1 and add the number for the paths of the same length. The complexity of this dynamic programming algorithm is $O(E \times k \times H)$ where E is the number of extant individuals, k is the number of direct children for each founder, H is the height of the pedigree. Therefore it is linear time with respect to the height of the pedigree.

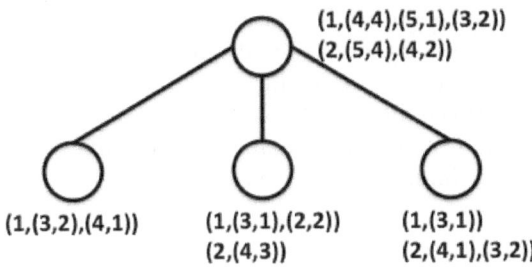

Fig. 2. An example of the dynamic programming algorithm

Once we compute the inheritance path pairs for each founder, we can calculate the number of meioses of any pair of extant individuals using Algorithm 1 and further compute the test score according to Formula 4.

2.4 Creating Parents

Once we determined the relationships of all the individuals of the current generation, we need to create parents for them. In order to guarantee that we create the same parents for all the individuals that are siblings, we create a graph for all the individuals at the current generation. Every individual is a node and there is an edge between a pair of nodes if they are determined as siblings according to the test. We call the graph *Sibling Graph*. Then we apply a Max-Clique algorithm [2] on the sibling graph for the current generation. We select the maximum clique where all the individuals in the clique are siblings to each other. We then create parents for them, and remove them from the sibling graph. We then select the next maximum clique from the remaining sibling graph and we repeat the procedure until all nodes are selected and all parents are created.

2.5 Performance Evaluation

Once we reconstructed the pedigree, we need to evaluate the accuracy of the reconstruction. We can not simply compare the reconstructed pedigree with the true pedigree directly due to graph polymorphism [8]. Therefore we consider the following metric:

$$accuracy(R, O) = \frac{\sum_{i \in E, j \in E} F(R_{i,j}, O_{i,j})}{|E|^2}$$

$$F(R_{i,j}, O_{i,j}) = \begin{cases} 1 & \text{if } R_{i,j} = O_{i,j} \\ 0 & \text{otherwise} \end{cases}$$

where R is the reconstructed pedigree, O is the original pedigree, E is the set of extant individuals, $|E|$ is the number of extant individuals, $R_{i,j}$ is the distance of individual i and j in pedigree R and $R_{i,j} = \infty$ if i, j are not connected in the pedigree graph. Notice if there are multiple paths between i and j in R, we select the shortest path. Therefore in this metric, we only compare the distance of extant individuals. If the distance between a pair of extant individuals in two pedigrees are the same (or two individuals are not connected in both pedigrees as the pedigrees are not high enough), we consider the reconstruction correct for this pair.

3   Experimental Results

We use the simulator from [7] to simulate the pedigrees. Instead of genotype data, we simulate haplotypes directly. The haplotypes of the individuals are generated according to the Wright-Fisher Model [11] with monogamy. The model takes parameters for a fixed populations size, a Poisson number of offspring and a number of generations (or the height of pedigree). We consider identical regions of length greater than 1Mb as IBD regions. We only compare our algorithm IPED with COP and CIP as the pedigree size in our simulation is relatively big and can not be handled by other algorithms. All the experiments are done on a 2.4GHz Intel Dual Core machine with 4G memory.

3.1   Outbreeding Simulation

We first test the outbreeding case. In the Wright-Fisher simulation, we fix the average number of children of each founder as 3, the individual of each generation is 20 and we vary the height of the pedigree. Notice according to the Wright-Fisher model, the number of individuals simulated each generation may not be 20. We compare the accuracy of COP and IPED. We randomly simulate 10 pedigrees for each parameter setting and show the averaged accuracy in Table 1. We can see that generally the accuracy drops as the generation and family size increase. IPED achieves slightly better results for outbreeding cases compared to COP. Also IPED is very fast, comparable to COP. For all different generations, IPED finishes in less than one second.

Next we show that COP algorithm is affected by the score threshold. As the empirically determined threshold is 0.7 in the work of [7], we vary the score threshold as 0.7 and 0.9. We show the results in Table 2. As we can see, the performance of COP varies with different thresholds. Our algorithm IPED, on the contrary has the advantage of not relying on any threshold.

Table 1. Outbreeding Accuracy for IPED and COP. Average number of children of each founder is 3. The number of individuals for each generation is 20. We vary the height of the pedigree.

Height	Family Size	IPED	COP
g = 3	52	0.966	0.955
g = 4	84	0.782	0.751
g = 5	144	0.831	0.836
g = 6	266	0.78	0.79
g = 7	384	0.706	0.655
g = 8	860	0.617	0.64

Table 2. Outbreeding Accuracy for COP with different test score thresholds. Average number of children of each family is 3. The number of individuals for each generation is 20. We vary the height of the pedigree.

Height	COP (0.7)	COP (0.9)
g = 4	0.905	0.89
g = 5	0.77	0.816
g = 6	0.874	0.895
g = 7	0.684	0.605

3.2   Inbreeding Simulation

Next we test the inbreeding case. As the CIP algorithm is very inefficient for inbreeding case, even for small pedigree it takes a long time and most often just simply crashes, we only compare our algorithm with CIP for pedigrees of height 3, with family size 40. IPED achieves an average accuracy of 0.91 while CIP achieves an average accuracy of 0.902 on 10 randomly simulated pedigrees.

Then we compare our algorithm with COP, which is aimed for outbreeding case, as it is able to finish fast on the simulated data sets. When COP is applied to a pedigree with inbreeding, it simply assumes there is only outbreeding in the pedigree.

We first fix the average number of children as 3, the individual of each generation is 20 and we vary the height of the pedigree. We show the averaged accuracy of IPED and COP in Table 3. We can see that for all generations, IPED achieves better results consistently. The accuracy generally drops for both methods. When the generation number is small, such as 3 and 4, the performances of IPED and COP are similar. However, as the pedigree gets bigger and more complicated, our algorithm significantly outperforms COP, which is reasonable as COP doesn't consider inbreeding. The algorithm CIP does consider inbreeding but it is not able to handle pedigrees of this size. IPED, on the contrary, is able to finish in just a few seconds for all parameter settings.

Next we show the performance of both algorithms for different family sizes. We vary the number of individuals of each generation as 20, 40 and 60. We set the generation number as 6. We show the averaged results from 10 random

Table 3. Inbreeding Accuracy of IPED and COP for different pedigree heights. Average number of children of each family is 3. The number of each generation is 20. We vary the height of the pedigree.

Height	Family Size	IPED	COP	improvement
g = 3	50	0.93	0.924	0.6%
g = 4	62	0.722	0.715	0.9%
g = 5	74	0.689	0.605	13.9%
g = 6	88	0.65	0.446	45.7%
g = 7	94	0.599	0.335	78.8%
g = 8	110	0.533	0.297	79.5%

Table 4. Inbreeding Accuracy of IPED and COP for different population size. Average number of children of each family is 3. We vary the number of individual for each generation used in the Wright-Fisher model as 20, 40, 60.

Number of Individual	Family Size	IPED	COP
S = 20	88	0.65	0.446
S = 40	156	0.66	0.55
S = 60	300	0.631	0.572

simulations in Table 4. We can see for all family sizes, our method achieves better accuracies, and the accuracies remain similar to each other, indicating the performance of our method is very stable w.r.t the size of the pedigree. Again, IPED is very fast and finishes in a few seconds for all datasets.

Finally we simulate a set of deep pedigrees and show the accuracy and running time of our algorithm in Table 5. As we can see, although the accuracy of IPED is relatively low, it is still a few times better than that of COP, the only existing algorithm that is able to handle such large pedigrees. In addition, IPED is faster than COP.

Table 5. Inbreeding Accuracy of IPED and COP for different family size. Average number of children of each family is 3.

Family Size	Generation	IPED	COP	IPED running time (.s)	COP running time (.s)
260	10	0.365	0.125	7	13
340	10	0.227	0.08	14	193

4   Conclusions

We proposed a very efficient algorithm IPED for pedigree reconstruction using genotype data. Our method is based on the idea of inheritance path where the time-consuming sampling can be avoided. A dynamic programming algorithm is developed to avoid redundant computation during the generation-by-generation reconstruction process. We show our method is much more efficient than the state-of-the-art methods especially when inbreeding is involved in the pedigree. To our knowledge it is the first algorithm that is able to reconstruct pedigrees with inbreeding containing hundreds of individuals with tens of generations. Our algorithm still does not consider all possible complicated cases in pedigrees, such as half-siblings. Also it reconstructs pedigree only from the extant individuals. When the genotype of the internal individuals are known, it is helpful to use all such information. We would like to address these problems in our future work.

Acknowledgement. The authors would like to thank Bonnie Kirkpatrick for her help on the pedigree simulation.

References

1. Abecasis, G.R., Cherny, S.S., Cookson, W.O., Cardon, L.R.: Merlin-rapid analysis of dense genetic maps using sparse gene flow trees. Nature Genetics 30(1), 97–101 (2002)
2. Bron, C., Kerbosch, J.: Algorithm 457: finding all cliques of an undirected graph. Communications of the ACM 16(9), 575–577 (1973)
3. Browning, B.L., Browning, S.R.: A fast, powerful method for detecting identity by descent. The American Journal of Human Genetics 88(2), 173–182 (2011)
4. Donnelly, K.P.: The probability that related individuals share some section of genome identical by descent. Theoretical Population Biology 23(1), 34–63 (1983)
5. Elston, R.C., Stewart, J.: A general model for the genetic analysis of pedigree data. Human Heredity 21(6), 523–542 (1971)
6. Fishelson, M., Dovgolevsky, N., Geiger, D.: Maximum likelihood haplotyping for general pedigrees. Human Heredity 59(1), 41–60 (2005)
7. Kirkpatrick, B., Li, S., Karp, R., Halperin, E.: Pedigree reconstruction using identity by descent. Journal of Computational Biology 18(3), 1181–1193 (2011)
8. Kirkpatrick, B., Reshef, Y., Finucane, H., Jiang, H., Zhu, B., Karp, R.M.: Comparing pedigree graphs. Arxiv preprint arXiv:1009.0909 (2010)
9. Lander, E.S., Green, P.: Construction of multilocus genetic linkage maps in humans. Proceedings of the National Academy of Sciences 84(8), 2363 (1987)
10. Li, X., Yin, X., Li, J.: Efficient identification of identical-by-descent status in pedigrees with many untyped individuals. Bioinformatics 26(12), i191–i198 (2010)
11. Press, W.H.: Wright-fisher models, approximations, and minimum increments of evolution (2011)
12. Sobel, E., Lange, K.: Descent graphs in pedigree analysis: applications to haplotyping, location scores, and marker-sharing statistics. American Journal of Human Genetics 58(6), 1323 (1996)
13. Thatte, B.D., Steel, M.: Reconstructing pedigrees: A stochastic perspective. Journal of Theoretical Biology 251(3), 440–449 (2008)
14. Thompson, E.A.: Pedigree analysis in human genetics. Johns Hopkins University Press, Baltimore (1986)

An Optimal Algorithm for Building
the Majority Rule Consensus Tree

Jesper Jansson[1,*], Chuanqi Shen[2], and Wing-Kin Sung[3,4]

[1] Laboratory of Mathematical Bioinformatics (Akutsu Laboratory),
Institute for Chemical Research,
Kyoto University, Gokasho, Uji, Kyoto 611-0011, Japan
jj@kuicr.kyoto-u.ac.jp
[2] Stanford University, 450 Serra Mall, Stanford, CA 94305-2004, U.S.A.
shencq@stanford.edu
[3] School of Computing, National University of Singapore, 13 Computing Drive,
Singapore 117417
ksung@comp.nus.edu.sg
[4] Genome Institute of Singapore, 60 Biopolis Street, Genome, Singapore 138672

Abstract. A deterministic algorithm for building the majority rule consensus tree of an input collection of conflicting phylogenetic trees with identical leaf labels is presented. Its worst-case running time is $O(nk)$, where n is the size of the leaf label set and k is the number of input phylogenetic trees. This is optimal since the input size is $\Omega(nk)$. Experimental results show that the algorithm is fast in practice.

1 Introduction

In the last 150 years, a vast number of phylogenetic trees [8,10,14,17,19] have been constructed and published in the literature. Existing phylogenetic trees may be based on different data sets or obtained by different methods, and do not always agree with each other; two trees can contain contradicting branching patterns even though their leaf label sets are identical. Also, when trying to infer a new, reliable phylogenetic tree from real data, heuristics for maximizing parsimony or resampling techniques such as bootstrapping may produce large collections of identically leaf-labeled phylogenetic trees having slightly different branching structures [2,3,7,8,19]. To deal with conflicts that arise between two or more such trees in a systematic manner, the concept of a *consensus tree* was invented [1,5]. Informally, a consensus tree is a phylogenetic tree which summarizes a given *collection* of phylogenetic trees. In addition to resolving conflicts, consensus trees may be employed to locate strongly supported groupings within a collection of trees [8] or as a basis for similarity measures between two given phylogenetic trees (measuring the similarity between phylogenetic trees is useful, e.g., when querying phylogenetic databases [3] or evaluating methods for phylogenetic reconstruction [12]).

---

* Funded by The Hakubi Project and KAKENHI grant number 23700011.

M. Deng et al. (Eds.): RECOMB 2013, LNBI 7821, pp. 88–99, 2013.

There are many ways to reconcile structural differences and remove inconsistencies in a collection of trees. Consequently, several alternative definitions of a "consensus tree" have been proposed since the 1970's.[1] In this paper, we concentrate on one particular type of consensus tree called the *majority rule consensus tree* [13], which is one of the most widely used consensus tree among practitioners, and present a new algorithm for constructing it. Our algorithm is fast in theory (it achieves optimal worst-case time complexity) and in practice. Furthermore, it is conceptually simple, relatively easy to implement, and deterministic, i.e., it does not use randomization or hash tables to keep track of clusters.

1.1   Definitions and Notation

We first give some basic definitions that will be used throughout the paper. A *phylogenetic tree* is a rooted, unordered, leaf-labeled tree in which every internal node has at least two children and all leaves have different labels. For short, phylogenetic trees will be referred to as "trees" from here on.

For any tree T, the set of all nodes in T is denoted by $V(T)$ and the set of all leaf labels in T by $\Lambda(T)$. Any subset of $\Lambda(T)$ is called a *cluster of $\Lambda(T)$*. For any $u \in V(T)$, $T[u]$ denotes the subtree of T rooted at the node u, so that $\Lambda(T[u])$ is the set of all leaf labels of leaves that are descendants of u.[2] The *cluster collection of T* is defined as $\mathcal{C}(T) = \bigcup_{u \in V(T)} \{\Lambda(T[u])\}$. See Fig. 1 for an example. If a cluster $C \subseteq \Lambda(T)$ belongs to $\mathcal{C}(T)$, we say that C *occurs in T*.

The *majority rule consensus tree* is defined next. Let $\mathcal{S} = \{T_1, T_2, \ldots, T_k\}$ be a set of trees satisfying $\Lambda(T_1) = \Lambda(T_2) = \cdots = \Lambda(T_k) = L$ for some leaf label set L. A cluster that occurs in more than $k/2$ of the trees in $\mathcal{S}$ is a *majority cluster of $\mathcal{S}$*, and the *majority rule consensus tree of $\mathcal{S}$* [13] is the unique tree T such that $\Lambda(T) = L$ and $\mathcal{C}(T)$ consists of all majority clusters of $\mathcal{S}$. The problem studied in this paper is:

> Given an input set $\mathcal{S}$ of trees with identical leaf label sets, compute the majority rule consensus tree of $\mathcal{S}$.

In the rest of the paper, we will use the following notation to refer to any input set of trees: $\mathcal{S} = \{T_1, T_2, \ldots, T_k\}$, $L = \Lambda(T_1) = \Lambda(T_2) = \cdots = \Lambda(T_k)$, and $k = |\mathcal{S}|$ and $n = |L|$. An example with $k = 3$ and $n = 6$ is provided in Fig. 1.

Finally, two clusters $C_1, C_2 \subseteq \Lambda(T)$ are said to be *pairwise compatible* if $C_1 \subseteq C_2$, $C_2 \subseteq C_1$, or $C_1 \cap C_2 = \emptyset$. Any cluster $C \subseteq \Lambda(T)$ is said to be *compatible with T* if C and $\Lambda(T[u])$ are pairwise compatible for every node $u \in V(T)$. For example, in Fig. 1, the cluster $\{a, c\}$ is compatible with T_1, but not compatible with any of the other trees. If T_1 and T_2 are two trees with $\Lambda(T_1) = \Lambda(T_2)$ such that every cluster in $\mathcal{C}(T_1)$ is compatible with T_2 then it follows that every cluster in $\mathcal{C}(T_2)$ is compatible with T_1, and we say that T_1 and T_2 are *compatible*.

---

[1] See reference [5], Chapter 30 in [8], or Chapter 8.4 in [19] for some surveys on consensus trees.

[2] For convenience, any node is considered to be a descendant of itself. This implies that if u is a leaf then $\Lambda(T[u])$ is a singleton set.

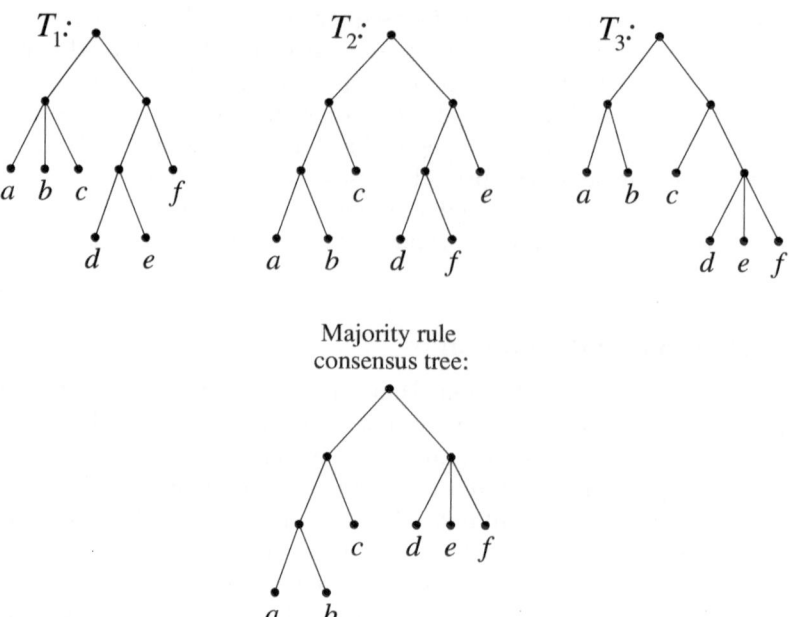

Fig. 1. In this example, $\mathcal{S} = \{T_1, T_2, T_3\}$ and $L = \Lambda(T_1) = \Lambda(T_2) = \Lambda(T_3) = \{a, b, c, d, e, f\}$. The cluster collections of T_1, T_2, and T_3 are:

$$\mathcal{C}(T_1) = \big\{\{a\}, \{b\}, \{c\}, \{d\}, \{e\}, \{f\}, \{a, b, c\}, \{d, e\}, \{d, e, f\}, L\big\},$$
$$\mathcal{C}(T_2) = \big\{\{a\}, \{b\}, \{c\}, \{d\}, \{e\}, \{f\}, \{a, b\}, \{a, b, c\}, \{d, e, f\}, \{d, f\}, L\big\},$$
$$\mathcal{C}(T_3) = \big\{\{a\}, \{b\}, \{c\}, \{d\}, \{e\}, \{f\}, \{a, b\}, \{c, d, e, f\}, \{d, e, f\}, L\big\},$$

The majority clusters of $\mathcal{S}$ are: $\{a\}, \{b\}, \{c\}, \{d\}, \{e\}, \{f\}, \{a, b\}, \{a, b, c\}, \{d, e, f\}, L$.

1.2   Previous Work

The majority rule consensus tree was introduced by Margush and McMorris [13] in 1981. In 1985, Wareham [21] published a deterministic algorithm for building the majority rule consensus tree with a worst-case running time of $O(n^2 + nk^2)$. This was the record until very recently; in [11], we developed a faster deterministic algorithm with $O(nk \log k)$ worst-case running time, based on recursion.

As for *randomized* methods, Amenta *et al.* [2] gave an algorithm with expected running time $O(nk)$ but unbounded worst-case running time. Here, randomization is used to count and store the number of occurrences of clusters from $\mathcal{S}$ in suitably constructed hash tables. We note that the implementations for computing majority rule consensus trees in existing software packages such as PHYLIP [9], MrBayes [16], SumTrees in DendroPy [18], COMPONENT [15], and PAUP* [20] also rely on randomization, and typically have unbounded worst-case running times as well.

1.3 New Results and Organization of the Paper

This paper presents a deterministic algorithm for computing the majority rule consensus tree. Its worst-case running time is $O(nk)$, which is optimal because the size of the input is $\Omega(nk)$. We thus resolve a long-standing open problem in Phylogenetics.

To ensure that our algorithm is practical, we implemented it and performed a series of experiments to compare its actual running time to that of the majority rule consensus tree method in PHYLIP [9]. (We chose PHYLIP's majority rule consensus tree method as a benchmark because it is freely available, frequently used in practice, and faster than many other methods such as SumTrees in DendroPy [18] and COMPONENT [15].) The experiments showed that our deterministic method is much faster than PHYLIP for certain types of large inputs, e.g., when $n \gg k$, implying that randomization may not be necessary in many cases.

The rest of the paper is organized as follows. Section 2 summarizes a few results from the literature that are needed later. To help us find an efficient solution to the majority rule consensus tree problem, Section 3 outlines a technique for identifying all majority elements in a list $\mathcal{W}$ of subsets of a fixed set, where a *majority element* is defined to be any element that occurs in more than half of the subsets in $\mathcal{W}$. This technique is subsequently employed in our new majority rule consensus tree algorithm, named `Fast_Maj_Rule_Cons_Tree`, which is described and analyzed in Section 4. Next, Section 5 reports the running times of our prototype implementation of `Fast_Maj_Rule_Cons_Tree` when applied to some simulated data sets. Finally, the availability of the prototype implementation is discussed in Section 6.

2 Preliminaries

We shall make use of the following results from the literature. (For further details, see the respective original references.)

2.1 Day's Algorithm [6]

Day's algorithm [6] takes two trees T_{ref} and T with identical leaf label sets as input. After linear-time preprocessing, the algorithm can check whether or not any specified cluster that occurs in T also occurs in T_{ref}, and each such check can be performed in constant time.

Theorem 1. *(Day [6]) Let T_{ref} and T be two given trees with $\Lambda(T_{ref}) = \Lambda(T) = L$ and let $n = |L|$. After $O(n)$ time preprocessing, it is possible to determine, for any $u \in V(T)$, if $\Lambda(T[u]) \in \mathcal{C}(T_{ref})$ in $O(1)$ time.*

2.2 Procedure `One-Way_Compatible` [11]

`One-Way_Compatible` is a linear-time procedure defined in Section 4.1 of [11]. Its input is two trees T_1 and T_2 with identical leaf label sets, and its output is a copy of T_1 in which every cluster that is not compatible with T_2 has been removed. The procedure is asymmetric; for example, if T_1 consists of n leaves attached to a root node and $T_2 \neq T_1$ then One-Way_Compatible$(T_1, T_2) = T_1$, while One-Way_Compatible$(T_2, T_1) = T_2$.

Theorem 2. *([11]) Let T_1 and T_2 be two given trees with $\Lambda(T_1) = \Lambda(T_2) = L$ and let $n = |L|$. Procedure One-Way_Compatible(T_1, T_2) returns a tree T with $\Lambda(T) = L$ such that $\mathcal{C}(T) = \{C \in \mathcal{C}(T_1) : C$ is compatible with $T_2\}$ in $O(n)$ time.*

2.3   Procedure Merge_Trees [11]

The procedure Merge_Trees from Section 2.4 in [11] combines all the clusters from two compatible trees into one tree in linear time.

Theorem 3. *([11]) Let T_1 and T_2 be two given trees with $\Lambda(T_1) = \Lambda(T_2) = L$ that are compatible and let $n = |L|$. Procedure Merge_Trees(T_1, T_2) returns a tree T with $\Lambda(T) = L$ and $\mathcal{C}(T) = \mathcal{C}(T_1) \cup \mathcal{C}(T_2)$ in $O(n)$ time.*

2.4   The *delete* and *insert* Operations on a Tree

Let T be a tree and let u be any non-root, internal node in T. Applying the *delete* operation on u modifies T as follows: First, all children of u become children of the parent of u, and then u and the edge between u and its parent are removed. See Fig. 2 for an illustration. Note that by applying the *delete* operation on node u, the cluster $\Lambda(T[u])$ is removed from the cluster collection $\mathcal{C}(T)$ while all other clusters are preserved. Also note that the time needed for this operation is proportional to the number of children of u.

The *insert* operation is the inverse of the *delete* operation. It inserts a new internal node into T, thereby creating an additional cluster in $\mathcal{C}(T)$.

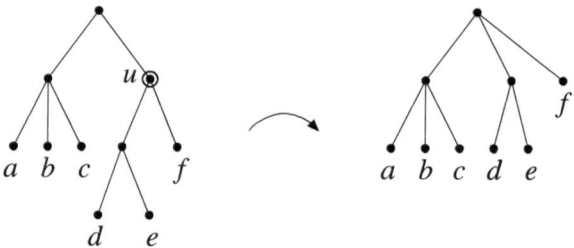

Fig. 2. Figure from [11]. Let T be the tree on the left and let u be the marked node. Then $\Lambda(T[u]) = \{d, e, f\}$ and applying the *delete* operation on u removes the cluster $\{d, e, f\}$ from $\mathcal{C}(T)$.

3   Finding All Majority Elements

In this section, we describe a technique for solving a problem closely related to the majority rule consensus tree problem: Given a list $\mathcal{W}$ of subsets of a set X, output all majority elements in $\mathcal{W}$, where a *majority element in* $\mathcal{W}$ is defined to be any element of X that occurs in more than half of the subsets in $\mathcal{W}$. It can be solved easily by using one counter for each element in X, but when $|X|$ is very large and many elements from X never occur in $\mathcal{W}$ at all, we need a method whose time complexity does not depend on $|X|$.

Denote $k = |\mathcal{W}|$, and for any $j \in \{1, 2, \ldots, k\}$, let $\mathcal{W}[j]$ be the jth subset in the list $\mathcal{W}$. For our purposes, it is sufficient to focus on the restriction of the problem in which X is an ordered set and each $\mathcal{W}[j]$ is specified as a sorted list. The following two-phase algorithm solves the restricted problem by maintaining a set of *current candidates*, which are certain elements belonging to X, along with a counter for each current candidate:

- Phase 1: Initialize the set of current candidates as the empty set. Sweep through $\mathcal{W}$, i.e., for each $j \in \{1, 2, \ldots, k\}$, consider $\mathcal{W}[j]$ and do the following. Firstly, for every current candidate x, increase x's counter by 1 if $x \in \mathcal{W}[j]$, or decrease it by 1 if $x \notin \mathcal{W}[j]$; if x's counter reaches 0 then remove x from the set of current candidates. Secondly, insert every $x \in \mathcal{W}[j]$ which is not a current candidate into the set of current candidates and initialize its counter to 1.
- Phase 2: Let X' be the set of current candidates. Sweep through $\mathcal{W}$ one more time to count the total number of occurrences in $\mathcal{W}$ of every element in X'. Output the ones that occur more than $\frac{k}{2}$ times.

As an example, let $X = \{a, b, c, d, e\}$ and $\mathcal{W} = (\mathcal{W}[1], \mathcal{W}[2], \mathcal{W}[3]) = (\{a, b, d\}, \{a, c\}, \{d, e\})$. Then the set of current candidates at the end of Phase 1 will be $\{a, d, e\}$. In Phase 2, the algorithm outputs a and d.

To prove the correctness of this method, observe that for any $x \in X$, if x occurs in more than $\frac{k}{2}$ subsets in $\mathcal{W}$, then x must be one of the current candidates at the end of Phase 1 because its counter is > 0. Hence, all majority elements in $\mathcal{W}$ (if any) belong to the set X'. However, as in the example above, some non-majority elements might also be included in X'. For this reason, Phase 2 is used to identify those elements that indeed occur more than $\frac{k}{2}$ times. To analyze the time complexity, since each $\mathcal{W}[j]$ is given as a sorted list, it is easy to maintain the set of current candidates in a sorted list and implement all operations for that value of j in time proportional to the number of current candidates. This yields:

Lemma 1. *Let X be an ordered set and let $\mathcal{W}$ be a list of sorted subsets of X. The above algorithm outputs all majority elements in $\mathcal{W}$ in $O(k \cdot y)$ time, where $k = |\mathcal{W}|$ and at most y elements from X belong to the set of current candidates at any point in time.*

Remark: Boyer and Moore's classical algorithm in [4] solves the special case of the problem where every subset in the list $\mathcal{W}$ has cardinality 1. The algorithm presented above can be viewed as an extension of [4].

4   An Optimal Algorithm for the Majority Rule Consensus Tree

This section presents the new algorithm Fast_Maj_Rule_Cons_Tree for building the majority rule consensus tree of an input collection $S = \{T_1, T_2, \ldots, T_k\}$ of identically leaf-labeled trees, where $\Lambda(T_1) = \Lambda(T_2) = \cdots = \Lambda(T_k) = L$. It uses the technique from Section 3 to locate all majority clusters in S by interpreting X as the set of all possible clusters of L (so that every element $x \in X$ is a subset of L) and the list $\mathcal{W}$ as the length-k sequence of cluster collections of the trees in S. In other words, $\mathcal{W} = (\mathcal{W}[1], \mathcal{W}[2], \ldots, \mathcal{W}[k]) = (\mathcal{C}(T_1), \mathcal{C}(T_2), \ldots, \mathcal{C}(T_k))$, and for every $j \in \{1, 2, \ldots, k\}$, it holds that $\mathcal{W}[j] \subseteq X$.

Algorithm Fast_Maj_Rule_Cons_Tree also consists of two phases. In Phase 1, it finds all clusters that might be majority clusters, and then, in Phase 2,

---

Algorithm   Fast_Maj_Rule_Cons_Tree

Input: A collection $S = \{T_1, T_2, \ldots, T_k\}$ of trees with $\Lambda(T_1) = \Lambda(T_2) = \cdots = \Lambda(T_k)$.

Output: The majority rule consensus tree of S.

/* Phase 1 */

1   $T := T_1$

2   for each $v \in V(T)$ do $count(v) := 1$

3   for $j := 2$ to k do

3.1   for each $v \in V(T)$ in top-down order do
          if $\Lambda(T[v])$ occurs in T_j then $count(v) := count(v) + 1$
          else $count(v) := count(v) - 1$; if $count(v)$ reaches 0 then delete node v.
        endfor

3.2   for every cluster C in T_j that is compatible with T but does not occur in T do
          Insert C into T.
          Initialize $count(v) := 1$ for the new node v satisfying $\Lambda(T[v]) = C$.
        endfor
      endfor

/* Phase 2 */

4   for each $v \in V(T)$ do $count(v) := 0$

5   for $j := 1$ to k do

5.1   for each $v \in V(T)$ do
          if $\Lambda(T[v])$ occurs in T_j then $count(v) := count(v) + 1$

6   for each $v \in V(T)$ in top-down order do
      if $count(v) \leq k/2$ then perform a delete operation on v.

7   return T

End Fast_Maj_Rule_Cons_Tree

Fig. 3. The pseudocode for Algorithm Fast_Maj_Rule_Cons_Tree

eliminates those candidates that do not occur in more than $\frac{k}{2}$ of the trees in $\mathcal{S}$. Whatever clusters that remain must be the majority clusters of $\mathcal{S}$. During the algorithm's execution, the current candidates are stored as nodes in a tree T, as explained below.

The pseudocode is summarized in Fig. 3. Phase 1 and Phase 2 are described in Sections 4.1 and 4.2, respectively. To achieve a good time complexity, some steps of the algorithm are implemented by applying Day's algorithm [6] and the procedures One-Way_Compatible and Merge_Trees mentioned in Section 2; the details are given in Section 4.3.

4.1 Description of Phase 1

Phase 1 of the algorithm examines the trees $T_1, T_2, \ldots, T_k$ in sequential order. As in Section 3, the algorithm maintains a set of current candidates, each equipped with its own counter. Every current candidate is some cluster of L and thus an element from X, like before. However, there are two crucial differences between Fast_Maj_Rule_Cons_Tree and the method in Section 3.

The first difference is that Fast_Maj_Rule_Cons_Tree does not store the set of current candidates in a sorted list as in Section 3, but encodes them as *nodes in a tree* T whose leaf label set equals L. (This is the key to getting an efficient algorithm.) To be precise, every node v in T represents a current candidate cluster $\Lambda(T[v])$ and has a counter $count(v)$. For any $j \in \{1, 2, \ldots, k\}$, when treating tree T_j, all clusters in $\mathcal{C}(T)$ that also belong to $\mathcal{C}(T_j)$ get their counters incremented by 1, while all clusters in $\mathcal{C}(T)$ that do not belong to $\mathcal{C}(T_j)$ get their counters decremented by 1. If this leads to some counter reaching 0 then the internal node in T corresponding to that cluster is deleted. Next, all other clusters in $\mathcal{C}(T_j)$ that are not current candidates but are compatible with T are upgraded to current candidate-status by inserting them into T and initializing their corresponding nodes' counters to 1.

The other important difference between this approach and the one in Section 3 is that for any $j \in \{2, \ldots, k\}$, a cluster C that occurs in T_j but is not a current candidate does not automatically become a current candidate; C will only be inserted into T if it is pairwise compatible with all the current candidates. We therefore need an additional lemma to guarantee the correctness of Phase 1:

Lemma 2. *For any $C \subseteq L$, if C is a majority cluster of $\mathcal{S}$ then $C \in \mathcal{C}(T)$ at the end of Phase 1.*

Proof. Suppose that C is a majority cluster of $\mathcal{S}$. During the execution of Phase 1, for any $j \in \{1, 2, \ldots, k\}$, say that C is *blocked in iteration j* if the following happens: C is not a current candidate, C occurs in tree T_j, and C is not allowed to become a current candidate because C is not compatible with the current T.

Let a denote the number of trees in $\mathcal{S}$ in which C occurs. By the definition of a majority cluster, $a > \frac{k}{2}$. Hence, there are $k - a < \frac{k}{2}$ trees in $\mathcal{S}$ in which C does not occur. We claim that each such tree T_x can cancel out the effect on C's counter of at most one of the a occurrences of C in $\mathcal{S}$. To prove the claim, let T_x be any tree in $\mathcal{S}$ in which C does not occur and consider the two possible cases:

- If C is a current candidate when T_x is considered, then C's counter will be decremented by 1.
- If C is not a current candidate when T_x is considered, then some clusters which are not pairwise compatible with C may get their counters incremented by 1. As a result, C may be blocked in another iteration.

Next, since $a - (k - a) > \frac{k}{2} - \frac{k}{2} = 0$, the counter for C will have a non-zero value at the end of Phase 1. By the definition of the tree T in the algorithm, $C \in \mathcal{C}(T)$ holds.   □

4.2   Description of Phase 2

Phase 2 of the algorithm is straightforward. It checks how many times every cluster in the tree T occurs among $T_1, T_2, \ldots, T_k$. Any clusters that do not occur more than $\frac{k}{2}$ times are removed from T. It follows immediately from Lemma 2 that the cluster collection of the remaining tree T equals the set of all majority clusters of $\mathcal{S}$. Hence, the output of the algorithm is the majority rule consensus tree.

Lemma 3. *The tree output by Algorithm* Fast_Maj_Rule_Cons_Tree *at the end of Phase 2 is the majority rule consensus tree of* $\mathcal{S}$.

4.3   Time Complexity Analysis

We now analyze the worst-case time complexity of Fast_Maj_Rule_Cons_Tree.

Theorem 4. *Algorithm* Fast_Maj_Rule_Cons_Tree *constructs the majority rule consensus tree of* $\mathcal{S}$ *in* $O(nk)$ *worst-case time, where* $n = |L|$ *and* $k = |\mathcal{S}|$.

Proof. We first show that in Phase 1, every iteration of the main loop in Step 3 takes $O(n)$ time. To perform Step 3.1 in $O(n)$ time, run Day's algorithm [6] with $T_{ref} = T_j$ and then check each $\Lambda(T[v])$ to see if it occurs in T_j. By Theorem 1, this requires $O(n)$ time for preprocessing, and each of the $O(n)$ nodes in $V(T)$ can be checked in $O(1)$ time. The *delete* operations take $O(n)$ time in total since the nodes are handled in top-down order (every node is moved at most once because if some node is deleted and its children moved then these children will not need to be moved again in the same iteration). Next, Step 3.2 can be implemented in $O(n)$ time by letting $P := $ One-Way_Compatible(T_j, T) and $Q := $ Merge_Trees(P, T), and then updating the structure of T to make T isomorphic to the obtained Q (and setting the counters of all new nodes to 1). This works because according to Theorem 2, P is a tree consisting of the clusters occurring in T_j that are compatible with the set of current candidates, and by Theorem 3, Q is the result of inserting each such cluster into T, if it did not already occur in T. There are $O(k)$ iterations in the main loop, so Phase 1 takes $O(nk)$ time.

In Phase 2, Step 5.1 is executed in $O(n)$ time, again by applying Day's algorithm [6] with $T_{ref} = T_j$ so that each $\Lambda(T[v])$ can be checked in $O(1)$ time.[3] Thus, the loop in Step 5 takes $O(nk)$ time. Step 6 can be carried out in $O(n)$ time by treating the nodes in top-down order as above. In total, Phase 2 also takes $O(nk)$ time.   □

5   Experimental Results

We implemented Fast_Maj_Rule_Cons_Tree in C++ and compared its worst-case and average running times to those of PHYLIP [9] and the previously fastest ($O(nk \log k)$ time) deterministic algorithm from [11] for some simulated data sets. The experiments were run on Ubuntu Nutty Narwhal, a 64-bit operating system with 8.00 GB RAM, and a 2.20 GHz CPU. Below, we refer to the majority rule consensus tree method in PHYLIP as "M-PHYLIP", the implementation of the algorithm in [11] as "M-Fast-v1", and the implementation of the new algorithm Fast_Maj_Rule_Cons_Tree presented in this paper as "M-Fast-v2".

We generated 10 data sets for various specified values of the parameters n and k with the method described in Section 6.2 of [11], applied the three majority consensus tree methods to each data set, and measured the running times. First, the following values of n and k were evaluated:

- (a) $n = 500$, $k = 1000$
- (b) $n = 1000$, $k = 500$
- (c) $n = 2000$, $k = 1000$
- (d) $n = 5000$, $k = 100$

The worst-case and average running times (in seconds) are reported below.

(a) $n = 500$, $k = 1000$:

	Worst-case	Average
M-PHYLIP	1.94	1.88
M-Fast-v1	8.10	8.00
M-Fast-v2	3.72	3.69

(b) $n = 1000$, $k = 500$:

	Worst-case	Average
M-PHYLIP	3.50	3.19
M-Fast-v1	7.54	7.38
M-Fast-v2	3.80	3.67

(c) $n = 2000$, $k = 1000$:

	Worst-case	Average
M-PHYLIP	34.07	30.03
M-Fast-v1	32.24	31.96
M-Fast-v2	16.09	14.86

(d) $n = 5000$, $k = 100$:

	Worst-case	Average
M-PHYLIP	93.25	90.04
M-Fast-v1	6.41	6.27
M-Fast-v2	4.40	4.28

---

[3] This way of counting occurrences of clusters has been used elsewhere in the literature, e.g., in [21] and on p. 217 of [19].

The experimental results indicate that `Fast_Maj_Rule_Cons_Tree` is exceptionally useful when n is large. For example, when $n = 5000$ and $k = 100$, it is about 20 times faster than M-PHYLIP. On the other hand, M-PHYLIP is faster in practice for inputs with $n \ll k$.

Next, we tried the methods on some even bigger inputs. M-PHYLIP returned "Error allocating memory" for $n = 2000$, $k \geq 2000$, whereas M-Fast-v2 worked fine and obtained the following worst-case and average running times.

(e) $n = 2000$, $k = \{2000, 3000, 4000, 5000\}$:

k	Worst-case	Average
2000	31.22	30.86
3000	47.42	46.23
4000	62.54	61.88
5000	78.96	77.78

This shows that `Fast_Maj_Rule_Cons_Tree` may come in handy when analyzing large phylogenetic data sets.

6   Concluding Remarks

We have proved that the majority rule consensus tree can be built in (optimal) $O(nk)$ time in the worst case, without using randomization. Although this might at first appear to be a purely theoretical result, it has practical implications as well. The experiments demonstrated that our deterministic algorithm `Fast_Maj_Rule_Cons_Tree` is much faster than randomized methods such as the one found in PHYLIP [9] when the input trees are very large, i.e., when $n \gg k$. In contrast to current practice, this suggests that it might not always be a good idea to use randomization and hashing when computing majority rule consensus trees.

We hope that the new algorithm will be a helpful tool for bioinformaticians working with huge phylogenetic trees in the future. We have included it in the FACT (Fast Algorithms for Consensus Trees) package [11] at:

http://compbio.ddns.comp.nus.edu.sg/~consensus.tree/

The C++ source code of our prototype implementation used in Section 5 can also be downloaded from the same webpage.

References

1. Adams III., E.N.: Consensus techniques and the comparison of taxonomic trees. Systematic Zoology 21(4), 390–397 (1972)
2. Amenta, N., Clarke, F., St. John, K.: A Linear-Time Majority Tree Algorithm. In: Benson, G., Page, R.D.M. (eds.) WABI 2003. LNCS (LNBI), vol. 2812, pp. 216–227. Springer, Heidelberg (2003)

3. Bansal, M.S., Dong, J., Fernández-Baca, D.: Comparing and aggregating partially resolved trees. Theoretical Computer Science 412(48), 6634–6652 (2011)
4. Boyer, R.S., Moore, J.S.: MJRTY – A Fast Majority Vote Algorithm. In: Boyer, R.S. (ed.) Automated Reasoning: Essays in Honor of Woody Bledsoe. Automated Reasoning Series, pp. 105–117. Kluwer Academic Publishers (1991)
5. Bryant, D.: A classification of consensus methods for phylogenetics. In: Janowitz, M.F., Lapointe, F.-J., McMorris, F.R., Mirkin, B., Roberts, F.S. (eds.) Bioconsensus. DIMACS Series in Discrete Mathematics and Theoretical Computer Science, vol. 61, pp. 163–184. American Mathematical Society (2003)
6. Day, W.H.E.: Optimal algorithms for comparing trees with labeled leaves. Journal of Classification 2(1), 7–28 (1985)
7. Degnan, J.H., DeGiorgio, M., Bryant, D., Rosenberg, N.A.: Properties of consensus methods for inferring species trees from gene trees. Systematic Biology 58(1), 35–54 (2009)
8. Felsenstein, J.: Inferring Phylogenies. Sinauer Associates, Inc., Sunderland (2004)
9. Felsenstein, J.: PHYLIP, version 3.6. Software package, Department of Genome Sciences, University of Washington, Seattle, U.S.A. (2005)
10. Gusfield, D.: Algorithms on Strings, Trees, and Sequences. Cambridge University Press, New York (1997)
11. Jansson, J., Shen, C., Sung, W.-K.: Improved algorithms for constructing consensus trees. In: Proceedings of the 24th Annual ACM-SIAM Symposium on Discrete Algorithms, SODA 2013, pp. 1800–1813. SIAM (2013)
12. Kuhner, M.K., Felsenstein, J.: A simulation comparison of phylogeny algorithms under equal and unequal evolutionary rates. Molecular Biology and Evolution 11(3), 459–468 (1994)
13. Margush, T., McMorris, F.R.: Consensus n-Trees. Bulletin of Mathematical Biology 43(2), 239–244 (1981)
14. Nakhleh, L., Warnow, T., Ringe, D., Evans, S.N.: A comparison of phylogenetic reconstruction methods on an Indo-European dataset. Transactions of the Philological Society 103(2), 171–192 (2005)
15. Page, R.: COMPONENT, version 2.0. Software package. University of Glasgow, U.K. (1993)
16. Ronquist, F., Huelsenbeck, J.P.: MrBayes 3: Bayesian phylogenetic inference under mixed models. Bioinformatics 19(12), 1572–1574 (2003)
17. Semple, C., Steel, M.: Phylogenetics. Oxford Lecture Series in Mathematics and its Applications, vol. 24. Oxford University Press (2003)
18. Sukumaran, J., Holder, M.T.: DendroPy: a Python library for phylogenetic computing. Bioinformatics 26(12), 1569–1571 (2010)
19. Sung, W.-K.: Algorithms in Bioinformatics: A Practical Introduction. Chapman & Hall/CRC (2010)
20. Swofford, D.L.: PAUP*, version 4.0. Software package. Sinauer Associates, Inc., Sunderland (2003)
21. Wareham, H.T.: An efficient algorithm for computing M_l consensus trees. B.Sc. Honours thesis, Memorial University of Newfoundland, Canada (1985)

UniNovo : A Universal Tool for *de Novo* Peptide Sequencing

Kyowon Jeong[1], Sangtae Kim[2], and Pavel A. Pevzner[2]

[1] Department of Electrical and Computer Engineering,
University of California, San Diego, CA
kwj@ucsd.edu
[2] Department of Computer Science and Engineering,
University of California, San Diego, CA
{sak008,ppevzner}@ucsd.edu

Abstract. Mass spectrometry (MS) instruments and experimental protocols are rapidly advancing, but *de novo* peptide sequencing algorithms to analyze tandem mass (MS/MS) spectra are lagging behind. While existing *de novo* sequencing tools perform well on certain types of spectra (e.g., Collision Induced Dissociation (CID) spectra of tryptic peptides), their performance often deteriorates on other types of spectra, such as Electron Transfer Dissociation (ETD), Higher-energy Collisional Dissociation (HCD) spectra, or spectra of non-tryptic digests. Thus, rather than developing a new algorithm for each type of spectra, we develop a *universal de novo* sequencing algorithm called UniNovo that works well for all types of spectra or even for spectral pairs (e.g., CID/ETD spectral pairs). The performance of UniNovo is compared with PepNovo+, PEAKS, and pNovo using various types of spectra. The results show that the performance of UniNovo is superior to other tools for ETD spectra and superior or comparable to others for CID and HCD spectra. UniNovo also estimates the probability that each reported reconstruction is correct, using simple statistics that are readily obtained from a small training dataset. We demonstrate that the estimation is accurate for all tested types of spectra (including CID, HCD, ETD, CID/ETD, and HCD/ETD spectra of trypsin, LysC, or AspN digested peptides). The appendix is available online at http://proteomics.ucsd.edu/Software/UniNovo.html.

1   Introduction

De novo peptide sequencing by tandem mass (MS/MS) spectrometry is a valuable alternative to MS/MS database search. In contrast to the database search approach that utilizes the information from proteome, the *de novo* sequencing approach attempts to identify peptides only using the information from the input spectrum. Hence, most *de novo* sequencing algorithms are based on the prior knowledge of the fragmentation characteristics (e.g., ion types and their propensities) of MS/MS spectra [27,12,11].

The fragmentation characteristics are highly dependent on the fragmentation method used to generate the spectrum. Among several fragmentation methods available, the collision induced dissociation (CID) is the most commonly

M. Deng et al. (Eds.): RECOMB 2013, LNBI 7821, pp. 100–117, 2013.
© Springer-Verlag Berlin Heidelberg 2013

used method. Accordingly, the fragmentation characteristics of CID have been well studied compared to recently introduced fragmentation methods, such as electron transfer dissociation (ETD) and higher-energy collisional dissociation (HCD) [18,36,4,35,15,2]. As a result, many *de novo* sequencing algorithms have been introduced for CID spectra; for example, PEAKS [27] and PepNovo+ [12,11], are the state of the art *de novo* sequencing tools for CID spectra.

Other fragmentation methods like ETD and HCD have a great potential for *de novo* sequencing. For example, for highly charged spectra, ETD provides better fragmentation and thus is better suited for *de novo* sequencing than CID [37,33]. Also, more complete fragmentation of peptide ions (especially in low mass regions) in HCD provides a better chance to obtain more accurate *de novo* reconstructions than CID [30,6]. Furthermore, modern mass spectrometers (e.g., LTQ-Orbitrap Velos) allow the generation of paired spectra (e.g., CID/ETD or HCD/ETD spectral pairs). Since CID (or HCD) and ETD spectra provide complementary information for peptide sequencing [32,8,14], such spectral pairs (or even triplets) enable more accurate *de novo* sequencing.

Several *de novo* sequencing algorithms were recently presented to take advantage of those new fragmentation methods. For instance, [25] proposed a *de novo* sequencing algorithm for ETD spectra, which is used by PEAKS. For HCD spectra, [6] introduced a *de novo* sequencing tool, pNovo, that not only takes advantage of the high precision peaks in HCD spectra but also uses the information of abundant immonium and internal ions. In case of spectral pairs, [32] proposed a greedy algorithm (for CID/ECD spectral pairs) that significantly boosts the performance of *de novo* sequencing. [8] presented Spectrum Fusion, a *de novo* sequencing algorithm for CID/ETD spectral pairs. Spectrum Fusion constructs a combined spectrum from the input CID/ETD spectral pair using a Bayesian Network. It generates multiple *de novo* sequences using the combined spectrum and score them by the scoring function in ByOnic [3]. [14] also presented a *de novo* sequencing algorithm, ADEPTS, for CID/ETD spectral pairs. Given a CID/ETD spectral pair, ADEPTS first finds 1,000 candidate *de novo* sequences from each spectrum, using PEAKS. The total 2,000 candidate sequences are then rescored against the input spectral pair, and the best-scoring peptide is reported.

While the above tools perform well for the spectra generated from the fragmentation method(s) that each tool targeted, they often generate inferior results for the spectra from other fragmentation methods. Moreover, if alternative proteases (e.g., LysC or AspN) are used for protein digestion, these tools may produce suboptimal results because different proteases often generate peptides with different fragmentation characteristics [24].

In case of the database search approach, [24] recently introduced a universal algorithm MS-GFDB that shows a significantly better peptide identification performance than other existing database search tools such as Mascot+Percolator [31,19]. However, a universal *de novo* sequencing tool is still missing.

We present UniNovo, a universal *de novo* sequencing tool that can be generalized for various *types* (i.e., the combinations of the fragmentation method and the protease used to digest sample proteins) of spectra. The scoring function of UniNovo is easily trainable using a training dataset consisting of thousands of annotated spectra. All information needed for *de novo* sequencing are learned from the training dataset, and the running time for training is less than 5 hours in a typical desktop environment. Currently UniNovo is trained for CID, HCD, and ETD spectra of trypsin, LysC, or AspN digested peptides. We show that the performance of UniNovo is better than or comparable to PepNovo+, PEAKS, and pNovo for various types of spectra.

One of the biggest challenges in *de novo* sequencing is to estimate the error rate of the resulting *de novo* reconstructions. Unlike MS/MS database search tools that commonly uses the *target-decoy approach* [9,28] to estimate the statistical significance of the peptide identifications, *de novo* reconstructions have rarely been subjected to a statistical significance analysis in the past.

Several *de novo* sequencing tools report the error rate of amino acid predictions (e.g. confidence scores in PEAKS), but this is often not sufficient because the overall quality of the sequence cannot be easily determined by the error rates of individual amino acid predictions. To our knowledge, only PepNovo+ reports the empirical probability that the output peptide is correct. PepNovo+ predicts the probabilities using logistic regression with multiple features of the reconstructions such as length and score, which are extracted from a training dataset consisting of hundreds of thousands of annotated spectra [11]. However, PepNovo+ does not include an automated training procedure (that would allow to easily extend PepNovo+ for newly emerging mass spectrometry approaches) and is currently trained only for CID.[1] Thus, in case of non-CID fragmentation methods, it remains unclear how to obtain accurate error rate estimation for *de novo* reconstructions.

UniNovo estimates the probability that each reported reconstruction is correct, using simple statistics that are readily obtained from a small training dataset. We demonstrate that the estimation is accurate for all tested types of spectra (including CID, HCD, ETD, CID/ETD, and HCD/ETD spectra of trypsin, LysC, or AspN digested peptides). This allows UniNovo to automatically filter out low quality spectra.

2   Methods

Similar to [23], we first describe the algorithm on a simplified model that assumes the following:

- the masses of amino acids are integers (e.g., the mass of Gly is 57).
- the m/z (mass to charge ratio) of peaks (in spectra) are integers.
- the intensity of all peaks is 1.

---

[1] Extending PepNovo+ beyond CID spectra requires training complex boosting-based re-ranking models for predicting peak ranks and rescoring peptide candidates. Pep-Novo+ training includes several manual steps and the availability of a very large corpus of training spectra (Ari Frank, personal communication, October 5, 2012).

- only N-terminal charge 1 ions are considered (e.g., b, c, or $b - H_2O$ ions, but not y-ion series).
- the *parent mass* (the mass of the precursor ion) of a spectrum equals to the mass of the peptide that generated the spectrum.

The algorithm on a more realistic model is described in the Appendix section A2.

Terminology and Definitions. Let A be the set of amino acids with (integer) masses $m(a)$ for $a \in A$. A *peptide* $a_1 a_2 \cdots a_k$ is a sequence of amino acids, and the mass of a peptide is the total mass of amino acids in the peptide. We represent a peptide $a_1 a_2 \cdots a_k$ with mass n by a Boolean vector $P = (P_1, \cdots, P_n)$, where $P_i = 1$ if $i = \sum_{t=1}^{j} a_t$ for $0 < j < k$, and $P_i = 0$ otherwise. If $P_i = 1$, we call a mass i a *fragmentation site*. For example, suppose there are two amino acids A and B with masses 2 and 3, respectively. Then, the peptide $ABBA$ has the mass of $2 + 3 + 3 + 2 = 10$ and is represented by a Boolean vector $(0, 1, 0, 0, 1, 0, 0, 1, 0, 0)$. The fragmentation sites of this peptide are, thus, 2, 5, and 8.

A *spectrum* is a list of peaks, where each peak is specified by an m/z. We represent a spectrum of parent mass n by a Boolean vector $S = (S_1, \cdots, S_n)$, where $S_i = 1$ if the peak of m/z i (or simply the peak i) is present and $S_i = 0$ otherwise.

A *peptide-spectrum match (PSM)* is a pair (P, S) formed by a peptide P and a spectrum S. Given an integer δ called an *ion type* and a PSM (P, S), we say a peak i is a δ-*ion peak* (with respect to P) if $i - \delta$ is a fragmentation site, that is, $P_{i-\delta} = 1$. In this model, the ion type can be any integer. In the connection to the experimental MS/MS spectra, ion types can represent common singly charged N-terminal ions; for example, the ion types 1 and -27 represent b and a ions, respectively.

Given an integer f called a *feature* and a spectrum S, we say that a peak i *satisfies* f if another peak $i + f$ is present in the spectrum, that is, $S_{i+f} = 1$. For instance, a peak 30 satisfies a feature $f = -18$, if $S_{30-18} = 1$. In experimental spectra, various ions are often observed along with neutral losses (e.g., b-ion and $b - H_2O$-ion) or with related ions (e.g., b-ion and a-ion). A feature describes the relation (the shift of m/z values in this simplified model) between two peaks that may correspond to a neutral loss or a mass gain/loss between related ions. For example, since we are dealing with only charge 1 ions, a water loss (from any ions) is represented by the feature $f = -18$, and the mass gain from a-ion to b-ion is represented by the feature $f = +27$.

Peptide-Spectrum Generative Model. We model how a peptide P (of mass n) generates a spectrum S. Departing from a 1-step generative model in [1] or [23], we introduce a more adequate 2-step probabilistic model in which the dependency between different ions can be described.

Assume that we are given the set of ion types (the *ion type set* Δ) and the set of features (the *feature set* F). For simplicity, we consider the case where only one ion type $\delta = 0$ is in Δ and one feature f is in F. Given a peptide

P, a *partial-spectrum* s is generated per each element P_i of P as follows: The probability that $s_i = 1$ is given by α if $P_i = 1$ or by β otherwise (the first generation step). This first step can be characterized by a 2×2 matrix called the *ion type matrix* (Figure 1). When $s_i = 1$, the probability that $s_{i+f} = 1$ (i.e., the peak i satisfies f) is given by μ if $P_i = 1$ or ν otherwise (the second generation step). The second step is characterized by the *feature-ion type matrix* (Figure 1).[2] The second step can describe the dependency between different ions (or an ion and its neutral loss) from the same fragmentation site. If multiple ion types and multiple features are considered, the ion type matrix should be defined per ion type, and the feature-ion type matrix per ion type and per feature. The spectrum S is generated by taking elementwise OR operation for the generated partial-spectra s.

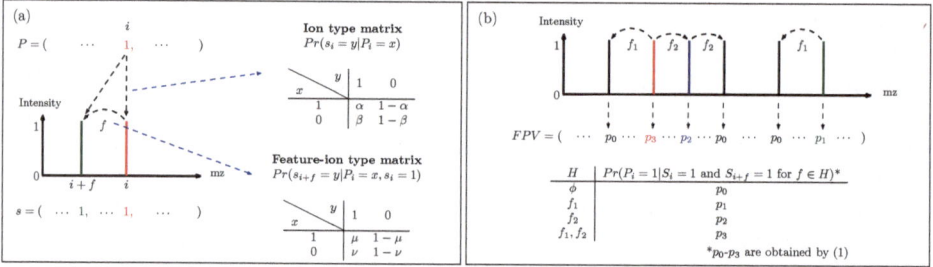

Fig. 1. (a) The generation of a partial-spectrum s for P_i. One ion type $\delta = 0$ and one feature f are considered. The probability that $s_i = 1$ is given by α if $P_i = 1$ or by β otherwise. When $s_i = 1$, the probability that $s_{i+f} = 1$ (i.e., the peak i satisfies f) is given by μ if $P_i = 1$ or by ν otherwise. The spectrum is generated by taking elementwise OR operation for generated partial-spectra for all elements of P.
(b) The calculation of the fragmentation probability vector FPV from a spectrum S (without knowing the peptide P that generated S). We consider one ion type $\delta = 0$ and two features f_1 and f_2. The events "a peak satisfies f_1" and "a peak satisfies f_2" are assumed to be independent. To derive FPV_i, first we examine which features the peak i satisfies in the spectrum S. Denote the features the peak i satisfies by H. Second, given H, we calculate the probability that $P_i = 1$ (using the probabilities given in ion type matrix and feature-ion type matrix - see the equation (1)).

Training UniNovo. Since the ion type matrices and feature-ion type matricies fully describe the generation of a spectrum, in the training step, UniNovo learns these matrices from the *training dataset* (a set of PSMs). To define these matrices, the ion type set Δ and the feature set F should be formed. Using the *offset frequency function* introduced in [7], we collect frequently observed ion types and form the ion type set Δ. Likewise, we collect frequently observed features and form the feature set F (see the Appendix section A1.2). From here on, we only consider ion types in the ion type set Δ and features in the feature set F.

---

[2] Given $s_i = 0$, the probability that $s_{i+f} = 1$ is assumed to be 0.

Next UniNovo learns the ion type and feature-ion type matrices that characterize the generative model of the PSMs in the training dataset. For example, $\alpha = Pr(s_i = 1|P_i = 1)$ can be empirically determined if partial-spectra s are given. However, it is not clear how to decompose a spectrum S into partial-spectra s (since partial spectra may share peaks in the spectrum). As a compromise, we learn $Pr(S_i = 1|P_i = 1)$ for estimation of α. Other probabilities are also empirically determined similarly by substituting the partial-spectra by the spectrum.

We emphasize that all the above probabilities can be learned from a small set of PSMs (e.g., 5000 PSMs per charge state are often sufficient to avoid overfitting; see the Appendix section A14) even if there are many ion types in Δ and features in F because each probability is associated to an individual ion type or a combination of an ion type and a feature, not a combination of multiple ion types and multiple features.

Lastly, we compute the probability that a random element of a peptide vector is a fragmentation site, i.e., $Pr(P_i = 1)$.[3] This probability is called the *prior fragmentation probability* and denoted by p. The detailed description of UniNovo training is given in the Appendix section A1.

How to Infer Fragmentation Sites from a Spectrum. Given a spectrum S of parent mass n, our goal is to predict the fragmentation sites of the (unknown) peptide P that generated S. For simplicity, assume that there exists a single ion type $\delta = 0$ is in the ion type set Δ (but multiple features in the feature set F). Given a peak i, define H as the set of features that the peak i satisfies. Then the fragmentation sites are predicted by solving the following Bayesian inference problem.

Fragmentation Inference Problem: Given the set of features H and P_i such that $Pr(P_i = 1) = p$ (the prior fragmentation probability), derive the posterior probability $Pr(P_i = 1|S_i = S_{i+f} = 1$ for $f \in H)$.

Since there is only one ion type, we have only one ion type matrix. On the other hand, per each feature we have a feature-ion type matrix. Let μ_f and ν_f denote μ and ν associated to the feature f, respectively. If we can assume that all features are independent (i.e., the events "$S_i = S_{i+f} = 1$ for f" are independent for $f \in H$), we obtain

$$Pr(P_i = 1|S_i = S_{i+f} = 1 \text{ for } f \in H) = \frac{\gamma \cdot \prod_{f \in H} \mu_f}{\gamma \cdot \prod_{f \in H} \mu_f + (1 - \gamma) \cdot \prod_{f \in H} \nu_f}. \quad (1)$$

where $\gamma = Pr(P_i = 1|S_i = 1) = \frac{p \cdot \alpha}{p \cdot \alpha + (1-p) \cdot \beta}$ (see the Appendix section A3 for derivation). Denote the obtained probability in (1) as π_i. We define a *fragmentation probability vector (FPV)* as a vector with n elements such that

---

[3] When masses of amino acids are rounded to integers, $Pr(P_i = 1) \approx \frac{1}{121.6}$. However, if we consider more accurate amino acid masses (for the spectra of high resolution), this probability should be learned from the training dataset.

$$FPV_i = \begin{cases} \pi_i & \text{if } S_i = 1 \\ 0 & \text{otherwise} \end{cases} \tag{2}$$

for $i = 1, \cdots, n - 1$, and $FPV_n := 1$ (see Figure 1 (b)). FPV_i is an estimated probability that $P_i = 1$ (see Figures A2 and A3, blue bars). We use FPV for the generation of *de novo* reconstructions.

The equation (1) is based on a simplified model in which a single one ion type and multiple independent features are used. However, some features are known to be strongly dependent each other (e.g., a feature describing a single water loss and a double water losses), and usually multiple ion types are present in the ion type set. Thus, in practice, per each peak, UniNovo automatically selects a small number of features (less than 10 out of thousands of features in the feature set) that are weakly correlated yet effective to determine the ion type of the peak. Assuming that the selected features are mutually independent, FPV is calculated per ion type using the equation (1), and then the final FPV is given by a weighted summation of the FPV's of different ion types. Note that there are many possible combinations of features due to the large number of all the features in the feature set (even if the number of the features to calculate FPV per peak is less than 10). Since different combinations of features are selected for different peaks, UniNovo is able to use more diverse relations between different ions as compared to other tools that typically use fixed dependencies between ions (e.g., PepNovo). The detailed description of the feature selection method and the calculation of FPV is given in the Appendix section A3.

Generating *de novo* Reconstructions. To generate *de novo* reconstructions, we first construct a *spectrum graph* [7]. Given a spectrum S of parent mass n from an unknown peptide P, the spectrum graph $G(V, E)$ is defined as a directed acyclic graph whose vertex set V consists of 0 (the source), n (the sink), and integers i such that $FPV_i > 0$. Two vertices i and j are connected by an edge (i, j) if $j - i$ equals to the mass of an amino acid or the total mass of multiple amino acids (*a mass gap*). Any path from 0 (the source) to n (the sink) in a spectrum graph corresponds to a peptide (possibly containing mass gaps). We say that a vertex i is *correct* if $P_i = 1$ and an edge (i, j) is *correct* if both vertices i and j are correct. We also say that a path r is *correct* if all vertices in r are correct. The *length* of a reconstruction is defined by the total number of amino acids and mass gaps in the reconstruction.

To score a *de novo* reconstruction, we use an additive (i.e., the score of a path is the sum of scores of vertices of the path) log likelihood ratio scoring (similar to [7]). Given a vertex i, let $FPV_i = x$. The likelihoods of the following two hypothesis for the outcome $FPV_i = x$ are tested: a) the vertex i is correct and b) the vertex i is incorrect. Let $Pr(P_i = 1|FPV_i = x) = x$. Then, we have

$$\frac{\mathcal{L}(P_i = 1|FPV_i = x)}{\mathcal{L}(P_i = 0|FPV_i = x)} = \frac{Pr(FPV_i = x|P_i = 1)}{Pr(FPV_i = x|P_i = 0)} = \frac{x}{1 - x} \cdot \frac{1 - p}{p}. \tag{3}$$

The score of the vertex i with $FPV_i = x$ is defined by $Score(i) := \left[\log \frac{x}{1-x} \cdot \frac{1-p}{p} \right]$ where $[\cdot]$ denotes the rounding to the nearest integer. Given a path r, the score of the path r is defined by $\sum\limits_{i \in r} Score(i)$.

Since an additive scoring is used, top scoring reconstructions can be efficiently generated using a dynamic programming as in [7]. We did not exclude symmetric paths in the spectrum graph that usually correspond to incorrect reconstructions. Considering only the antisymmetric paths would further enhance the performance of UniNovo [5].

After generating the reconstructions, a probability that each reconstruction is correct (termed the *accuracy* of the reconstruction) is predicted, using Hunter's bound [16] (see the Appendix section A4 for the definition of the accuracy of reconstructions). Hunter's bound can be calculated from relatively simple statistics that are readily learned from a small set of PSMs (about 5,000 PSMs). Figures A2 and A3 (green bars) in the Appendix section A9 show that the accuracy of a reconstruction is a conservative estimate of the empirical probability of the reconstruction being correct.

3 Results

Datasets. To benchmark UniNovo, we used 13 different datasets with diverse fragmentation methods (CID/ETD/HCD), digested with diverse proteases (trypsin, LysC, and AspN), and having diverse charge states (see Table 1). We re-analyzed the spectral datasets (*original datasets*) from Albert Heck's and Joshua Coon's laboratories that were previously analyzed in [24], [34], and [13]. The CID and ETD spectra in these original datasets were acquired in a hybrid linear ion trap/Orbitrap mass spectrometers (high MS1 resolution and low MS2 resolution). The HCD spectra have high MS1 and MS2 resolution. All spectra in the original datasets were identified by MS-GFDB (ver. 01/06/2012) [24] at 1% peptide-level FDR without allowing any modification except the carbamidomethylation of Cys (C+57) as a fixed modification.[4] Out of all identified spectra, we selected 1,000 spectra (or pairs of spectra) from distinct peptides randomly and formed the 13 datasets listed in Table 1. The unselected identified spectra (about 5,000-20,000 spectra depending on the type of spectra) were used for the training of UniNovo. The peptide contained in the training dataset were not contained in the above 13 datasets. See the Appendix section A10 for the detailed description of these datasets.

Benchmarking UniNovo. We benchmarked UniNovo, PepNovo+ (ver. 3.1 beta) [11], PEAKS (ver. 5.3, online) [27], and pNovo (ver. 1.1) [6] using the datasets in Table 1. For each tool, we generated N *de novo* reconstructions per each spectrum for $N = 1, 5$, and 20. We say that a spectrum is *correctly sequenced*

---

[4] In the Appendix section A12, we also re-analyzed the dataset reported in [22] that contains doubly charged CID spectra identified using Sequest [10] and PeptideProphet [20].

Table 1. Summary of the datasets used for benchmarking. Number of spectra (or spectral pairs) is 1,000 for each dataset. While UniNovo is applicable to all datasets, other tools are only applicable to (or optimized for) datasets marked by '*'. PEAKS was not tested for HCD datasets.

Dataset	CID2	CIDL2	CIDA2	ETD2	ETD3	ETDL3	ETDL4	ETDA3	ETDA4	HCD2	HCD3	CID/ETD2	CID/ETD3
Fragmentation	CID	CID	CID	ETD	ETD	ETD	ETD	ETD	ETD	HCD	HCD	CID/ETD	CID/ETD
Charge	2	2	2	2	3	3	4	3	4	2	3	2	3
Enzyme	Tryp	LysC	AspN	Tryp	Tryp	LysC	LysC	AspN	AspN	Tryp	Tryp	Tryp	Tryp
Avg. pep. length	12.6	11.4	12.3	12.5	16.4	12.5	18.7	12.8	18.9	10.5	14.5	12.3	17.1
UniNovo	*	*	*	*	*	*	*	*	*	*	*	*	*
PepNovo+	*	*	*	N/A	N/A	N/A	N/A	N/A	N/A	*	*	N/A	N/A
PEAKS	*	*	*	*	*	*	*	*	*			N/A	N/A
pNovo	N/A	N/A	N/A	N/A	N/A	N/A	N/A	N/A	N/A	*	*	N/A	N/A

if at least one of N reconstructions generated from the spectrum is correct. To evaluate the performance of each tool, the number of correctly sequenced spectra and the average length of correct reconstructions were measured for each tool.[5]

For UniNovo, the maximum number of mass gaps in a reconstruction was set to 2. UniNovo was tested for all datasets. For PepNovo+, also N top scoring reconstructions were generated per spectrum. PepNovo+ was used for CID2, CIDL2, CIDA2, HCD2, and HCD3 datasets. In case of PEAKS, we first generated 500 top scoring reconstructions per each spectrum. Then, for each reconstruction we converted amino acids with the local confidence lower than 30% into mass gaps. Such conversion is adopted because PEAKS generates reconstruction without mass gaps while UniNovo and PepNovo+ generate reconstructions with up to two mass gaps. In this procedure, multiple reconstructions without mass gaps were often converted into the same reconstruction with mass gaps. The score of a converted reconstruction is defined as the highest score of the reconstructions before conversion. Out of the converted reconstructions, N top high scoring (distinct) ones were chosen and used for further analysis. PEAKS was tested for all datasets except for HCD2 and HCD3 datasets. For pNovo, N top scoring reconstructions were generated per a spectrum.[6] Only HCD2 and HCD3 datasets were analyzed by pNovo. The parameters of each tool for each dataset is provided in the Appendix section A8.

We also indirectly compared UniNovo with MS-GFDB [24] as both tools were developed to analyze diverse types of spectra. We replaced the scoring function of UniNovo with that of MS-GFDB and generated reconstructions using the replaced scoring method. More precisely, the spectrum graph was generated by MS-GFDB per each spectrum, and the reconstructions were generated by UniNovo on that spectrum graph (instead of the spectrum graph generated by UniNovo). This generation method is specified by MS-GFDBScore. All experimental parameters for MS-GFDBScore were the same as for UniNovo.

---

[5] Since mass gaps are allowed for reconstructions, often multiple correct reconstructions were reported for a spectrum. To calculate the average length of correct reconstructions, only the top scoring correct reconstruction was counted per a spectrum.

[6] pNovo also generates reconstructions without mass gaps. However, the conversion of reconstructions as in PEAKS could not be applied to pNovo because pNovo does not report any local score.

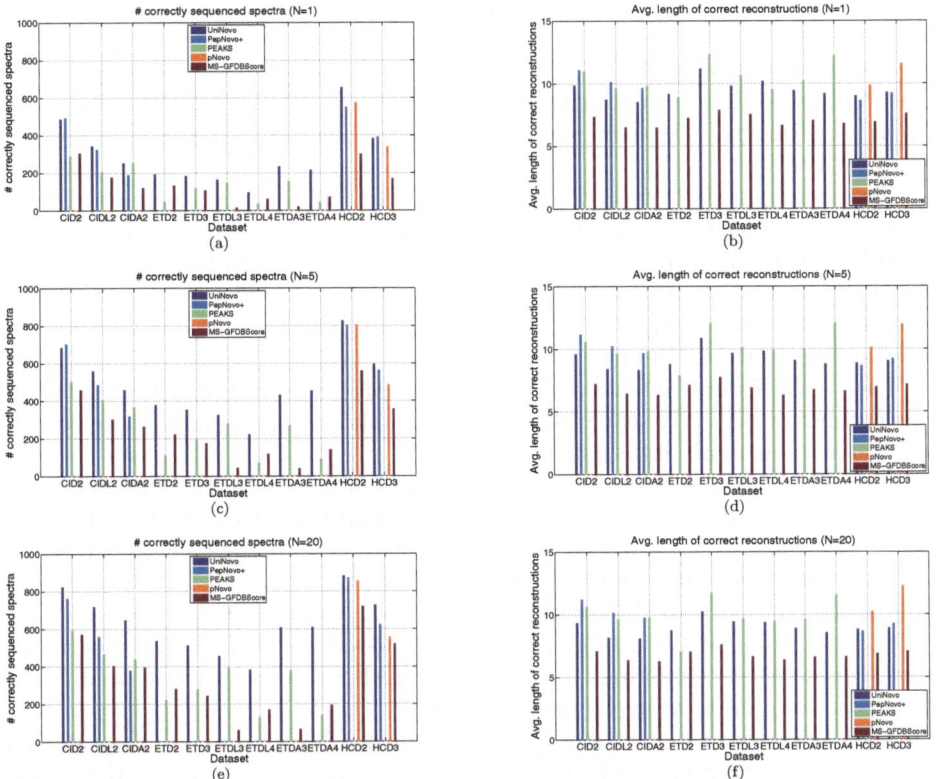

Fig. 2. Comparison of *de novo* sequencing tools (as well as a database search tool MS-GFDB [24] tweaked for *de novo* sequencing). Per each spectrum, N top scoring reconstructions were generated by UniNovo, PepNovo+ [12,11], PEAKS [27], pNovo [6], and MS-GFDBScore. MS-GFDBScore provides UniNovo with MS-GFDB's scoring function. The number of reported reconstructions per a spectrum (N) is set to 1, 5, and 20. A reconstruction is correct if all the fragmentation sites of the reconstruction are correct, and a spectrum is classified as correctly sequenced if at least one of the reconstructions generated from the spectrum is correct. Figures on the left side ((a), (c), and (e)) show the number of correctly sequenced spectra in each dataset, and figures on the right side ((b), (d), and (f)) show the average length of the correct reconstructions.

Figure 2 shows the comparison results for different datasets. UniNovo found the largest number of correctly sequenced spectra among all the tested tools in most datasets. In particular, for ETD spectra, UniNovo reported significantly more correctly sequenced spectra than PEAKS. For example, in case of ETD2 or ETDL4 dataset, the number of correctly sequenced spectra was more than twice for UniNovo than for PEAKS.

For CID spectra, UniNovo and PepNovo+ showed similar results. When $N = 1$, UniNovo and PepNovo+ found about the same number of correctly sequenced spectra in CID2 and CIDL2 datasets, but UniNovo found about 35% more correctly sequenced spectra than PepNovo+ in CIDA2 dataset.

While trypsin and LysC digested peptides generate the spectra of similar fragmentation characteristics, AspN digested peptides generate spectra with distinct fragmentation propensities. UniNovo worked well with AspN digested peptides, but PepNovo+ showed suboptimal results for the spectra of AspN digested peptides.[7] The length of correct reconstructions for PepNovo+ was slightly longer than for UniNovo.

The results on HCD spectra also demonstrate that UniNovo finds the largest number of correctly sequenced spectra in general. The reconstructions reported by pNovo were, however, longer than those by UniNovo (and PepNovo+) by 2-3 amino acids. This suggests that UniNovo still has room for improvement for HCD spectra (e.g., introducing features better reflecting the high mass resolution and information from immonium or internal ions).

The results from UniNovo were superior to MS-GFDBScore in both terms of the number of correctly sequenced spectra and the average length of the correct reconstructions in all datasets.

For each dataset, we drew the Venn diagrams of the correctly sequenced spectra (Figure 3 and Figure A4-A11) to see the overlaps of the spectra between different tools. For all datasets, the overlaps between different tools increase as N grows, as expected. Relatively small overlaps are observed for ETD spectra (as compared to CID or HCD spectra). It indicates that UniNovo may have been using some valuable features of ETD spectra missed by PEAKS (and vice versa) and suggests that combining UniNovo and PEAKS results may potentially lead to a promising *de novo* sequencing approach.

While the above results measure the sequence level accuracy, they do not directly show the amino acid level precision or recall. To measure the amino acid level precision and recall, the top scoring reconstruction was generated per spectrum for each tool (i.e., $N = 1$). For this experiment, MS-GFDB was not tested, and the reconstructions of PEAKS were not converted using the local confidence. From the generated reconstructions, the number of (predicted) fragmentation sites and the number of correct fragmentation sites are counted. Also, since the spectra are annotated, we can count the number of all fragmentation sites in test sets. The precision and recall are defined by

$$\text{precision} = \frac{\# \text{ correct fragmentation sites}}{\# \text{ predicted fragmentation sites}} \tag{4}$$

$$\text{recall} = \frac{\# \text{ correct fragmentation sites}}{\# \text{ all fragmentation sites in test sets.}} \tag{5}$$

Figure 4 shows the precision and recall values of the tested tools for different datasets. For all datasets, UniNovo showed the highest precision value. But the recall values of UniNovo tended to be lower than others in particular for CID

---

[7] Training of the parameters for the Bayesian network of PepNovo [12] for the CID spectra of AspN or LysC digested peptides would lead to better results; however, as mentioned above, the re-ranking models of PepNovo+ [11], which are crucial for the suprior performance of PepNovo+ for CID tryptic spectra (see the Appendix section A13), cannot be readily trained.

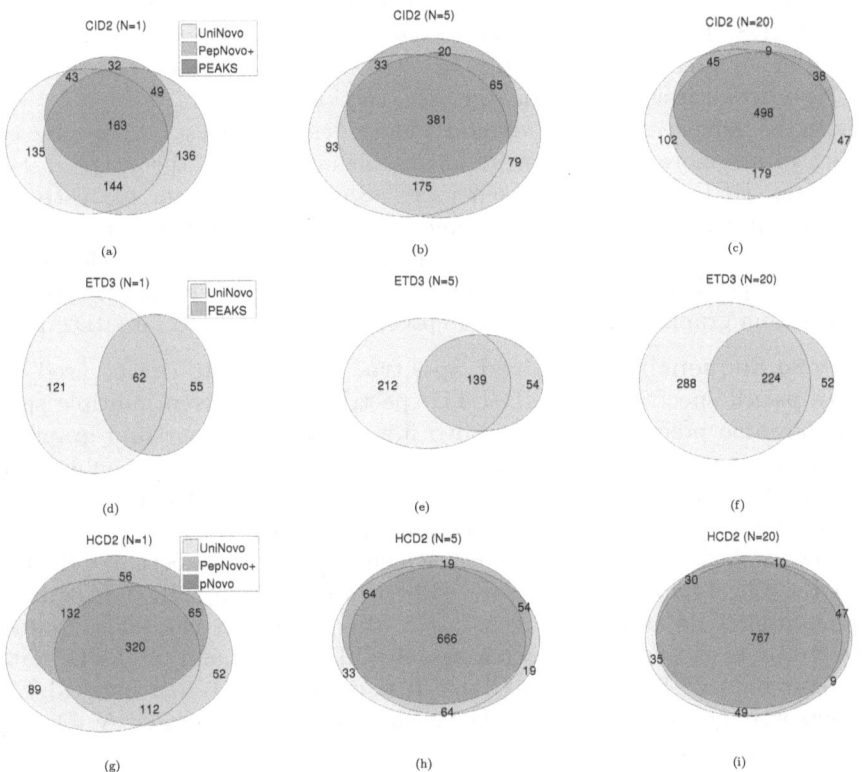

Fig. 3. The Venn diagrams of the correctly sequenced spectra for CID2 (a)-(c), ETD3 (d)-(f), and HCD2 (g)-(i) datasets. For all datasets, the overlaps between different tools increase as N grows, as expected. Relatively small overlaps are observed for ETD spectra when compared to CID or HCD spectra. The Venn diagrams for other datasets are found in Figure A4-A11 in the Appendix section A11.

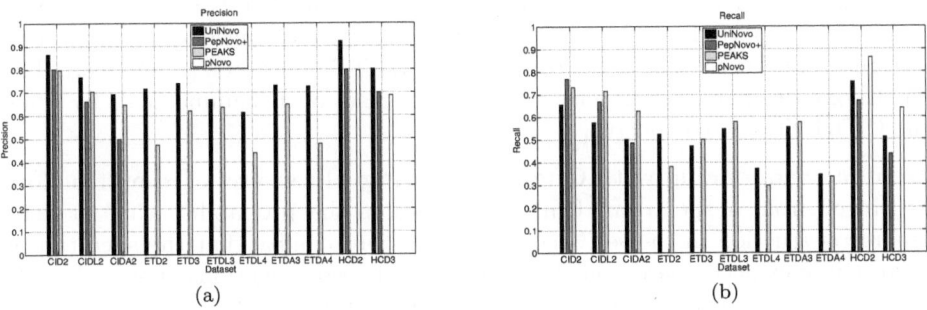

Fig. 4. Comparison of *de novo* sequencing tools in terms of amino acid level precision (a) and recall (b). The definitions of precision and recall are given in (4) and (5), respectively.

spectra. For ETD2 and ETDL4 datasets, UniNovo had higher precision and recall than PEAKS. These observations are consistent with the sequence level results above; higher precision of UniNovo resulted in more accurate reconstructions, and lower recall resulted in shorter reconstructions.

Both the sequence level and amino acid level results suggest that specific types of spectra are more suitable for *de novo* sequencing than others. For instance, in general, HCD spectra generated more accurate and longer reconstructions (or higher precision and recall in amino acid level) than ETD spectra. Further evaluation of the scoring function (i.e., spectrum graph) of UniNovo for different spectrum types is found in the Appendix section A13, where we also compared the spectrum graphs from UniNovo, PepNovo, and MS-GFDB for CID2 dataset.

De novo Sequencing of Paired Spectra. UniNovo also can be used to sequence paired spectra (e.g., CID/ETD spectral pairs). Given multiple spectra from the same precursor ion, UniNovo first generates a spectrum graph from each of the spectra and next merges the spectrum graphs into a combined spectrum graph, on which the reconstructions are generated (refer to the Appendix section A5 for the spectrum graph merging algorithm).

To benchmark UniNovo in *de novo* sequencing of paired spectra, CID/ETD2 and CID/ETD3 datasets were analyzed by UniNovo. From CID/ETD2 dataset, two additional datasets were generated: CID/etd2 and cid/ETD2 datasets. CID/etd2 dataset was formed by taking only CID spectra, and cid/ETD2 dataset by taking only ETD spectra in CID/ETD2 dataset. CID/etd3 and cid/ETD3 datasets were generated similarly. For each dataset, we generated $N = 1$, 5, and 20 top scoring reconstructions.

The results are shown in Figure 5. When precursor ions were doubly charged, the performance boost from the paired spectra was very modest. For $N = 1, 5$, and 20, UniNovo reported 5% more correctly sequenced spectral pairs in CID/ETD2 datasets than in CID/etd2 dataset. The average length of correct reconstructions for CID/ETD2 dataset was slightly longer than for CID/etd2 dataset.

In contrast, for triply charged spectra, the use of paired spectra was highly beneficial for generating more accurate reconstructions. For example, when $N = 1$, UniNovo reported 100% and 50% more correctly sequenced spectral pairs in CID/ETD3 dataset than in CID/etd3 and cid/ETD3 datasets, respectively. The length of correct reconstructions typically increases by 1-2 amino acids by using the CID/ETD paired spectra.

De novo Sequencing with Quality Filtering. Given a set of reconstructions generated from a spectrum, UniNovo estimates the probability that at least one reconstruction in the set is correct (i.e., a probability that the spectrum is correctly sequenced) based on the accuracies of reconstructions. The estimated probability is called the *set accuracy*.[8] Denote the set of reconstructions by $R = \{r_1, \cdots, r_N\}$. If the events "r_i is correct" for $i = 1, \cdots, N$ are independent,

---

[8] When multiple *de novo* reconstructions are reported, it is important to guarantee that one of them is correct.

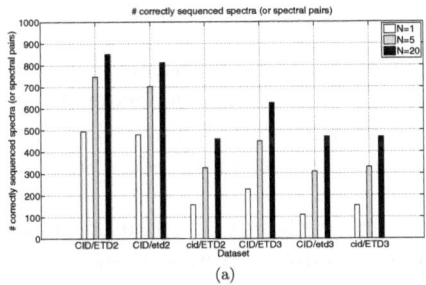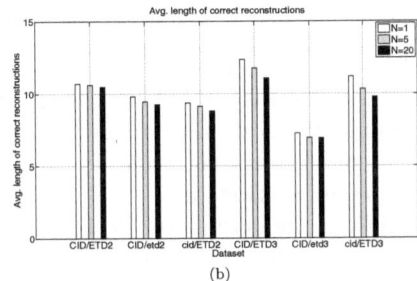

(a)   (b)

Fig. 5. *De novo* sequencing of paired spectra. CID/ETD spectral pairs were analyzed by UniNovo (in CID/ETD2 and CID/ETD3 datasets). To see if the spectral pairs are beneficial for *de novo* sequencing, CID/etd2 (cid/ETD2) dataset was generated from CID/ETD2 dataset by collecting only CID (ETD) spectra in CID/ETD2 dataset. Likewise, CID/etd3 and cid/ETD3 datasets were generated from CID/ETD3 dataset. (a) the number of correctly sequenced spectra (or spectral pairs), (b) the average length of correct reconstructions for each dataset. The spectral pairs resulted in more accurate and longer reconstructions, in particular for triply charged spectral pairs.

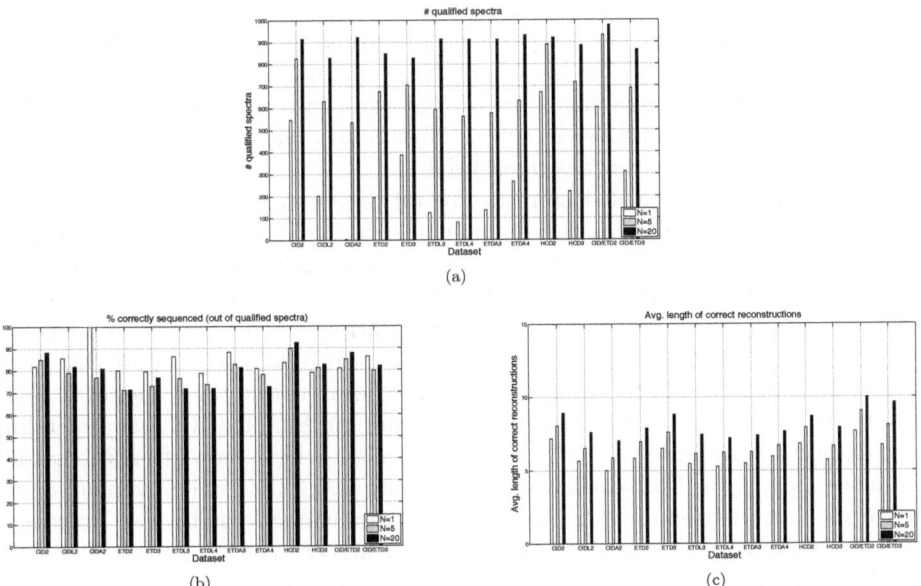

Fig. 6. *De novo* sequencing with qualify filtering of spectra. Given a spectrum, if the parameter *SetAccuracyThreshold* is set, UniNovo attempts to achieve set accuracy (an estimated probability of the spectrum being correctly sequenced) exceeding *SetAccuracyThreshold*. If it fails to generate such a set, the spectrum is filtered out. An unfiltered spectrum is called a *qualified spectrum*. We set *SetAccuracyThreshold* = 0.8. (a) the number of qualified spectrum, (b) the percentage of qualified spectra that were correctly sequenced, (c) the average length of correct reconstructions.

the set accuracy is simply given by $1 - \prod_{i=1}^{N}(1 - Accuracy(r_i))$. However, since the reconstructions are often similar to each other, the dependency between reconstructions should be taken into account. To model this dependency, we assume Markov property between the events "r_i is correct" for $i = 1, \cdots, N$ and compute the set accuracy. The derivation of the set accuracy is given in the Appendix section A6.

When the parameter N is set, one may want to choose N reconstructions with the highest accuracies to maximize the set accuracy. However, such a selection often results in a set of short reconstructions (because short reconstructions have relatively high accuracies). Since short reconstructions are not very useful in many cases (e.g., in follow-up homology searches), UniNovo uses a greedy algorithm to select long and accurate reconstructions. The inputs to the algorithm are the parameters $SetAccuracyThreshold$ and N. The algorithm tries to form an output set of N reconstructions of set accuracy higher than $SetAccuracyThreshold$ while maximizing the minimum length of the reconstructions (see the Appendix section A7 for the description of the algorithm). If UniNovo fails to generate a set of N reconstructions with the set accuracy higher than $SetAccuracyThreshold$, it filters out the query spectrum.

We set $SetThreshold = 0.8$ and reanalyzed the datasets in Table 1. The maximum number of mass gaps per each reconstruction was set to 10. For each dataset, we measured the number of unfiltered spectra (termed *qualified* spectra) and the percentage of qualified spectra that were correctly sequenced (which is expected to be 80% since $SetAccuracyThreshold = 0.8$). The average length of correct reconstructions was also measured.

The results are given in Figure 6. For all datasets, the number of qualified spectra increases sharply as the number of reconstructions N grows (Figure 6 (a)). For example, UniNovo reported only few qualified spectra (less than 5) from CIDA2 dataset when $N = 1$. When $N = 20$, it reported more than 900 qualified spectra from the same dataset. In contrast to the dramatic changes in the number of qualified spectra, the percentage of qualified spectra that were correctly sequenced hardly changed across the datasets and the values of N (Figure 6 (b)). As expected, the percentage was around 80% for all cases (including the datasets containing CID/ETD spectral pairs), which shows that the set accuracy reported by UniNovo is reliable. Figure 6 (c) shows the average length of correct reconstructions. As N decreases, the average length also decreases. This is because shorter reconstructions (with higher accuracies) are chosen by UniNovo when N is small to achieve high set accuracy.

4 Conclusion

We presented a universal *de novo* sequencing tool UniNovo that works well for various types of spectra. UniNovo can be easily trained for different types of spectra using only thousands of PSMs that typically can be obtained from a single MS/MS run. The experimental results show that UniNovo generates accurate and long *de novo* reconstructions from spectra of CID, ETD, HCD, and

CID/ETD fragmentation methods and spectra of trypsin, LysC, or AspN digested peptides. We also showed that UniNovo is better than or comparable to other state of the art tools.

As pointed out by [26], *de novo* sequences not only are valuable for the analysis of the novel peptides that are not present in proteome databases but also can facilitate the homology-based database searches. Since the reconstructions reported by UniNovo contain mass gaps representing the total mass of multiple amino acids (termed *gapped peptides* [22,17]), MS-BPM algorithm [29] can be used for fast exact or homology searches (UniNovo⊕MS-BPM). MS-BPM enables searches against a sequence database using gapped peptides as queries. Currently MS-BPM takes gapped peptides generated by MS-GappedDictionary [17] (MS-GappedDictionary⊕MS-BPM). However, the reconstructions from UniNovo are usually longer than those from MS-GappedDictionary (8-9 vs. 5-6). Since the search time of MS-BPM strongly depends on the length of gapped peptides - the longer gapped peptides, the shorter search time - the running time of UniNovo⊕MS-BPM is smaller than MS-GappedDictionary⊕MS-BPM by an order of magnitude in a blind search against the IPI Human proteome database ver.3.87 [21] (data not shown).

Acknowledgement. We are grateful to Albert Heck and Joshua Coon for making their spectral datasets available. We are also thankful to Ari Frank, Sunghee Woo, and Nuno Bandeira for helpful discussion.

References

1. Bandeira, N., Olsen, J.V., Mann, M., Pevzner, P.A.: Multi-spectra peptide sequencing and its applications to multistage mass spectrometry. Bioinformatics 24(13), i416–i423 (2008)
2. Barton, S.J., Whittaker, J.C.: Review of factors that influence the abundance of ions produced in a tandem mass spectrometer and statistical methods for discovering these factors. Mass Spectrometry Reviews 28(1), 177–187 (2009)
3. Bern, M., Cai, Y., Goldberg, D.: Lookup peaks: a hybrid of de novo sequencing and database search for protein identification by tandem mass spectrometry. Anal. Chem. 79(4), 1393–1400 (2007)
4. Breci, L.A., Tabb, D.L., Yates, J.R., Wysocki, V.H.: Cleavage n-terminal to proline: analysis of a database of peptide tandem mass spectra. Analytical Chemistry 75(9), 1963–1971 (2003)
5. Chen, T., Kao, M.Y., Tepel, M., Rush, J., Church, G.M.: A dynamic programming approach to de novo peptide sequencing via tandem mass spectrometry. Journal of Computational Biology: A Journal of Computational Molecular Cell Biology 8(3), 325–337 (2001)
6. Chi, H., Sun, R., Yang, B., Song, C., Wang, L., Liu, C., Fu, Y., Yuan, Z., Wang, H., He, S., Dong, M.: pNovo: de novo peptide sequencing and identification using HCD spectra. J. Proteome Res. 9(5), 2713–2724 (2010)
7. Dancik, V., Addona, T.A., Clauser, K.R., Vath, J.E., Pevzner, P.A.: De novo peptide sequencing via tandem mass spectrometry. Journal of Computational Biology 6(3-4), 327–342 (1999)

8. Datta, R., Bern, M.: Spectrum fusion: using multiple mass spectra for de novo peptide sequencing. Journal of Computational Biology 16(8), 1169–1182 (2009)
9. Elias, J.E., Gygi, S.P.: Target-decoy search strategy for increased confidence in large-scale protein identifications by mass spectrometry. Nature Methods 4(3), 207–214 (2007)
10. Eng, J.K., McCormack, A.L., Yates, J.R.: An approach to correlate tandem mass spectral data of peptides with amino acid sequences in a protein database. Journal of the American Society for Mass Spectrometry 5(11), 976–989 (1994)
11. Frank, A.: A ranking-based scoring function for peptide-spectrum matches. Journal of Proteome Research 8(5), 2241–2252 (2009)
12. Frank, A., Pevzner, P.: PepNovo: de novo peptide sequencing via probabilistic network modeling. Anal. Chem. 77(4), 964–973 (2005)
13. Frese, C.K., Altelaar, A.F.M., Hennrich, M.L., Nolting, D., Zeller, M., Griep-Raming, J., Heck, A.J.R., Mohammed, S.: Improved peptide identification by targeted fragmentation using CID, HCD and ETD on an LTQ-Orbitrap velos. J. Proteome Res. 10(5), 2377–2388 (2011)
14. He, L., Ma, B.: ADEPTS: advanced peptide de novo sequencing with a pair of tandem mass spectra. Journal of Bioinformatics and Computational Biology 8(6), 981–994 (2010)
15. Huang, Y., Triscari, J.M., Tseng, G.C., Pasa-Tolic, L., Lipton, M.S., Smith, R.D., Wysocki, V.H.: Statistical characterization of the charge state and residue dependence of low-energy CID peptide dissociation patterns. Analytical Chemistry 77(18), 5800–5813 (2005)
16. Hunter, D.: An upper bound for the probability of a union. Journal of Applied Probability 13(3), 597–603 (1976)
17. Jeong, K., Kim, S., Bandeira, N., Pevzner, P.A.: Gapped spectral dictionaries and their applications for database searches of tandem mass spectra. Molecular & Cellular Proteomics 10(6), M110.002220 (2011)
18. Johnson, R.S., Martin, S.A., Biemann, K., Stults, J.T., Watson, J.T.: Novel fragmentation process of peptides by collision-induced decomposition in a tandem mass spectrometer: differentiation of leucine and isoleucine. Anal. Chem. 59(21), 2621–2625 (1987)
19. Käll, L., Canterbury, J.D., Weston, J., Noble, W.S., MacCoss, M.J.: Semi-supervised learning for peptide identification from shotgun proteomics datasets. Nature Methods 4(11), 923–925 (2007)
20. Keller, A., Nesvizhskii, A., Kolker, E., Aebersold, R.: Empirical statistical model to estimate the accuracy of peptide identifications made by ms/ms and database search. Anal. Chem. 74, 5383–5392 (2002)
21. Kersey, P.J., Duarte, J., Williams, A., Karavidopoulou, Y., Birney, E., Apweiler, R.: The international protein index: an integrated database for proteomics experiments. Proteomics 4(7), 1985–1988 (2004)
22. Kim, S., Bandeira, N., Pevzner, P.A.: Spectral profiles, a novel representation of tandem mass spectra and their applications for de novo peptide sequencing and identification. Molecular & Cellular Proteomics 8(6), 1391–1400 (2009)
23. Kim, S., Gupta, N., Bandeira, N., Pevzner, P.A.: Spectral dictionaries. Molecular & Cellular Proteomics 8(1), 53–69 (2009)
24. Kim, S., Mischerikow, N., Bandeira, N., Navarro, J.D., Wich, L., Mohammed, S., Heck, A.J.R., Pevzner, P.A.: The generating function of CID, ETD, and CID/ETD pairs of tandem mass spectra: Applications to database search. Molecular & Cellular Proteomics 9(12), 2840–2852 (2010)

25. Liu, X., Shan, B., Xin, L., Ma, B.: Better score function for peptide identification with ETD MS/MS spectra. BMC Bioinformatics 11(suppl. 1), S4 (2010)
26. Ma, B., Johnson, R.: De novo sequencing and homology searching. Molecular & Cellular Proteomics, O111.014902 (2011)
27. Ma, B., Zhang, K., Hendrie, C., Liang, C., Li, M., Doherty-Kirby, A., Lajoie, G.: PEAKS: powerful software for peptide de novo sequencing by tandem mass spectrometry. Rapid Communications in Mass Spectrometry: RCM 17(20), 2337–2342 (2003)
28. Nesvizhskii, A.I.: A survey of computational methods and error rate estimation procedures for peptide and protein identification in shotgun proteomics. J. Proteomics 73(11), 2092–2123 (2010)
29. Ng, J., Amir, A., Pevzner, P.A.: Blocked Pattern Matching Problem and Its Applications in Proteomics. In: Bafna, V., Sahinalp, S.C. (eds.) RECOMB 2011. LNCS, vol. 6577, pp. 298–319. Springer, Heidelberg (2011)
30. Olsen, J.V., Macek, B., Lange, O., Makarov, A., Horning, S., Mann, M.: Higher-energy c-trap dissociation for peptide modification analysis. Nature Methods 4(9), 709–712 (2007)
31. Perkins, D.N., Pappin, D.J., Creasy, D.M., Cottrell, J.S.: Probability-based protein identification by searching sequence databases using mass spectrometry data. Electrophoresis 20(18), 3551–3567 (1999)
32. Savitski, M.M., Nielsen, M.L., Kjeldsen, F., Zubarev, R.A.: Proteomics-Grade de novo sequencing approach. J. Proteome Res. 4(6), 2348–2354 (2005)
33. Swaney, D.L., McAlister, G.C., Coon, J.J.: Decision tree-driven tandem mass spectrometry for shotgun proteomics. Nature Methods 5(11), 959–964 (2008)
34. Swaney, D.L., Wenger, C.D., Coon, J.J.: Value of using multiple proteases for Large-Scale mass Spectrometry-Based proteomics. J. Proteome Res. 9(3), 1323–1329 (2010)
35. Tabb, D.L., Huang, Y., Wysocki, V.H., Yates, J.R.: Influence of basic residue content on fragment ion peak intensities in Low-Energy Collision-Induced dissociation spectra of peptides. Anal. Chem. 76(5), 1243–1248 (2004)
36. Wysocki, V.H., Tsaprailis, G., Smith, L.L., Breci, L.A.: Mobile and localized protons: a framework for understanding peptide dissociation. Journal of Mass Spectrometry 35(12), 1399–1406 (2000)
37. Zubarev, R.A., Zubarev, A.R., Savitski, M.M.: Electron Capture/Transfer versus collisionally Activated/Induced dissociations: Solo or duet? Journal of the American Society for Mass Spectrometry 19, 753–761 (2008)

Efficiently Identifying Significant Associations in Genome-Wide Association Studies

Emrah Kostem and Eleazar Eskin

Computer Science Department, University of California, Los Angeles,
California 90095, USA
{ekostem,eeskin}@cs.ucla.edu

Abstract. Over the past several years, genome wide association studies (GWAS) have implicated hundreds of genes in common disease. More recently, the GWAS approach has been utilized to identify regions of the genome which harbor variation affecting gene expression or expression quantitative trait loci (eQTLs). Unlike GWAS applied to clinical traits where only a handful of phenotypes are analyzed per study, in (eQTL) studies, tens of thousands of gene expression levels are measured and the GWAS approach is applied to each gene expression level. This leads to computing billions of statistical tests and requires substantial computational resources, particularly when applying novel statistical methods such as mixed-models. We introduce a novel two-stage testing procedure that identifies all of the significant associations more efficiently than testing all the SNPs. In the first-stage a small number of informative SNPs, or proxies, across the genome are tested. Based on their observed associations, our approach locates the regions which may contain significant SNPs and only tests additional SNPs from those regions. We show through simulations and analysis of real GWAS datasets that the proposed two-stage procedure increases the computational speed by a factor of 10. Additionally, efficient implementation of our software increases the computational speed relative to state of the art testing approaches by a factor of 75.

1 Introduction

Research in complex diseases has progressed rapidly in the last decade with the advent of genomic technologies [13,17,19,31]. In genome-wide association studies (GWAS), information on millions of single nucleotide polymorphisms (SNPs) across the genome is collected from thousands of case and control individuals. Typically, each SNP is statistically tested for disease association by comparing the minor allele frequency (MAF) between the cases and controls. The significant associations are used to gain insight into the genetic basis of disease, and hundreds of GWASs have been performed on dozens of complex diseases and successfully discovered many novel loci involved in disease susceptibility [18].

More recently, there has been great interest in applying the GWAS approach to genomic data such as gene expression. In these studies, the goal is to identify

M. Deng et al. (Eds.): RECOMB 2013, LNBI 7821, pp. 118–131, 2013.

regions of the genome harboring genetic variation which affect gene expression levels or expression quantitative trait loci (eQTL) [3, 10, 32].

A challenge in applying GWAS to genomic data is that these technologies typically obtain tens of thousands of measurements for each sample resulting in a tremendous computational burden when performing the analysis, including computing billions of tests and requires substantial computational resources. This challenge is compounded for novel statistical approaches such as linear mixed models, which account for population structure [20, 25, 46], yet themselves are computationally intensive.

eQTL studies are already very popular [4, 5, 21] and with rapidly decreasing costs of RNA-seq technologies [26, 45] will likely become more popular in the future. These include, several major efforts collecting expression from multiple-tissues in human [1, 9, 14, 35, 39] and mouse [6, 8]. More broadly, application of the GWAS approach to phenotypes measured by other genomic technologies such as those reported by the ENCODE consortium [41–44] will face similar computational challenges.

In this paper, we introduce a novel two-stage method which can be applied to reduce the computational burden of a wide range of association studies including those employ case-control, quantitative trait and mixed-model statistical testing methodologies. In each trait, typically only a small percentage of the SNPs are significantly associated and the SNPs neighboring a significant association have elevated statistics. Intuitively, one can first test an informative subset of the SNPs, termed proxy SNPs, across the genome to quickly locate these regions and test the SNPs therein. This way, many of the regions with no associations can be discarded from the analysis to reduce the computational burden.

Our novel method for genome-wide rapid association testing (GRAT), guarantees to identify all of the significant associations with high-probability while reducing the total number of tests. The proposed method chooses the proxy SNPs and determines which additional SNPs to test based on the observed proxy SNP statistics and the patterns of linkage disequilibrium (LD) in the region. The key insight underlying GRAT is that by taking advantage of how the statistics at SNPs in LD with each other behave, we can estimate the probability that an untested SNP has a significant association and use this probability to only eliminate SNPs from consideration if they are highly unlikely to have significant associations. We have selected a set of proxy SNPs for the 1000 Genomes Project and any study which imputes to the 1000 Genomes Project SNPs can readily use our approach. We also provide our method for choosing proxy SNPs, which can be applied to any reference dataset.

We show through simulations and analysis of real eQTL datasets that the proposed two-stage procedure identifies the significant associations while only testing approximately 10% of the SNPs. GRAT's efficient software implementation reduces the computational time for computing large-scale association studies by a factor of 30 compared to currently used state of the art methods. When our method is applied to association studies that utilize linear mixed models, the speed-up is cumulative with recent efforts that decrease the computational burden of computing the actual association statistic such as EMMAX, FaST-LMM and GEMMA [20, 25, 46].

2   Material and Methods

2.1   Genome-Wide Association Studies

For the simplicity of description, we consider a balanced case-control genome-wide association study (GWAS) with $N/2$ individuals (N copies of each chromosome) per panel. For our actual experiments, we will use association statistics for quantitative phenotypes, but the approach assuming case-control phenotypes is equivalent. For SNP m_i, p_i denotes its population minor allele frequency (MAF); p_i^+ and p_i^- denote its population case and control MAFs; $\hat{p}_i^+$ and $\hat{p}_i^-$ denote its observed case and control MAFs in the GWAS. Given the relative risk of the SNP, γ_i, in the disease and the prevalence of the disease, F, in the population, it can be shown that the case and control MAFs of the SNP follows,

$$p_i^+ = \frac{\gamma_i p_i}{(1 - \gamma_i)p_i + 1}, \ p_i^- = \frac{p_i - F p_i^+}{1 - F}. \tag{1}$$

A SNP is defined as *not* associated if $p_i^+ = p_i^-$.

In case-control GWASs the following statistic is widely used, which is normally distributed for large N with mean $\lambda_i \sqrt{N}$ (the non-centrality parameter), and unit variance,

$$S_i = \hat{s}_i = \frac{\hat{p}_i^+ - \hat{p}_i^-}{\sqrt{2\hat{p}_i(1-\hat{p}_i)}} \sqrt{N} \sim \mathcal{N}\left(\lambda_i \sqrt{N}, 1\right), \text{where } \lambda_i = \frac{p_i^+ - p_i^-}{\sqrt{2p_i(1-p_i)}} \text{ and } \hat{p}_i = \frac{\hat{p}_i^+ + \hat{p}_i^-}{2}. \tag{2}$$

Given the significance level α and the observed value of the test statistic $\hat{s}_i$, the SNP is deemed as significant, or statistically associated, if $|\hat{s}_i| > \Phi^{-1}\left(1 - \frac{\alpha}{2}\right)$, where $\Phi^{-1}(.)$ is the quantile function of the standard normal distribution. For simplicity, we use the notation: $t_\alpha \equiv \Phi^{-1}\left(1 - \frac{\alpha}{2}\right)$. Typically, in a GWAS the significance level is chosen as $\alpha = 10^{-8}$.

2.2   A Two-Stage Approach for Identifying the Significant Associations

We propose the following two-stage testing procedure for identifying the significant associations within a set of SNPs $\mathcal{M}$. Given a subset of the SNPs $\mathcal{T} \subset \mathcal{M}$, referred to as the proxy SNPs, for each proxy SNP, $m_t \in \mathcal{T}$, its association statistic, $\hat{s}_t$, is computed. In the second stage, a decision rule is exercised for each of the remainder SNP, $m_i \in \mathcal{M} \backslash \mathcal{T}$, in order to determine whether or not to compute the association statistic of the remainder SNP. The decision rule for a remainder SNP m_i is defined using a proxy SNP, $m_t \in \mathcal{T}$, and a threshold, s_t^*, for its observed statistic $\hat{s}_t$. If the observed statistic of the proxy SNP is more extreme than the threshold value, $\hat{s}_t > s_t^*$ the remainder SNP is tested.

2.3 Performance of the Two-Stage Approach

In a GWAS, the performance of the two-stage approach can be summarized by the total number of SNPs tested (NT), and the percentage of the significant SNPs identified, or the recall rate (RR). The total number of tests is the sum of the tests performed on the proxy SNPs, plus the remainder SNPs that are tested as a result of the decision rules. We use a standard GWAS simulation model [22] to evaluate a given set of proxy SNPs and decision rules based on their *expected* performance within the simulated data.

The simulation model considers the probability of each SNP being causal, c_i, and the non-centrality parameter (NCP) of the causal SNP, $\lambda_c \sqrt{N}$. For simplicity, we give a brief explanation of the simulation procedure for a single causal SNP using a genomic reference dataset such as the HapMap. Using the given probabilities of each SNP being causal, at most a single causal SNP is randomly selected. Given the disease prevalence F and the NCP of the causal SNP $\lambda_c \sqrt{N}$, the case and control MAFs, p_c^+ and p_c^- are determined. Next, the HapMap haplotypes are divided into two pools according to the minor and major allele of the causal SNP, and case-control panels are sampled using p_c^+ and p_c^-.

For each simulation dataset, each association statistic is computed to identify which SNPs are significant in the dataset. We then apply the two-stage method to observe the NT and RR. The expected recall rate (ERR) and the expected number of SNPs to be tested (ENT) then can be computed by repeatedly simulating datasets, applying the two-stage approach and averaging the observed NT and RR value.

2.4 Finding the Optimal Decision Rules for Given Proxy SNPs

For a given set of proxy SNPs, one can determine the decision rules empirically by evaluating the performance of using different threshold values on the remainder SNPs in the simulated data. The empirical approach can be cumbersome and instead we derive an analytical framework for estimating the expected performance, which eliminates the need for generating simulated data and saves time. Furthermore, using this analytical framework we show how to determine the optimal decision rules for the remainder SNPs given a set of proxy SNPs.

A SNP that is disease-associated can be either causal in the disease or in LD with the causal SNP. Given that SNP m_i is the causal SNP, the non-centrality parameter (NCP) of a correlated SNP m_t, $\lambda_t \sqrt{N}$, is proportional to the NCP of the causal SNP, $\lambda_c \sqrt{N}$, by their correlation coefficient, r, where $\lambda_t = r\lambda_c$. It can be shown that the joint distribution of the association statistics of the causal SNP m_i and the non-causal SNP m_t follows a bivariate normal distribution [16].

We follow a conservative approach in which each remainder SNP m_i is paired with the proxy SNP that is most strongly correlated, referred to as the *best*-proxy, and denoted by $m_{b(i)}$. For each remainder SNP m_i, we denote the association statistic of its best-proxy $m_{b(i)}$ with $s_{b(i)}$ and test SNP m_i if its best-proxy SNP association statistic is more extreme than a given threshold, $s_{b(i)} > s_{b(i)}^*$. For simplicity, we assume only the remainder SNP can be causal and express the density function of the joint distribution, $f(s_i, s_{b(i)})$,

$$f\left(s_i, s_{b(i)}\right) = c_i \phi\left(\begin{bmatrix} s_i \\ s_{b(i)} \end{bmatrix} ; \begin{bmatrix} \lambda_c\sqrt{N} \\ r\lambda_c\sqrt{N} \end{bmatrix}, \begin{bmatrix} 1 & r \\ r & 1 \end{bmatrix}\right) + (1 - c_i)\phi\left(\begin{bmatrix} s_i \\ s_{b(i)} \end{bmatrix} ; \begin{bmatrix} 0 \\ 0 \end{bmatrix}, \begin{bmatrix} 1 & r \\ r & 1 \end{bmatrix}\right),$$
(3)

where $\phi(\boldsymbol{x} ; \boldsymbol{\mu}, \boldsymbol{\Sigma})$ denotes the density of a multivariate normal distribution with mean vector $\boldsymbol{\mu}$ and covariance matrix $\boldsymbol{\Sigma}$. The first term corresponds to having the remainder SNP as causal, with probability c_i, and the second term to not casual with probability $1 - c_i$.

Assume we are given K proxy SNPs, where $\mathcal{T} = \{m_1, \ldots, m_K\}$. The expected number of SNPs to be tested (ENT) can be expressed as the fixed cost of testing K proxy SNPs, plus the expected number of decision rules that are triggered,

$$\text{ENT}(s^*_{b(K+1)}, \ldots, s^*_{b(M)}) = K + \sum_{i=K+1}^{M} \Pr\left(|S_{b(i)}| > s^*_{b(i)}\right).$$
(4)

We *approximate* the expected recall rate (ERR) as the ratio of the expected number of significant SNPs that the two-stage approach discovers, to the expected number of significant SNPs in a GWAS,

$$\text{ERR}(s^*_{b(K+1)}, \ldots, s^*_{b(M)}) = \frac{\sum_{t=1}^{K} \Pr\left(|S_t| > t_\alpha\right) + \sum_{i=K+1}^{M} \Pr\left(|S_i| > t_\alpha, |S_{b(i)}| > s^*_{b(i)}\right)}{\sum_{i=1}^{M} \Pr\left(|S_i| > t_\alpha\right)},$$
(5)

where the first and the second terms in the numerator correspond to the expected number of significant SNPs obtained from testing the proxy SNPs and the remainder SNPs, respectively. Further, we refer to the second term as the expected recall function, which can be computed using the joint distribution,

$$\text{ER}(s^*_{b(K+1)}, \ldots, s^*_{b(M)}) = \sum_{i=K+1}^{M} \Pr\left(|S_i| > t_\alpha, |S_{b(i)}| > s^*_{b(i)}\right),$$

$$\Pr\left(|S_i| > t_\alpha, |S_{b(i)}| > s^*_{b(i)}\right) = \iint_{\Omega_i} f\left(s_i, s_{b(i)}\right) ds_i \, ds_{b(i)},$$
(6)

where $\Omega_i = \left\{(s_i, s_{b(i)}) \mid |s_i| > t_\alpha, |s_{b(i)}| > s^*_{b(i)}\right\}$.

We are interested in determining the decision rules that lead to the least expected number of SNPs to be tested (ENT), while the expected recall rate (ERR) satisfies a given target value, ρ, which can be expressed as an optimization problem,

$$\text{minimize } \text{ENT}(s^*_{b(K+1)}, \ldots, s^*_{b(M)}),$$
$$\text{such that } \text{ERR}(s^*_{b(K+1)}, \ldots, s^*_{b(M)}) = \rho.$$
(7)

2.5   Choosing the Optimal Proxy SNPs

The expected number of SNPs to be tested (ENT) in the two-stage approach depends on the number of proxy SNPs and which SNPs are chosen as proxies.

It can be shown that the problem of finding the optimal set of proxy SNPs, among all possible sets of proxy SNPs the set that gives the minimum ENT, is an NP-Hard problem. Therefore, we propose a heuristic algorithm for choosing the proxy SNPs using a greedy approach, which incrementally builds the set of proxy SNPs.

Starting with an empty set, let $\mathcal{T}_k$ denote the current set of proxy SNPs with size k, where ENT_k and ERR_k denote the values of its ENT and ERR. ($\mathrm{ENT}_0 = +\infty$ and $\mathrm{ERR}_0 = -\infty$). Each remainder SNP m_i is a candidate to extend the current set of proxy SNPs to become $\{\mathcal{T}_k \cup m_i\}$, which performs $\mathrm{ENT}_{k+1}^{(i)}$. The remainder SNP with the least $\mathrm{ENT}_{k+1}^{(i)}$ is chosen for extending the current set of proxy SNPs:

$$\mathcal{T}_{k+1} = \mathcal{T}_k \cup \operatorname*{argmin}_{m_i \in \mathcal{M} \setminus \mathcal{T}_k} \left(\mathrm{ENT}_{k+1}^{(i)} \right). \tag{8}$$

While the extended set $\mathcal{T}_{k+1}$ improves the ENT, i.e., $\mathrm{ENT}_{k+1} < \mathrm{ENT}_k$, the algorithm continues.

For each candidate set of proxy SNPs, the algorithm solves the optimization problem (7) to compute $\mathrm{ENT}_{k+1}^{(i)}$. This leads to a quadratic computational complexity in the order of the number of the collected SNPs and in practice makes it hard to scale to large numbers. We further introduce a heuristic extension to the above greedy-approach to reduce this complexity. While extending the current set of proxy SNPs $\mathcal{T}_k$ to $\mathcal{T}_{k+1}$, the optimization problem (7) is solved $M - k$ times. In particular, solving the optimization problem (7) corresponds to finding the gradient, g^*, at which the ENT function is minimized while satisfying the constraints. We assume that for $\mathcal{T}_k$ and $\mathcal{T}_{k+1}$ the gradient values of their ENT functions are close enough, $g_k^* \approx g_{k+1}^*$. Therefore, while extending the current proxy set, we compute the ENT of each candidate set, $\mathrm{ENT}_{k+1}^{(i)}$, using the gradient value from the previous step, g_k^*. This way, rather than solving the optimization problem $M - k$ times for each possible proxy SNP at each step k, the gradient is updated once after the new set $\mathcal{T}_{k+1}$ is determined. Using this approach the optimization problem (7) is solved a total of K times, where K is the size of the final set of proxy SNPs.

2.6   Updating the Remainder SNP Thresholds in Linear Mixed Models

We consider the following linear mixed model (LMM) formulation,

$$y = X\beta + g + e, \tag{9}$$

where y is the $(n \times 1)$ vector of phenotypic values, X is the $(n \times p)$ matrix of fixed-effects, which includes the mean, covariates and the SNP to be tested, β is the $(p \times 1)$ vector of fixed-effect weights, g is the variance component accounting for the population structure and e is the iid noise. We assume the random effects, g and e, follow multivariate normal distribution, $g \sim \mathcal{N}\left(0, \sigma_g^2 K\right)$, $e \sim \mathcal{N}\left(0, \sigma_e^2 I\right)$, where K is the known, $(n \times n)$, genetic similarity matrix, I is the $(n \times n)$ identity matrix with unknown magnitudes σ_g^2 and σ_e^2. We follow the approach taken in EMMAX [20] and estimate σ_g^2 and σ_e^2 in the null model, with no SNP effect, and use these parameters while testing the SNPs. That is, when each SNP is tested, the covariance of y is kept fixed, $\mathrm{Cov}(y) = \Sigma = \hat{\sigma}_g^2 K + \hat{\sigma}_e^2 I$, where $\hat{\sigma}_g^2$ and $\hat{\sigma}_e^2$ are the restricted log likelihood (REML) estimates [20, 25].

In GRAT, the threshold value for each remainder SNP is computed after the covariance matrix Σ is estimated and the alternate model is transformed by the inverse square root of this matrix,

$$\Sigma^{-1/2} y \sim \mathcal{N}\left(\Sigma^{-1/2} X\beta, \ \sigma^2 I\right), \tag{10}$$

where the residuals are iid. For two SNPs m_i and m_j, let x_i and x_j be their $(n \times 1)$ allelic indicator vectors. When the SNPs are tested individually in the above model, the same transformation is applied to the genotype vectors, which may moderately change the pairwise correlation between the SNPs. The transformed genotype vectors are $\tilde{x}_i = \Sigma^{-1/2} x_i$ and $\tilde{x}_j = \Sigma^{-1/2} x_j$ and their correlation coefficient is,

$$\tilde{r}_{ij} = \frac{\mathrm{Cov}(\tilde{x}_i, \tilde{x}_j)}{\sqrt{\mathrm{Var}(\tilde{x}_i)}\sqrt{\mathrm{Var}(\tilde{x}_j)}}. \tag{11}$$

3 Results

3.1 Genome-Wide Rapid Association Testing (GRAT)

In Figure 1, we consider two possible scenarios for a genomic region in a GWAS. In (a) the region contains no significant associations and in (b) the region contains a causal SNP. In (a) and (b), the statistics for each SNP are shown, denoting what could have been observed in each scenario had all the SNPs in the region been tested. Let m_2 be the proxy SNP for this region to decide whether or not to test the rest of the SNPs. We refer to the SNPs other than the proxy SNP (m_1, m_3, m_4, m_5, m_6 and m_7) as the "remainder SNPs". If the observed statistic of the proxy SNP is stronger than a threshold value, which in this example is 3.0, the remainder SNPs are tested.

In the first-stage, only the proxy SNP is tested and its association statistic is observed. In (a), where the region contains no associations, the statistic of the proxy SNP is 0.7. The observed statistic of the proxy is less than the threshold value ($0.7 < 3.0$) and hence none of the remainder SNPs within the region are tested. In (b), the region contains associations and the proxy SNP captures this information. The observed statistic of the proxy SNP is stronger than the threshold value ($5.0 > 3.0$), which leads to testing each of the remainder SNPs in the region. This results in identifying all the significant SNPs (m_3, m_4 and m_5).

In Methods, we introduce a novel approach for choosing the proxy SNPs and the threshold values, which provide guarantees that all statistically significant associations will be discovered while computing the least amount of association tests. Due to the complexity of linkage disequilibrium (LD) across the genome, we use a separate threshold value for each remainder SNP rather than using

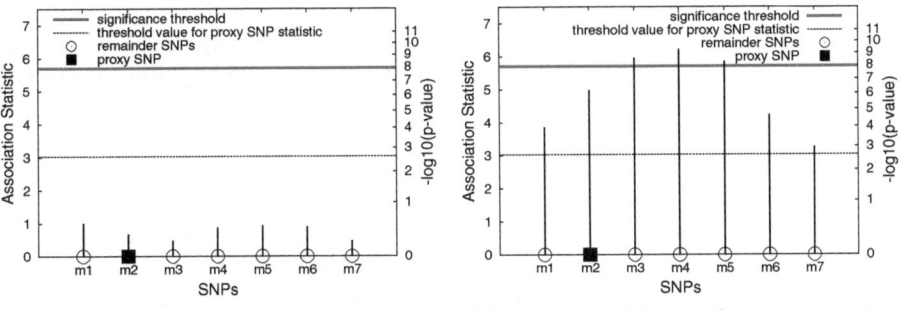

(a) A region with no associations.   (b) A region with significant associations.

Fig. 1. An example of applying GRAT in two hypothetical regions. First, the proxy SNP (rectangle) is tested and its statistics is compared to the threshold (dashed line). If the statistic is above the threshold, the remaining SNPs in the region are tested.

a common threshold value for all the remainders SNPs in an LD region. This is performed by pairing each remainder SNP with its most strongly correlated proxy SNP and a threshold value is used for the pair to decide whether or not to test the remainder SNP. We have precomputed the proxy SNPs for the 1000 Genomes Project and studies imputing to SNPs in this reference can benefit from our method. Even though the LD structure among the SNPs in the study and the reference dataset may be different, our method guarantees to discover all significant associations with high-probability. This is achieved by updating the threshold values using the LD structure observed in the study. We term our novel two-stage testing procedure as Genome-wide Rapid Association Testing (GRAT).

GRAT can be applied to a wide range of statistical models, such as case-control studies, quantitative traits and linear mixed models (LMM). In particular, the LMM approach has recently become popular due to its effective control of population structure. Computing the LMM association statistic is computationally expensive and recently its efficient computation has attracted great interest [20, 25, 46]. The speed-up due to GRAT is cumulative with these efforts.

3.2 Application to a Large-Scale eQTL Study

We compared the performance of GRAT to the standard approach of testing all the SNPs using a large-scale eQTL study [38] that contains 47, 292 gene expression traits on 80 HapMap ASN (East Asian ancestry) individuals that are fully sequenced in the 1000 Genomes Project. We obtained the genotype data from the MACH website [23] and retained approximately 5.9 million SNPs that are filtered for Hardy-Weinberg equilibrium (HWE) and minor allele frequency (MAF) greater than 5%. We eliminated SNPs with lower MAF frequency since they could not be genome-wide significant due to the sample size.

We performed the standard analysis using PLINK [29] which took approximately 2600 hours. We used a conservative genome-wide significance threshold level, $\alpha = 10^{-8}$, to label the significant SNPs and observed 85, 219 significant associations. We repeated the association analysis by applying GRAT using the proxy SNPs precomputed for the 1000 Genomes Project ASN population SNPs. The number of proxies is 276, 702, which means GRAT tests approximately 5% of the SNPs in the first stage.

Applying GRAT to the whole eQTL dataset took 35 hours using the same computational resources (single core of an Opteron CPU). In addition to the proxies, GRAT tested 8.5% of the SNPs in the second stage, reducing the computational cost down to analyzing 13.5% of all the SNPs with the rest of the speedup coming from a faster implementation compared to PLINK. GRAT identified all of the significant associations and speeded up the computation by a factor of 75.

3.3   GRAT Applied to Linear Mixed Model Association

We applied GRAT to a linear mixed model (LMM) association of the eQTL dataset. A challenge in applying GRAT to LMMs is that GRAT utilizes the fact that the joint distribution of traditional association statistics for correlated markers is directly dependent on the correlation between the markers as shown in Pritchard & Przeworski [28]. Unfortunately, when applying LMMs, this relation no longer holds. We derive an analogous relationship between LMM statistics that takes into account both the correlation between the markers and the kinship matrix. Utilizing this relationship, we apply GRAT to LMMs using an efficient implementation [25].

We performed the standard analysis, testing each SNP in each expression trait, which identified 66, 818 significant associations ($\alpha = 10^{-8}$). We applied GRAT using the proxy SNPs precomputed for the 1000 Genomes Project ASN population. In two-stages, GRAT statistically tested a total of 9.1% of the SNPs, identifying all of the significant associations, demonstrating that GRAT can speed up LMM association by a factor of 10.

3.4   Simulations Using the 1000 Genomes Project

To obtain a more robust estimate of the performance, we applied GRAT to thousands of simulated GWAS studies. We simulated the studies using common SNPs (minor allele frequency > 5%) available from the 1000 Genomes Project [40] using the phased SNP genotypes obtained from the MACH website [23] on four populations: African (AFR), East Asian (ASN), Ad Mixed American (AMR) and European (EUR) ancestries.

We divided each chromosome into panels of 1000 SNPs and simulated case-control GWASs by randomly selecting 5% of the panels as the alternate panels, in which we simulated a causal SNP, and the remaining panels as the null panels, without any causal SNPs. In each alternate panel, we randomly selected the causal SNP and set its statistical power to be $\mathcal{P}_c = 50\%$ at the significance level $\alpha = 10^{-8}$. Using this procedure, we simulated 500 GWASs in each population.

We applied GRAT to each simulated GWAS and recorded the recall rate of the significant SNPs and total number of tests performed. In Table 1, we show the performance of GRAT in each population averaged over the simulations. GRAT practically identified all significant associations and reduced the number of tests by 10 folds. Across the simulations, from the total $3,718,126$ significant associations GRAT only missed 1052 significant associations.

Table 1. The average performance of GRAT in 500 simulated GWASs using 1000 Genomes Project data in four populations. GRAT identified practically all significant associations by only testing 10% of the SNPs.

Population	Number of SNPs	Recall Rate	Reduction
AFR	8.5×10^6	$> 99.9\%$	88.2%
AMR	6.7×10^6	$> 99.9\%$	92.4%
ASN	6.1×10^6	$> 99.9\%$	92.8%
EUR	6.6×10^6	$> 99.9\%$	92.6%

3.5   Comparison to Tradition Tag-SNP Based Association Testing

Choosing an informative subset of the SNPs, termed tag-SNPs, under various criteria has been extensively investigated [2, 7, 11, 12, 15, 24, 27, 30, 33, 34, 36, 37]. The main goal of these methods is to reduce the cost of GWASs by genotyping a subset of the SNPs, yet collect as much information as possible on the remaining SNPs.

We mimic a two-stage association testing approach using a traditional tag-SNP selection method and compare its performance to GRAT. In the first stage, we test all the tag-SNPs and use a p-value threshold, α_{tag}, to choose which of the tag-SNPs to follow. If the p-value of a tag-SNP is more stronger than the threshold, the remainder SNPs tagged by this tag-SNP are tested.

We simulated association studies using the 10 HapMap ENCODE regions, which are densely genotyped for four HapMap populations [41]. In each simulation study, we used the ENCODE regions to generate null regions that harbor no causal SNPs and alternate regions each harboring a causal SNP with 50% statistical power at the genome-wide significance level of $\alpha = 10^{-8}$. Following this approach, we generated 500 association studies in each population.

In each region and in each population, we identified the tag-SNPs using the widely utilized tag-SNP selection method Tagger [2]. Given a set of SNPs and information on their minor allele frequencies and pairwise correlation coefficients, Tagger selects the minimum number of tag-SNPs such that each of the remaining SNPs correlates to a tag-SNP with a minimum r^2 pairwise correlation value. In our evaluations, we have used the default value of $r^2 = 0.8$. In order to perform a comparison, we also applied GRAT to identify the proxy SNPs and the statistic threshold rules for testing the remainder SNPs to achieve 99% target recall rate on the significant associations.

In Table 2 the performance of GRAT is compared to Tagger in four HapMap populations using various p-value threshold values, $\alpha_{\text{tag}} = \{10^{-8}, 10^{-7}, 10^{-6}, 10^{-5}\}$.

In each population, GRAT achieved more than 99% recall rate, while testing approximately 10% of all SNPs. Among all the p-value threshold values used, the traditional tag-SNPs led to testing more than twice the number of SNPs tested by GRAT and only achieved the target recall rate in all populations when the p-value threshold value was $\alpha_{tag} = 10^{-5}$. Unfortunately, Tagger, unlike GRAT does not guarantee a recall rate so it is not clear how to set the threshold and be certain that no associations are missed.

Table 2. In each HapMap population, the average performance of GRAT and Tagger in 500 simulated GWASs are shown. GRAT guarantees to achieve the 99% target recall rate, while reducing the number of tests by 90%. Using Tagger, we test the remainder SNPs that are tagged by the tag-SNPs that exceed a p-value cut-off threshold, α_{tag}. GRAT outperforms the traditional tag-SNPs in all populations.

Method	CEU			CHB		
	Recall	Reduction	Speedup	Recall	Reduction	Speedup
GRAT	99.89%	89.7%	9.7×	99.73%	89.6%	9.6×
Tagger $\alpha_{tag=1e\text{-}8}$	86.25%	78.9%	4.7×	87.78%	79.7%	4.9×
Tagger $\alpha_{tag=1e\text{-}7}$	95.74%	78.6%	4.7×	97.70%	79.4%	4.8×
Tagger $\alpha_{tag=1e\text{-}6}$	98.40%	78.3%	4.5×	99.62%	79.0%	4.8×
Tagger $\alpha_{tag=1e\text{-}5}$	99.30%	77.8%	4.5×	99.97%	78.4%	4.6×

Method	JPT			YRI		
	Recall	Reduction	Speedup	Recall	Reduction	Speedup
GRAT	99.63%	90.2%	10.2×	99.72%	88.4%	8.6×
Tagger $\alpha_{tag=1e\text{-}8}$	88.53%	80.5%	5.1×	87.62%	65.3%	2.9×
Tagger $\alpha_{tag=1e\text{-}7}$	98.10%	80.1%	5.0×	97.55%	65.3%	2.9×
Tagger $\alpha_{tag=1e\text{-}6}$	99.52%	79.6%	4.9×	99.39%	65.1%	2.9×
Tagger $\alpha_{tag=1e\text{-}5}$	99.92%	79.1%	4.8v	99.94%	65.0%	2.9×

4   Discussion

In the genome-wide association study (GWAS), information on single-nucleotide polymorphisms (SNPs) across the genome is collected from thousands of case and control individuals. Typically, each SNP is tested individually for disease association and the significant SNPs provide insight into the genetics of the disease. Association studies attempt to collect information on as many SNPs as possible to cover the whole genome. However, as the number of collected SNPs increases so does the computational burden to identify the significant associations.

We introduced a novel method, GRAT, for genome-wide rapid association testing to identify all significant associations by testing a small subset of the SNPs. Due to the correlation, or linkage disequilibrium (LD), testing a SNP provides information about the associations of its neighboring SNPs. Using this intuition, the procedure first tests a subset of the SNPs, referred to as the proxy SNPs, across the genome to locate the regions that may contain the significant associations. Once located, additional SNPs are tested from those regions to identify the significant SNPs. Each unobserved, or remainder, SNP is paired with its most strongly correlated proxy SNP, termed best-proxy, and a threshold value is used for the best-proxy's statistic to decide whether or not to test the unobserved SNP. We introduced a novel approach to choose the proxy SNPs and determine the threshold values for each best-proxy SNP. Through simulations and real GWAS data we showed that the proposed approach can identify more than 99% of the significant SNPs by reducing the number of tests by a factor of 10. Furthermore, GRAT can also be applied to association studies that utilize linear mixed models, where the speed-up is cumulative with recent efforts that decrease the computational burden of computing the actual association statistic. GRAT is implemented in C++ for high performance and is available at http://genetics.cs.ucla.edu/GRAT.

Acknowledgments. E.K is supported by training grant 2T32NS048004-06A1. E.K. and E.E. are supported by National Science Foundation grants 0513612, 0731455, 0729049, 0916676 and 1065276, and National Institutes of Health grants K25-HL080079, U01-DA024417, P01-HL30568 and PO1-HL28481.

References

1. Baker, M.: Biorepositories: Building better biobanks. Nature 486(7401), 141–146 (2012)
2. de Bakker, P.I.W., Yelensky, R., Pe'er, I., Gabriel, S.B., Daly, M.J., Altshuler, D.: Efficiency and power in genetic association studies. Nature Genetics 37(11), 1217–1223 (2005)
3. Bochner, B.R.: Innovations: New technologies to assess genotype-phenotype relationships. Nature Rev. Genet. 4(4), 309–314 (2003)
4. Brem, R.B., Kruglyak, L.: The landscape of genetic complexity across 5,700 gene expression traits in yeast. Proc. Natl. Acad. Sci. U S A 102(5), 1572–1577 (2005)
5. Brem, R.B., Yvert, G., Clinton, R., Kruglyak, L.: Genetic dissection of transcriptional regulation in budding yeast. Science 296(5568), 752–755 (2002)
6. Bystrykh, L., Weersing, E., Dontje, B., Sutton, S., Pletcher, M.T., Wiltshire, T., Su, A.I., Vellenga, E., Wang, J., Manly, K.F., Lu, L., Chesler, E.J., Alberts, R., Jansen, R.C., Williams, R.W., Cooke, M.P., de Haan, G.: Uncovering regulatory pathways that affect hematopoietic stem cell function using 'genetical genomics'. Nat. Genet. 37(3), 225–232 (2005)
7. Carlson, C.S., Eberle, M.A., Rieder, M.J., Yi, Q., Kruglyak, L., Nickerson, D.A.: Selecting a maximally informative set of single-nucleotide polymorphisms for association analyses using linkage disequilibrium. The American Journal of Human Genetics 74(1), 106–120 (2004)
8. Chesler, E.J., Lu, L., Shou, S., Qu, Y., Gu, J., Wang, J., Hsu, H.C., Mountz, J.D., Baldwin, N.E., Langston, M.A., Threadgill, D.W., Manly, K.F., Williams, R.W.: Complex trait analysis of gene expression uncovers polygenic and pleiotropic networks that modulate nervous system function. Nat. Genet. 37(3), 233–242 (2005)
9. Cheung, V.G., Spielman, R.S., Ewens, K.G., Weber, T.M., Morley, M., Burdick, J.T.: Mapping determinants of human gene expression by regional and genome-wide association. Nature 437(7063), 1365–1369 (2005)
10. Cookson, W., Liang, L., Abecasis, G., Moffatt, M., Lathrop, M.: Mapping complex disease traits with global gene expression. Nature Rev. Genet. 10(3), 184–194 (2009)

11. Cousin, E., Deleuze, J.F., Genin, E.: Selection of SNP subsets for association studies in candidate genes: comparison of the power of different strategies to detect single disease susceptibility locus effects. BMC Genetics 7 (2006)
12. Cousin, E., Genin, E., Mace, S., Ricard, S., Chansac, C., del Zompo, M., Deleuze, J.F.: Association studies in candidate genes: strategies to select SNPs to be tested. Human Heredity 56(4), 151–159 (2003)
13. Devlin, B., Risch, N.: A comparison of linkage disequilibrium measures for fine-scale mapping. Genomics 29(2), 311–322 (1995)
14. Emilsson, V., Thorleifsson, G., Zhang, B., Leonardson, A.S., Zink, F., Zhu, J., Carlson, S., Helgason, A., Walters, G.B., Gunnarsdottir, S., Mouy, M., Steinthorsdottir, V., Eiriksdottir, G.H., Bjornsdottir, G., Reynisdottir, I., Gudbjartsson, D., Helgadottir, A., Jonasdottir, A., Jonasdottir, A., Styrkarsdottir, U., Gretarsdottir, S., Magnusson, K.P., Stefansson, H., Fossdal, R., Kristjansson, K., Gislason, H.G., Stefansson, T., Leifsson, B.G., Thorsteinsdottir, U., Lamb, J.R., Gulcher, J.R., Reitman, M.L., Kong, I., Schadt, E.E., Stefansson, K.: Genetics of gene expression and its effect on disease. Nature 452(7186), 423–428 (2008)
15. Halperin, E., Kimmel, G., Shamir, R.: Tag SNP selection in genotype data for maximizing SNP prediction accuracy. Bioinformatics 21(suppl. 1) (2005)
16. Han, B., Kang, H.M., Eleazar, E.: Rapid and accurate multiple testing correction and power estimation for millions of correlated markers. PLoS Genet 5(4) (2009)
17. Hardy, J., Singleton, A.: Genomewide association studies and human disease. N. Engl. J. Med. 360(17), 1759–1768 (2009)
18. Hindorff, L.A., Sethupathy, P., Junkins, H.A., Ramos, E.M., Mehta, J.P., Collins, F.S., Manolio, T.A.: Potential etiologic and functional implications of genome-wide association loci for human diseases and traits. PNAS 106(23), 9362–9367 (2009)
19. International HapMap Consortium: A haplotype map of the human genome. Nature 437(7063), 1299–1320 (2005)
20. Kang, H.M., Sul, J.H., Service, S.K., Zaitlen, N.A., Kong, S.Y., Freimer, N.B., Sabatti, C., Eskin, E.: Variance component model to account for sample structure in genome-wide association studies. Nature Genet. 42(4), 348 (2010)
21. Keurentjes, J.J.B., Fu, J., Terpstra, I.R., Garcia, J.M., van den Ackerveken, G., Snoek, L.B., Peeters, A.J.M., Vreugdenhil, D., Koornneef, M., Jansen, R.C.: Regulatory network construction in arabidopsis by using genome-wide gene expression quantitative trait loci. Proc. Natl. Acad. Sci. U S A 104(5), 1708–1713 (2007)
22. Kostem, E., Lozano, J.A., Eskin, E.: Increasing power of genome-wide association studies by collecting additional single-nucleotide polymorphisms. Genetics 188(2), 449–460 (2011)
23. Li, Y., Willer, C.J., Ding, J., Scheet, P., Abecasis, G.: Mach: using sequence and genotype data to estimate haplotypes and unobserved genotypes. Genet. Epidemiol. 34(8), 816–834 (2010)
24. Lin, Z., Altman, R.B.: Finding haplotype tagging SNPs by use of principal components analysis. The American Journal of Human Genetics 75(5), 850–861 (2004)
25. Lippert, C., Listgarten, J., Liu, Y., Kadie, C.M., Davidson, R.I., Heckerman, D.: Fast linear mixed models for genome-wide association studies. Nature Methods 8(10), 833 (2011)
26. Majewski, J., Pastinen, T.: The study of eQTL variations by RNA-seq: from snps to phenotypes. Trends Genet. 27(2), 72–79 (2011)
27. Pardi, F., Lewis, C.M., Whittaker, J.C.: SNP selection for association studies: Maximizing power across SNP choice and study size. Annals of Human Genetics 69(6), 733–746 (2005)

28. Pritchard, J.K., Przeworski, M.: Linkage disequilibrium in humans: models and data. Am. J. Hum. Genet. 69(1), 1–14 (2001)
29. Purcell, S., Neale, B., Todd-Brown, K., Thomas, L., Ferreira, M.A.R., Bender, D., Maller, J., Sklar, P., de Bakker, P.I.W., Daly, M.J., Sham, P.C.: Plink: a tool set for whole-genome association and population-based linkage analyses. Am. J. Hum. Genet. 81(3), 559–575 (2007)
30. Qin, Z.S., Gopalakrishnan, S., Abecasis, G.R.: An efficient comprehensive search algorithm for tag SNP selection using linkage disequilibrium criteria. Bioinformatics 22(2), 220–225 (2006)
31. Risch, N., Merikangas, K.: The future of genetic studies of complex human diseases. Science 273(5281), 1516–1517 (1996)
32. Rockman, M.V., Kruglyak, L.: Genetics of global gene expression. Nature Rev. Genet. 7(11), 862–872 (2006)
33. Saccone, S.F., Rice, J.P., Saccone, N.L.: Power-based, phase-informed selection of single nucleotide polymorphisms for disease association screens. Genetic Epidemiology 30(6), 459–470 (2006)
34. Santana, R., Mendiburu, A., Zaitlen, N., Eskin, E., Lozano, J.A.: Multi-marker tagging single nucleotide polymorphism selection using estimation of distribution algorithms. Artificial Intelligence in Medicine 50(3), 193–201 (2010)
35. Spielman, R.S., Bastone, L.A., Burdick, J.T., Morley, M., Ewens, W.J., Cheung, V.G.: Common genetic variants account for differences in gene expression among ethnic groups. Nat. Genet. 39(2), 226–231 (2007)
36. Stram, D.O.: Tag SNP selection for association studies. Genetic Epidemiology 27(4), 365–374 (2004)
37. Stram, D.O.: Software for tag single nucleotide polymorphism selection. Human Genomics 2(2), 144–151 (2005)
38. Stranger, B.E., Montgomery, S.B., Dimas, A.S., Parts, L., Stegle, O., Ingle, C.E., Sekowska, M., Smith, G.D., Evans, D., Gutierrez-Arcelus, M., Price, A., Raj, T., Nisbett, J., Nica, A.C., Beazley, C., Durbin, R., Deloukas, P., Dermitzakis, E.T.: Patterns of cis regulatory variation in diverse human populations. PLoS Genet. 8(4), e1002639 (2012)
39. Stranger, B.E., Nica, A.C., Forrest, M.S., Dimas, A., Bird, C.P., Beazley, C., Ingle, C.E., Dunning, M., Flicek, P., Koller, D., Montgomery, S., Tavaré, S., Deloukas, P., Dermitzakis, E.T.: Population genomics of human gene expression. Nat. Genet. 39(10), 1217–1224 (2007)
40. The 1000 Genomes Project Consortium: A map of human genome variation from population-scale sequencing. Nature 467(7319), 1061 (2010)
41. The ENCODE Project Consortium: The ENCODE (ENCyclopedia Of DNA Elements) project. Science 306(5696), 636–640 (2004)
42. The ENCODE Project Consortium: Identification and analysis of functional elements in 1% of the human genome by the ENCODE pilot project. Nature 447(7146), 799–816 (2007)
43. The ENCODE Project Consortium: A user's guide to the encyclopedia of DNA elements (ENCODE). PLoS Biol. 9(4), e1001046 (2011)
44. The ENCODE Project Consortium: An integrated encyclopedia of DNA elements in the human genome. Nature 489(7414), 57–74 (2012)
45. Wang, Z., Gerstein, M., Snyder, M.: RNA-seq: a revolutionary tool for transcriptomics. Nature Rev. Genet. 10(1), 57–63 (2009)
46. Zhou, X., Stephens, M.: Genome-wide efficient mixed-model analysis for association studies. Nature Genet. 44(7), 821–824 (2012)

Identification of Ultramodified Proteins Using Top-Down Spectra

Xiaowen Liu[1,3], Shawna Hengel[2], Si Wu[2], Nikola Tolić[2],
Ljiljana Pasa-Tolić[2], and Pavel A. Pevzner[3]

[1] School of Informatics, Indiana University-Purdue University Indianapolis
[2] EMSL, Pacific Northwest National Laboratory
[3] Department of Computer Science and Engineering,
University of California, San Diego

Abstract. Post-translational modifications (PTMs) play an important
role in various biological processes through changing protein structure
and function. Some *ultramodified* proteins (like histones) have multiple
PTMs forming *PTM patterns* that define the functionality of a protein.
While bottom-up mass spectrometry (MS) has been successful in identifying *individual* PTMs within short peptides, it is unable to identify
PTM patterns spread along entire proteins in a coordinated fashion. In
contrast, top-down MS analyzes intact proteins and reveals PTM patterns along the entire proteins. However, while recent advances in instrumentation have made top-down MS accessible to many laboratories,
most computational tools for top-down MS focus on proteins with few
PTMs and are unable to identify complex PTM patterns. We propose
a new algorithm, MS-Align-E, that identifies both expected and unexpected PTMs in ultramodified proteins. We demonstrate that MS-Align-
E identifies many protein forms of histone H4 and benchmark it against
the currently accepted software tools.

1   Introduction

Post-translational modifications (PTMs) affect protein structure and function.
In some proteins, the function of the protein is determined by a *combination* of
multiple PTM sites (*PTM pattern*) rather than individual PTMs at specific sites.
We refer to proteins with many modifications sites as *ultramodified* proteins. For
example, histones often have multiple PTM sites with various PTM types such
as acetylation, methylation, and phosphorylation. Specifically for histones, the
PTM patterns define their gene regulatory functions [1, 2] through the "combinatorial histone code" [3, 4]. PTM patterns in histones are part of the epigenetic
mechanisms that are now being linked to several human diseases. However, revealing PTM patterns in histones has proven to be a challenge. As Garcia and
colleagues wrote in a recent review: "The ability to detect combinatorial histone
PTMs is now much easier than it has been before, but the most difficult issue
with these analyses still remains: deconvolution of the data" [5]. Highly complex
top-down spectra of histones feature multiple ion series that are either shared
and unique to the multiple protein forms. These spectra have to be decoded for

M. Deng et al. (Eds.): RECOMB 2013, LNBI 7821, pp. 132–144, 2013.

revealing the histone PTM space and deriving rules governing the combinatorial histone code.

Bottom-up database search tools offer a variety of algorithms to search for both expected [6] and unexpected [7, 8] PTMs. However, while bottom-up mass spectrometry (MS) has been successful in identifying some PTM sites, it is not well suited for identification of complex PTM patterns. Because bottom-up MS is based on digesting proteins into short peptides, PTMs identified are restricted to *individual* peptides, lacking information on how many protein isoforms are present (i.e. how the combination of modified/unmodified peptide sequences are put back together). Even if all peptides within a protein, and all PTMs within each peptide were identified, the ability to identify PTM patterns would still be lacking because the correlations between PTMs located on different peptides are lost (Fig. 1 in the Appendix). Moreover, bottom-up MS rarely provides full coverage of proteins by identified peptides: a typical shotgun proteomics study (with a single protease like trypsin) provides on average about 25% coverage for proteins [9]. It implies that many PTMs may remain below the radar of bottom-up proteomics. Middle-down proteomics [10, 11] identifies PTM sites on longer peptides and thus takes an intermediate position between bottom-up and top-down approaches with respect to identifying PTM patterns, however there is still a gap between intact protein forms and digestion products.

Over the last several years, applications of top-down MS have significantly expanded due to the recent progress in MS instrumentation and protein separation. The widely available commercial mass spectrometers are now capable of analyzing short proteins with molecular weight up to 30 kDa [12]. However, software tools for analyzing ultramodified proteins by top-down MS have not kept pace with rapid developments in top-down technology.

PTMs are often classified into *expected* and *unexpected* referring to the types of PTMs that are commonly and rarely observed (on specific proteins). For example, with respect to histones, acetylation, methylation, and phosphorylation represent expected PTMs, while carbamylation may represent an unexpected PTM. We emphasize that by expected PTMs we mean expected PTM *types* rather than PTM *sites*. Expected PTM types are often referred to as "variable PTMs" in peptide identification tools.

Existing top-down protein identification tools score Protein-Spectrum-Matches (PrSMs) using various scoring functions $Score(P, S)$, where (P, S) refers to a PrSM formed by a protein P and a spectrum S. The simplest scoring function (called the "shared peak count") counts the number of peaks in the spectrum S "explained" by the protein P, i.e., the number of shared monoisotopic peaks between S and the *theoretical spectrum* of P. Given a PrSM (P^*, S) between a modified form P^* of a protein P and a spectrum S, the shared peak count is the number of shared monoisotopic peaks between S and the theoretical spectrum of P^*.

Given an unmodified protein P, a set of expected PTM types Ω, and an integer F, we define $ProteinDB(P, \Omega, F)$ as the set of *all* modified forms of P with exactly F expected PTM sites. Since the size of $ProteinDB(P, \Omega, F)$ increases exponentially with an increase in F, exploring all protein forms in this database becomes computationally intractable, particularly when the set of expected PTM types is large. This motivates the following *Expected PTM Identification* (EPI) problem: given a top-down spectrum S, an unmodified protein P, an integer F, and a set of expected PTM types Ω, find a modified form P^* of the protein P with F expected PTM sites such that $Score(P^*, S)$ is maximized among all protein forms in $ProteinDB(P, \Omega, F)$. Below we describe two existing approaches to solving the EPI problem.

The "virtual database" approach (proposed by Neil Kelleher's group and implemented in ProSightPC [13]) compares each spectrum against the "virtual database" $ProteinDB(P, \Omega, F)$ with the goal to find the best scoring PrSM [11, 13]. This approach faces a combinatorial explosion when the number of PTM sites F is large and thus is not well suited for ultramodified proteins. For instance, based on the UniProt [14] flat file, histone H4 has more than 26 billion potential modified forms making it impractical to generate a "virtual database" containing all its modified forms. The number of modified forms explodes even further in searches for both expected and unexpected PTMs. Another limitation of the "virtual database" approach is its inability to find unexpected modifications that are not included in the set Ω.

To avoid combinatorial explosion, the *spectral alignment* algorithms for top-down protein identification find the best-scoring PrSM without explicitly exploring all protein forms in the virtual database in the case-by-case fashion [15,16]. However, the existing spectral alignment approaches, while working well for identification of proteins with a relatively small number of PTM sites (e.g., up to 3-4), were not designed for identification of ultramodified proteins like histones. First, they are primarily aimed at unexpected PTMs and the capabilities remain limited in the case of searches for both expected and unexpected PTMs. For example, due to limitations of the scoring functions, they tend to interpret two closely located expected PTM sites with masses a and b as a single unexpected PTM with mass $a+b$. Another limitation of the existing spectral alignment tools is that they require evidence for each PTM in the form of a "diagonal" in the spectral alignment matrix (See [16]). In the case when there are no fragmentation sites between two consecutive PTM sites along the protein, such diagonals may not exist, preventing the spectral alignment algorithms from solving the EPI problem. This situation is quite common for histones since PTM sites in histones are often closely located to each other.

Acknowledging that the "virtual database" approach is useful for identification of known protein forms, we emphasize that it also promotes erroneous identifications when known protein forms are used to explain spectra originating from unknown protein forms. Since such erroneous assignments turn out to be quite common (See the Appendix), they may severely limit our ability to construct the comprehensive list of PTM patterns for ultramodified proteins. Fig. 1 illustrates the case when ProSightPC reports an erroneous erroneous PrSM between a spectrum and a known protein form (from ProSightPC database). However, while this PrSM is high-scoring, the correct PrSM (found by MS-Align-E) explains

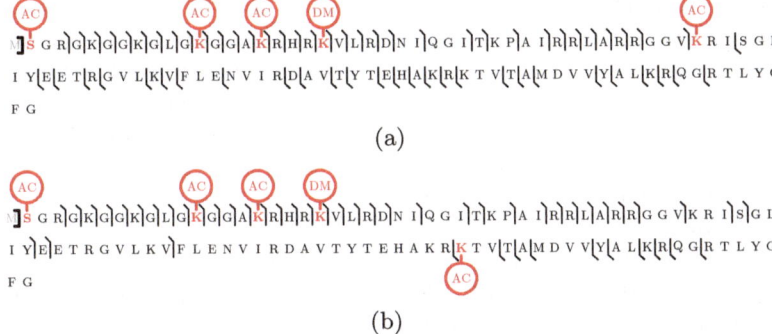

(a)

(b)

Fig. 1. MS-Align-E and ProSightPC reported two different protein forms for a spectrum from histone H4 spectral dataset. (a) The protein form reported by MS-Align-E has 62 matched fragment ions. (b) The protein form reported by ProSightPC has 49 matched fragment ions. The ']' symbol right to the first methionine residue represents N-terminal methionine excision. Residues with PTMs are shown in red. AC and DM stand for acetylation and di-methylation, respectively.

many more fragmentation sites and has a much higher score. The general problem with identification of ultramodified proteins using the "virtual database" approach is that high-scoring PrSMs often turn out to be incorrect because there are even higher scoring PrSMs that the "virtual database" approach has no ability to detect.

MS-Align-E (Mass Spectral ALIGNment for Expected PTMs) addresses this limitation of ProSightPC since it does not rely on a virtual database. It solves the EPI problem and is further extended for identifying proteins with both expected and unexpected PTMs. Even in the case of closely located sites of expected PTMs, MS-Align-E is capable of identifying correct PTM patterns. We tested MS-Align-E on a top-down MS/MS data set from histone H4 and identified 199 protein forms. The large number of reported protein forms illustrates the complexity of the combinatorial histone code. We further compared the resulting PrSMs with those reported by ProSightPC and found that in many cases, MS-Align-E finds protein forms explaining many more fragmentation sites than those reported by ProSightPC. Similarly to the case illustrated in Fig. 1, such PrSM may represent erroneous protein forms reported by ProSightPC.

2   Methods

MS-Align-E uses the spectral alignment to find PrSMs and the generating function approach to compute the E-values of these PrSMs. The key part of the generating function approach is the assumption that amino acids have integer masses [17]. However, rounding amino acid masses into integers introduces errors. These rounding errors reduce after rescaling by 0.9995 as described in [18–20]. While the scaling constant 0.9995 proved to be useful for bottom-up peptide identification, the rounding errors remain too large, even after rescaling, for highly accurate top-down spectra. MS-Align-E uses a scaling constant 274.335215 (e.g. mass(G) = 57.021464 × 274.335215 = 15642.995586 ≈ 15643) that reduces the rounding error to 2.5 parts per million (ppm). We thus assume that masses of all amino acids are integers (the mass of an amino acid r is referred to as $mass(r)$).

A protein $B = r_1 r_2 \ldots r_m$ is a sequence of amino acids. The mass of a protein prefix $r_1 r_2 \ldots r_k$ is $b_k = \sum_{i=1}^{k} mass(r_i)$. We will find it convenient to represent a protein B as a sequence of its prefix masses $b_0 < b_1 < \ldots < b_m$ (we assume $b_0 = 0$). The molecular mass of protein B is $\sum_{i=1}^{m} mass(r_i) + mass(H_2O)$, where $mass(H_2O)$ is the (rounded) mass of a water molecule.

A tandem mass spectrum (MS/MS) generated from a protein is represented by a precursor mass and a list of peaks. The precursor mass corresponds to the molecular mass of the protein and each peak, represented as (m/z, *intensity*), corresponds to a fragment ion of the protein. The values m/z and *intensity* are the mass-to-charge ratio and the abundance of the fragment ion, respectively. In preprocessing of top-down spectra, m/z values are usually converted into neutral masses of fragment ions by deconvolution algorithms [21,22]. Most of the neutral masses correspond to either protein prefixes or protein suffixes. The list of neutral masses can be further converted to a list of *prefix residue masses (PRMs)* corresponding to the masses of protein prefixes [23]. For a collision-induced dissociation (CID) spectrum with a precursor mass M, the PRM spectrum is generated as follows: (1) two masses 0 and $M - mass(H_2O)$ are added to the PRM spectrum (the mass $M - mass(H_2O)$ equals to the sum of the masses of all residues in the protein); (2) for each neutral mass x extracted from the experimental spectrum, two masses x and $M - x$ are added to the PRM spectrum. If mass x corresponds to a protein suffix (prefix), then mass $M - x$ corresponds to a protein prefix (suffix). Similar to discretization of amino acid masses, the precursor masses and the PRMs are discretized resulting in PRM spectra with integer mass values.

In contrast to bottom-up peptide identification tools that benefit from information about peak intensities, the existing top-down protein identification algorithms hardly use information about peak intensities (except for filtering out low intensity peaks). While in this paper we also ignore peak intensities, all proposed algorithms can be easily generalized to incorporate peak intensities. We represent a PRM spectrum A with a precursor mass M simply as a list of ordered integers $a_0 < a_1 < \ldots < a_n$, where $a_0 = 0$ and $a_n = M - mass(H_2O)$.

The mass difference between the modified and unmodified residues is the *mass shift* of the PTM. A PTM with a mass shift s on the ith residue in B transforms it into $b_0, b_1, \ldots, b_i + s, \ldots, b_m + s$. The mass shifts of all PTMs are discretized in the same way as PRMs are discretized. Let $S_1 = \{s_1, s_2, \ldots, s_k\}$ be the set of mass shifts corresponding to the expected PTM types. The (composite) *mass shift* of several expected PTM sites is the sum of their mass shifts. The set of mass shifts of all combinations of f expected PTM sites is defined recursively as $S_f = \{s | s = u + v, u \in S_1 \text{ and } v \in S_{f-1}\}$, for $f = 2, 3, \ldots$. For example, if $S_1 = \{14, 42\}$, then $S_2 = \{28, 56, 84\}$ and $S_3 = \{42, 70, 98, 126\}$. The *modification number* of an integer s is the minimum number f satisfying $s \in S_f$. For example, when

$S_1 = \{14, 42\}$, the composite mass shift 84 is present in three sets S_2, S_4, and S_6 since $84 = 42 + 42 = 42 + 14 + 14 + 14 = 14 + 14 + 14 + 14 + 14 + 14$. The modification number of 84 is 2. We also define $mod(0) = 0$ and $mod(s) = \infty$ if s cannot be partitioned into a sum of integers from S_1.

Typically, a PTM type modifies only several types of amino acids rather than all 20 standard amino acids. For example, phosphorylation is observed on amino acids S, T, and Y, but not on A. To simplify the presentation, we first consider a rather unrealistic case when each expected PTM type can modify all 20 amino acids. We will later describe how MS-Align-E restricts each expected PTM type to some specific amino acids that can be modified by this PTM.

2.1 Spectral Alignment

Given sequences of integers $A = a_0, a_1, \ldots, a_n$ and $B = b_0, b_1, \ldots, b_m$, the *grid* of A and B is defined as a two dimensional grid within a rectangle formed by four points $(0, 0)$, $(b_m, 0)$, $(0, -a_n)$, $(b_m, -a_n)$ [15]. The grid has $(n + 1)(m + 1)$ *matching points* $p_{i,j} = (b_j, -a_i)$. We refer to the upper leftmost matching point $(0, 0)$ and the lower rightmost matching point $(b_m, -a_n)$ as the *source* and the *sink*, respectively. Given matching points $p_{i,j}$ and $p_{i',j'}$, we say $p_{i',j'} < p_{i,j}$ if $i' < i$ and $j' < j$. We construct a *grid graph* with vertices corresponding to matching points and directed edges from matching points $p_{i',j'}$ to $p_{i,j}$ if $p_{i',j'} < p_{i,j}$. The grid graph has $O(n \cdot m)$ vertices and $O(n^2 \cdot m^2)$ edges.

The *mass shift* of an edge from vertex (matching point) $p_{i',j'}$ to vertex $p_{i,j}$ is defined as $(a_i - b_j) - (a_{i'} - b_{j'})$. An edge is called a *diagonal edge* if its mass shift is zero, and a *shift edge* otherwise. The diagonal edges are represents by $(-45°)$ diagonal segments. An *alignment* between sequences A and B is a path from the source to the sink in the grid graph. We assign *scores* to the vertices in the grid graph and define the score of an alignment (path) as the total score of its vertices. Below we assume that every vertex in the grid graph has score 1. An *optimal alignment* is an alignment with the maximum score.

As an example, consider a protein B =GSTGRTK and its modified version B^* =GS[+160]T[-30]GRT[-30]K with 3 PTMs. The grid for these proteins (represented as sequences $B = \{0, 57, 144, 245, 302, 458, 559, 687\}$ and $B^* = \{0, 57, 304, 375, 432, 588, 659, 787\}$) is shown in Fig. 2(a). The alignment shown in Fig. 2(a) represents every unmodified (modified) amino acid as a diagonal (shift) edge. The score of the alignment is simply the number of vertices in the alignment path (length of the protein plus 1).

Fig. 2(b) shows the grid in the case when the protein B^* is substituted by its spectrum A. As compared to B^*, the spectrum A has two missing masses 304 and 432, and a noise mass 482. As a result, the optimal alignment in Fig. 2(b) differs from the alignment in Fig. 2(a): the missing mass 384 results in two consecutive shift edges substituted by a single one, while the missing mass 432 results in two consecutive diagonal edges substituted by a single one.

When A and B correspond to a spectrum and a peptide, we refer to the grid and alignment between them as their *spectral grid* and *spectral alignment*, correspondingly. Diagonal edges in a spectral alignment correspond to segments of B matched to spectrum A without PTMs; shift edges correspond to segments of B with PTMs. The *modification number* of an edge is defined as the modification number of its mass shift (e.g., diagonal edges have modification number 0). The modification number of an edge from $p_{i',j'}$ to $p_{i,j}$ is denoted by $mod(p_{i',j'} \rightarrow p_{i,j})$. A shift edge from $p_{i',j'}$ to $p_{i,j}$ is *valid* if its modification number $x \leq F$ and $x \leq j - j'$. The condition $x \leq j - j'$ guarantees that for a shift edge with modification number x, there exist at least x modified residues in the protein supporting the mass shift. A spectral alignment is *valid* if all its shift edges are valid. The *modification number* of a spectral alignment is the sum of the modification numbers of its shift edges. A spectral alignment between A and B with modification number F is *optimal* if it has the maximum score among all alignments with modification number F. It is easy to check that a path shown in Fig. 2(b) is an optimal valid alignment with modification number 3. Since a valid spectral alignment with a modification number F corresponds to a modified protein form with F PTM sites [24], the EPI problem is reduced to the following graph-theoretical problem:

Expected PTM spectral alignment (EPSA) problem. Given a spectrum $A = \{a_0, a_1, \ldots, a_n\}$, a protein $B = \{b_0, b_1, \ldots, b_m\}$, an integer F, a set of mass shifts S_1 corresponding to expected PTMs, find an optimal valid spectral alignment of A and B with the modification number F.

To solve the EPSA problem one can use the *parametric dynamic programming* algorithm (similar to the generating function approach in [17]) for finding a longest path in a spectral grid graph with a given number of modifications. However, the running time of the longest path algorithm is proportional to the number of edges in the spectral grid graph (equal to $O(n^2 \cdot m^2)$) making this algorithm prohibitively time consuming. Pevzner *et al.*, 2000, 2001 [24,25] described an *equivalent transformation* of the spectral grid graph that greatly reduces the number of edges in the graph while preserving an optimal spectral alignment path. Below we develop similar approaches for top-down spectra.

EPSA Algorithm. We modify the spectral alignment approach [15,16,24] for solving the EPSA problem. To trace the number of modifications along a path in a spectral grid graph, we recursively fill an $(n+1) \times (m+1) \times (F+1)$ array D in which the value $D_{i,j}(f)$ is the highest score among all valid paths with a modification number f from the source $p_{0,0}$ to a vertex $p_{i,j}$. Below we show how to greatly reduce the number of edges in the spectral grid graph to make the spectral alignment algorithm efficient.

Let $\mathcal{S} = \{0\} \cup S_1 \cup \ldots \cup S_F$. We remove all invalid shift edges from the spectral grid graph and further reduce the number of edges as follows. An edge between vertices x and y is called *dispensable* if there is a vertex z such that $x < z < y$ and at least one of the edges (x, z) and (z, y) is diagonal. Since every dispensable edge (x, y) can be substituted by a path formed by edges (x, z) and (z, y) with a higher score, optimal spectral alignments do not include dispensable edges. We thus can safely remove all dispensable edges from the spectral grid graph.

Consider all edges entering into $p_{i,j}$ and denote the set of all vertices where these edges originate as $\mathcal{M}_{i,j}$. It is easy to see that no two vertices in $\mathcal{M}_{i,j}$ are located on the same $-45°$ line since otherwise an edge from one of these vertices to $p_{i,j}$ would be dispensable. Therefore, no two vertices in $\mathcal{M}_{i,j}$ have the same mass shift implying that the size of $\mathcal{M}_{i,j}$ is small (does not exceed $|\mathcal{S}|$). The recurrence function for $D_{i,j}(f)$ can be rewritten as follows:

$$D_{i,j}(f) = \begin{cases} \max\limits_{p_{i',j'} \in \mathcal{M}_{i,j}} D_{i',j'}(f - mod(p_{i',j'} \to p_{i,j})) + 1 & \text{if } \mathcal{M}_{i,j} \neq \phi; \\ -\infty & \text{otherwise,} \end{cases} \qquad (1)$$

where $D_{i',j'}(f - mod(p_{i',j'} \to p_{i,j})) = -\infty$ when $f - mod(p_{i',j'} \to p_{i,j}) < 0$.

The algorithm using the recurrence (1) is referred to as EPSA algorithm. The total number of matching points is $(n+1)(m+1)$, thus $|\mathcal{M}_{i,j}| \leq (n+1)(m+1)$. Also, as we showed before, $|\mathcal{M}_{i,j}| \leq |\mathcal{S}|$. Since the set $\mathcal{M}_{i,j}$ is easy to compute and $|\mathcal{M}_{i,j}| \leq T = \min\{|\mathcal{S}|, (n+1)(m+1)\}$, the time complexity of the EPSA algorithm is $O(n \cdot m \cdot T \cdot F)$.

2.2   From Spectral Grid to Diagonal Grid

A mass spectrum A of protein B contains *some* but not necessarily all fragmentation points of a protein B. As a result, the spectral alignment in Fig. 2(b) deteriorates as compared to Fig. 2(a). However, given the set of (composite) mass shifts $\mathcal{S}$, one can construct a set containing *all* putative fragmentation points of protein B (and to "restore" the quality of spectral alignment) as follows.

A $-45°$ line l passing the spectral grid at point (x, y) is called a *diagonal line* with *offset* equal to $offset(l) = -x - y$. For example, a diagonal line starting at the left vertical border of the grid at $(0, -10)$ has offset 10. Similarly to the standard grid formed by crossing $(n + 1)$ horizontal lines with $(m + 1)$ vertical lines (originated from spectrum $A = \{a_0, \ldots, a_n\}$ and protein $B = \{b_0, \ldots, b_m\}$), we form a *diagonal grid* by crossing $|\mathcal{S}|$ diagonal lines with $(m+1)$ vertical lines. For each $s \in \mathcal{S}$, there exists a diagonal line with offset s contributing to the diagonal grid (Fig. 2(c)). The intersection of a diagonal line and a vertical line is called a *diagonal point* (there are $|\mathcal{S}| \cdot (m + 1)$ diagonal points in the diagonal grid). Let $l_0, l_1, \ldots, l_{|\mathcal{S}|-1}$ be the diagonal lines ordered in the increasing order of $offset(l_0) < offset(l_1) < \ldots < offset(l_{|\mathcal{S}|-1})$. The diagonal point of a crossing line l_i and a vertical line corresponding to mass b_j is denoted by $q_{i,j}$.

The *diagonal grid graph* (or simply *diagonal graph*) is defined similarly to the grid graph. The vertex set of the diagonal graph consists of all diagonal points. Score 1 is assigned to vertices in the diagonal grid if they are present in the spectral grid (all other vertices are assigned score 0). The set of edges in the diagonal graph is redefined (as compared to the spectral grid graph) by only connecting vertices located on *consecutive* vertical lines in the diagonal grid. Specifically, a vertex (diagonal point) $q_{i,j}$ is connected with a vertex $q_{i+1,j'}$ by an edge if the difference between the offsets of diagonal lines l_j and $l_{j'}$ is either 0 (i.e., connecting consecutive vertices on the same diagonal line) or in set S_1.

A *diagonal alignment* is defined as an alignment (path) in the diagonal graph (Fig. 2(c)). Each valid path in the spectral grid graph has a corresponding path

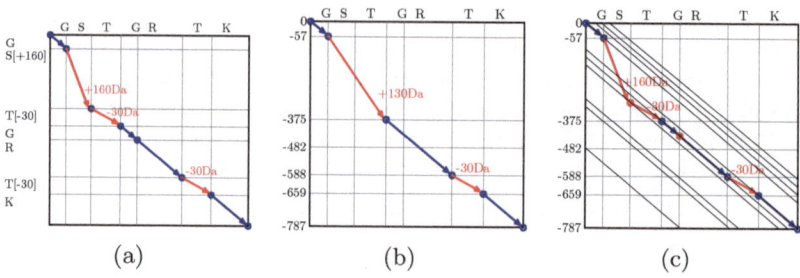

Fig. 2.   Spectral alignment. (a) A spectral alignment between the theoretical spectrum $B = \{0, 57, 144, 245, 302, 458, 559, 687\}$ of a protein GSTGRTK and the theoretical spectrum $B^* = \{0, 57, 304, 375, 432, 588, 659, 787\}$ of a modified protein GS[+160]T[-30]GRT[-30]K. The path from the top left corner (source) to the bottom right corner (sink) represents the alignment of B and B^* with three PTMs: +160 Da on the first S and −30 Da on the two T's. Diagonal and shift edges are shown in blue and red, respectively. The circles along the path denote the matching points in the alignment path. (b) A spectral alignment between a spectrum $A = \{0, 57, 375, 482, 588, 659, 787\}$ generated from GS[+160]T[-30]GRT[-30]K and the theoretical spectrum B. Because mass 304 is missing in A, the PTM on the first S and the PTM on the first T are represented by a single shift edge (+130 Da) with a modification number 2. Another missing mass 432 in A results in replacing two consecutive diagonal edges by one diagonal edge. In addition, mass 482 is a noise mass. (c) A diagonal alignment between the spectrum A and the theoretical spectrum B (for a set of mass shifts $S_1 = \{-30, 160\}$ and $F = 3$). The diagonal grid of A and B has 10 diagonal lines with offsets -90, -60, -30, 0, 100, 130, 160, 290, 320, and 480. The path from the source to the sink represents a diagonal alignment of spectrum A and protein B. The circles along the path denote diagonal points: blue ones have weight 1 and red ones have weight 0.

in the diagonal grid graph (all shift edges have a modification number 1). Edges with modification number larger than 1 in the spectral grid graph correspond to paths (formed by edges with modification number 1) in the diagonal graph. As Fig. 2(c) illustrates, the diagonal alignment improves as compared to the spectral alignment in Fig. 2(b) and now looks like the protein-protein alignment in Fig. 2(a). The EPSA problem in the spectral grid graph is reduced to the following problem in the diagonal graph:

Expected PTM diagonal alignment (EPDA) problem. Given a spectrum $A = \{a_0, a_1, \ldots, a_n\}$, a protein $B = \{b_0, b_1, \ldots, b_m\}$, an integer F, a set of mass shifts S_1 corresponding to expected PTMs, find an optimal diagonal alignment of A and B with F shift edges in the diagonal graph.

EPDA Algorithm. To solve the EPDA problem, we recursively fill an $|S| \times (m+1) \times (F+1)$ array D in which the value $D_{i,j}(f)$ is the highest score among all paths with a modification number f from the source $q_{0,0}$ to a diagonal point $q_{i,j}$. Consider all edges entering into vertex $q_{i,j}$ in the diagonal graph and denote the set of all vertices where these edges originate as $\mathcal{N}_{i,j}$. The recurrence for $D_{i,j}(f)$ can be rewritten as follows:

$$D_{i,j}(f) = \begin{cases} \max\limits_{q_{i',j'} \in \mathcal{N}_{i,j}} D_{i',j'}(f - mod(q_{i',j'} \to q_{i,j})) + w_{i,j} & \text{if } \mathcal{N}_{i,j} \neq \phi; \\ -\infty & \text{otherwise,} \end{cases} \tag{2}$$

where $w_{i,j}$ is the score of $q_{i,j}$ and $D_{i',j'}(f - mod(q_{i',j'} \to q_{i,j})) = -\infty$ when $f - mod(q_{i',j'} \to q_{i,j}) < 0$.

The algorithm using the recurrence (2) is referred to as EPDA algorithm. Similar to the analysis of $\mathcal{M}_{i,j}$ in Section 2.2, one can prove that $|\mathcal{N}_{i,j}| \leq (|S_1|+1)$. Therefore, the complexity of EPDA algorithm is $O(m \cdot F \cdot |\mathcal{S}|)$, a significant speed-up compared to the EPSA Algorithm.

In many cases, the size of the diagonal graph can be further reduced. Every alignment path starts at the source (with offset 0) and ends at the sink (with $offset(sink) = a_n - b_m$). Let v be a vertex in the spectral grid graph with offset s. If a diagonal alignment path passes through v then this path has the modification number that is greater than or equal to $mod(s) + mod(offset(sink) - mod(s))$. Therefore, diagonal alignments with modification number F do not include vertices with offset s if $mod(s) + mod(offset(sink) - mod(s)) > F$. Thus, every $s \in S$ that satisfies the condition $mod(s) + mod(offset(sink) - mod(s)) > F$, should not contribute a diagonal line to the construction of the diagonal graph.

Typically, a PTM type modifies only several types of amino acids rather than all 20 amino acids. Restricting PTMs to a subset of amino acids can be naturally modeled in the framework of the diagonal graph. Since every shift edge in the diagonal graph corresponds to a specific amino acid in the protein, we simply remove shift edges whose shift values are not present in the list of allowed PTMs for the amino acid.

2.3   Identifying Spectra with Both Expected and Unexpected PTMs

The spectral alignment algorithms can be modified to identify proteins with both expected and unexpected PTMs [15]. However, the complexity of the resulting algorithm is $O(n \cdot m \cdot T \cdot F_e \cdot F_u)$, where $T = \min\{(n+1)(m+1), |\mathcal{S}|\}$, and F_e/F_u are the numbers of expected and unexpected PTM sites, respectively. Since this algorithm is too slow in practice, we propose a fast heuristic algorithm for identifying proteins with both expected and unexpected PTMs (See the Appendix for details). To identify protein isoforms truncated at N- or C-terminus, the local alignment algorithm described in [16] is used. E-values of identified PrSMs are computed using a method described in [16].

3   Results

We implemented MS-Align-E (CPDA algorithm in the Appendix) in Java and tested it on a top-down MS/MS data set of histone H4. The data set contains $1,626$ CID and $1,626$ ETD spectra (See the Appendix for details). The experiments were run on a desktop PC with 3.4 GHz CPU (Intel Core i7-3770) and 16 GB memory.

3.1   Identification of Protein Forms from Ultramodified Histone H4

All MS/MS spectra were deconvoluted using MS-Deconv [22]. MS-Align-E was used to align the deconvoluted spectra with the histone H4 protein sequence. The error tolerances for precursor ions and fragment ions were set as 15 ppm. Five PTM types were treated as expected ones (Table 1 in the Appendix); maximum 10 expected PTM sites and 1 unexpected PTM sites were allowed. The running time of MS-Align-E was $\approx$ 505 minutes (with computing E-values). With E-value cut off 0.01^1, MS-Align-E identified 624 spectra from 199 protein forms. These results can provide hints to help identify and functionally characterize differentially modified protein forms of histone H4. Many identified protein forms have more than 3 expected PTM sites (Fig. 2 in the Appendix). When one unexpected PTM site is allowed, several expected or unexpected PTM sites might be combined to an unexpected PTM site with a large mass shift. Thus, the protein forms with one unexpected PTM sites tend to have less expected PTM sites compared with those without unexpected PTM sites.

3.2   Comparison with ProSightPC

ProSightPC computes E-values of identified PrSMs based on the size of the target protein database and a Poisson distribution of three parameters: the number of fragment ions, the number of matched fragment ions, and the probability of an observed fragment ion matching a random theoretical fragment ion. Because the distribution fails to consider the peak positions in spectra, the estimation of E-values of identified PrSMs is not accurate [16]. MS-Align-E uses a more accurate generation function approach [17] to estimate E-values of identified PrSMs. ProSightPC and MS-Align-E often report different E-values for the same PrSM. Thus, it is not fair to compare the number of PrSMs identified by the two tools using the same cutoff for E-values. The number of matched fragment ions was used to rank PrSMs identified by the two tools, and all PrSMs with at least 10 matched fragment ions were reported and compared. ProSightPC identified 1, 034 PrSMs from 114 protein forms with at least 10 matched fragment ions.

Using used the parameter setting in Section 3.1, MS-Align-E identified 1, 081 PrSMs from 434 protein forms with at least 10 matched fragment ions (Fig. 3(b) in the Appendix). ProSightPC and MS-Align-E may report two different protein forms for the same spectrum. While the numbers of PrSMs identified by MS-Align-E and ProSightPC are similar (Fig. 3(b) in the Appendix), the PrSMs reported by MS-Align-E have more matched fragment ions than those reported by ProSightPC for many spectra (See the Appendix for details).

4   Conclusion

We proposed MS-Align-E algorithm for identifying ultramodified proteins from top-down MS data. Since MS-Align-E identifies more protein forms that other

---

[1] The target/decoy approach was used to estimate false discovery rate of the identified PrSMs, but no PrSMs with an E-value ≤ 0.01 were reported from the shuffled decoy protein database.

top-down database search tools, it has a potential to become a method of choice for analyzing ultramodified proteins and to contribute to studies of the combinatorial histone code.

Acknowledgment. This work was supported by National Institutes of Health Grant P-41-RR024851 and a startup fund provided by Indiana University-Purdue University Indianapolis. Portions of this work were performed in the William R. Wiley Environmental Molecular Sciences Laboratory (EMSL), a DOE BER national scientific user facility located on the campus of PNNL in Richland, Washington. PNNL is a multi-program national laboratory operated by Battelle for the DOE under Contract DE-AC05-76RLO1830. We thank Dr. David Stenoien for providing the core histone mixtures.

Appendix

http://mypage.iu.edu/~xwliu/msaligne/

References

1. Cosgrove, M.S., Wolberger, C.: How does the histone code work? Biochemistry and Cell Biology 83, 468–476 (2005)
2. Strahl, B.D., Allis, C.D.: The language of covalent histone modifications. Nature 403, 41–45 (2000)
3. Garcia, B.A., Pesavento, J.J., Mizzen, C.A., Kelleher, N.L.: Pervasive combinatorial modification of histone H3 in human cells. Nature Methods 4, 487–489 (2007)
4. Young, N.L., DiMaggio, P.A., Plazas-Mayorca, M.D., Baliban, R.C., Floudas, C.A., Garcia, B.: High throughput characterization of combinatorial histone codes. Molecular & Cellular Proteomics 8, 2266–2284 (2009)
5. Britton, L.M.P., Gonzales-Cope, M., Zee, B.M., Garcia, B.A.: Breaking the histone code with quantitative mass spectrometry. Expert Review of Proteomics 8, 631–643 (2011)
6. Aebersold, R., Mann, M.: Mass spectrometry-based proteomics. Nature 422, 198–207 (2003)
7. Tsur, D., Tanner, S., Zandi, E., Bafna, V., Pevzner, P.A.: Identification of post-translational modifications by blind search of mass spectra. Nature Biotechnology 23, 1562–1567 (2005)
8. Na, S., Bandeira, N., Paek, E.: Fast multi-blind modification search through tandem mass spectrometry. Molecular & Cellular Proteomics 11, M111.010199 (2012)
9. de Godoy, L.M.F., Olsen, J.V., de Souza, G.A., Li, G., Mortensen, P., Mann, M.: Status of complete proteome analysis by mass spectrometry: SILAC labeled yeast as a model system. Genome Biology 7, R50 (2006)
10. Baliban, R.C., DiMaggio, P.A., Plazas-Mayorca, M.D., Young, N.L., Garcia, B.A., Floudas, C.A.: A novel approach for untargeted post-translational modification identification using integer linear optimization and tandem mass spectrometry. Molecular & Cellular Proteomics 9, 764–779 (2010)

11. DiMaggio, P.A., Young, N.L., Baliban, R.C., Garcia, B.A., Floudas, C.A.: A mixed integer linear optimization framework for the identification and quantification of targeted post-translational modifications of highly modified proteins using multiplexed electron transfer dissociation tandem mass spectrometry. Molecular & Cellular Proteomics 8, 2527–2543 (2009)

12. Tran, J.C., Zamdborg, L., Ahlf, D.R., Lee, J.E., Catherman, A.D., Durbin, K.R., Tipton, J.D., Vellaichamy, A., Kellie, J.F., Li, M., Wu, C., Sweet, S.M.M., Early, B.P., Siuti, N., Leduc, R.D., Compton, P.D., Thomas, P.M., Kelleher, N.L.: Mapping intact protein isoforms in discovery mode using top-down proteomics. Nature 480, 254–258 (2011)

13. Zamdborg, L., LeDuc, R.D., Glowacz, K.J., Kim, Y.B., Viswanathan, V., Spaulding, I.T., Early, B.P., Bluhm, E.J., Babai, S., Kelleher, N.L.: ProSight PTM 2.0: improved protein identification and characterization for top down mass spectrometry. Nucleic Acids Research 35, W701–W706 (2007)

14. Consortium, U.: Reorganizing the protein space at the Universal Protein Resource (UniProt). Nucleic Acids Research 40(Database issue) D71–D75 (2012)

15. Frank, A.M., Pesavento, J.J., Mizzen, C.A., Kelleher, N.L., Pevzner, P.A.: Interpreting top-down mass spectra using spectral alignment. Analytical Chemistry 80, 2499–2505 (2008)

16. Liu, X., Sirotkin, Y., Shen, Y., Anderson, G., Tsai, Y.S., Ting, Y.S., Goodlett, D.R., Smith, R.D., Bafna, V., Pevzner, P.A.: Protein identification using top-down spectra. Molecular & Cellular Proteomics, M111.008524 (2012)

17. Kim, S., Gupta, N., Pevzner, P.A.: Spectral probabilities and generating functions of tandem mass spectra: a strike against decoy databases. Journal of Proteome Research 7, 3354–3363 (2008)

18. Kim, S., Gupta, N., Bandeira, N., Pevzner, P.A.: Spectral dictionaries: Integrating de novo peptide sequencing with database search of tandem mass spectra. Molecular & Cellular Proteomics 8, 53–69 (2009)

19. Taylor, J.A., Johnson, R.S.: Sequence database searches via de novo peptide sequencing by tandem mass spectrometry. Rapid Communications in Mass Spectrometry 11, 1067–1075 (1997)

20. Bern, M., Cai, Y., Goldberg, D.: Lookup peaks: a hybrid of de novo sequencing and database search for protein identification by tandem mass spectrometry. Analytical Chemistry 79, 1393–1400 (2007)

21. Horn, D.M., Zubarev, R.A., McLafferty, F.W.: Automated reduction and interpretation of high resolution electrospray mass spectra of large molecules. Journal of the American Society for Mass Spectrometry 11, 330–332 (2000)

22. Liu, X., Inbar, Y., Dorrestein, P.C., Wynne, C., Edwards, N., Souda, P., Whitelegge, J.P., Bafna, V., Pevzner, P.A.: Deconvolution and database search of complex tandem mass spectra of intact proteins: A combinatorial approach. Molecular & Cellular Proteomics 9, 2772–2782 (2010)

23. Tanner, S., Shu, H., Frank, A., Wang, L.C., Zandi, E., Mumby, M., Pevzner, P.A., Bafna, V.: InsPecT: Identification of posttranslationally modified peptides from tandem mass spectra. Analytical Chemistry 77, 4626–4639 (2005)

24. Pevzner, P.A., Dančík, V., Tang, C.L.: Mutation-tolerant protein identification by mass spectrometry. Journal of Computational Biology 7, 777–787 (2000)

25. Pevzner, P.A., Mulyukov, Z., Dancik, V., Tang, C.L.: Efficiency of database search for identification of mutated and modified proteins via mass spectrometry. Genome Research 11, 290–299 (2001)

Distinguishing between Genomic Regions Bound by Paralogous Transcription Factors

Alina Munteanu[1] and Raluca Gordân[2]

[1] Faculty of Computer Science,
Alexandru I. Cuza University, Iasi, Romania
`alina.munteanu@info.uaic.ro`
[2] Institute for Genome Sciences and Policy,
Departments of Biostatistics & Bioinformatics,
Computer Science, and Molecular Genetics and Microbiology,
Duke University, Durham, NC 27708, USA
`raluca.gordan@duke.edu`

Abstract. Transcription factors (TFs) regulate gene expression by binding to specific DNA sites in cis regulatory regions of genes. Most eukaryotic TFs are members of protein families that share a common DNA binding domain and often recognize highly similar DNA sequences. Currently, it is not well understood why closely related TFs are able to bind different genomic regions *in vivo*, despite having the potential to interact with the same DNA sites. Here, we use the Myc/Max/Mad family as a model system to investigate whether interactions with additional proteins (co-factors) can explain why paralogous TFs with highly similar DNA binding preferences interact with different genomic sites *in vivo*. We use a classification approach to distinguish between targets of c-Myc versus Mad2, using features that reflect the DNA binding specificities of putative co-factors. When applied to c-Myc/Mad2 DNA binding data, our algorithm can distinguish between genomic regions bound uniquely by c-Myc versus Mad2 with 87% accuracy.

Keywords: Transcription factors, protein binding microarray, ChIP-seq, co-factors, support vector machine, random forrest.

1 Introduction

Transcription factors (TFs) regulate gene expression by binding to specific, short DNA sites in in the promoters or enhancers of the regulated genes. Determining the DNA sequences recognized by TFs is essential for understanding how these proteins achieve their DNA binding specificities and exert their specific regulatory roles in the cell. The DNA binding site motifs of hundred of eukaryotic TFs have been determined thus far using high-throughput *in vivo* techniques such as ChIP-chip [1] or ChIP-seq [2], as well as *in vitro* assays such as protein binding microarrays (PBMs [3]). A close examination of the available TF-DNA binding motifs from databases such as UniPROBE [4], Transfac [5], and Jaspar [6] reveals that many eukaryotic TFs have highly similar DNA binding properties. This is

M. Deng et al. (Eds.): RECOMB 2013, LNBI 7821, pp. 145–157, 2013.

not surprising given that most TFs are members of protein families that share a common DNA binding domain and thus have very similar sequence preferences [7]. However, it is surprising that, despite having the potential to bind the same genomic sites, individual members of TF families (*i.e.*, paralogous TFs) often function in a non-redundant manner by binding different sets of target genes and controlling different regulatory programs. For example, among TFs in the E2F family, E2F1 has specific target genes [8] and it is the only factor equipped with an ability to induce apoptosis [9], despite the fact that all E2F family members have the same DNA binding specificity [10]. Similarly, ETS1 and ELK1, members of the ETS family of TFs, each have unique target genes not bound by other ETS factors [11], despite the fact that their DNA binding motifs are virtually identical [12]. Currently, it is not well understood how closely related TFs achieve their differential DNA binding specificity *in vivo*. In some cases, intrinsic differences in DNA binding preferences contribute to the observed functional differences between paralogous TFs [13]. However, in other cases, the core DNA motifs are virtually identical [10], and still the proteins interact differently with putative genomic binding sites *in vivo*, as revealed by genome-wide ChIP-chip and ChIP-seq data [14, 15]. In such cases, it has been hypothesized that interactions with specific protein partners (henceforth referred to as co-factors) may contribute to the differential DNA binding *in vivo* [16].

Here, we use the Myc/Max/Mad family of TFs a model system to investigate whether interactions with putative co-factors can explain why paralogous TFs with seamingly identical DNA binding preferences interact with different genomic sites *in vivo*. Myc, Max, and Mad proteins are members of the basic helix-loop-helix leucine zipper (bHLH/Zip) family and they play essential roles in cell proliferation, differentiation, and death. Myc proteins are transcriptional activators that promote cell growth and proliferation, and are often overexpressed in cancer cells [17]. Proteins of the Mad family act as transcriptional repressors, they inhibit cell proliferation and are typically expressed at lower levels in human cancers [17]. In order to bind DNA, both Myc and Mad must heterodimerize with Max, a bHLH/Zip TF with little transcriptional activity [17]. Mad factors compete with Myc for dimerization with Max and for binding to genomic regions containing the E-box motif (CAnnTG), with both Myc and Mad having a strong preference for the E-box site CACGTG. Thus, it is not surprising that there is a high degree of overlap between the sets of targets bound by Myc and Mad factors *in vivo*, as illustrated by ChIP-seq data available from ENCODE [15]. However, despite a significant overlap in their sets of ChIP-bound regions, Myc and Mad also have unique targets, as illustrated in Fig. 1A for c-Myc and Mad2 (Mxi1), representatives members of the Myc and Mad subfamilies, respectively. Here, we focus on c-Myc and Mad2 because high-quality *in vivo* TF-DNA binding data is available for both these factors as part of the ENCODE project [15].

We show that the intrinsic DNA binding preferences of c-Myc and Mad2 cannot explain why the two factors bind distinct sets of targets *in vivo*. High-quality DNA binding site motifs have been previously reported for c-Myc [5], but not Mad2 (nor other Mad factors). Therefore, we use PBM assays [3] to thoroughly characterize the sequence preferences of c-Myc and Mad2. Then, we use Support Vector Machines (SVM) [18] and Random Forrests (RF) [19] to identify sets of putative co-factors that can successfully distinguish between the genomic regions bound uniquely by c-Myc or Mad2, with an accuracy of ~87%.

Our classification-based approach is not restricted to c-Myc and Mad2. Instead, we implemented this approach in a general framework named **COUGER** (**co**-factors associated with **u**niquely-bound **ge**nomic **r**egions). Our framework can be applied to any two sets of genomic regions bound by paralogous TFs to identify the uniquely-bound targets and to determine the sets of co-TFs that best distinguish between the two sets of unique targets. Compared to related tools for analyzing ChIP-seq data, **COUGER** has several advantages, as detailed in the Discussion section: it uses state-of-the-art classification algorithms (SVM and RF) that are robust even when the feature set is large and some of the features are highly correlated; it makes use of high-quality TF-DNA binding data (from PBM experiments) to generate the features used in the classification; it takes into account the fact that TF binding sites may occur in clusters (while other tools only consider the highest affinity TF binding sites). Furthermore, given the large amount of ChIP-seq data available from ENCODE, we have implemented **COUGER** to accept as input ChIP-seq files in the narrowPeak format; such files can be downloaded directly from the ENCODE website. We anticipate that our framework will be extremely useful in analyzing ChIP-seq data to understand how interactions with specific co-factors contribute to differences in the *in vivo* DNA binding specificities of paralogous TFs.

COUGER is available at: www.genome.duke.edu/labs/gordan/COUGER. The PBM data for c-Myc and Mad2 is available at www.genome.duke.edu/labs/gordan/DATA.

2   Intrinsic DNA Binding Preferences of c-Myc and Mad2 Cannot Explain Their Differential *in vivo* DNA Binding

We combined *in vitro* and *in vivo* TF-DNA binding data for c-Myc and Mad2 to determine whether subtle differences in their intrinsic sequence preferences can explain, at least in part, the unique genomic targets bound by only one of the two factors *in vivo*. As evidence of *in vivo* binding we used ChIP-seq data from the ENCODE project [15]. We focused on the Hela S3 and K562 cell lines because ChIP-seq data is available for both c-Myc and Mad2, from the same laboratory. For both c-Myc and Mad2 we downloaded the ChIP-seq data in narrowPeak format from the UCSC Genome Browser [20]. For the HeLa S3 cell line, 7,440 binding regions (i.e., ChIP-seq peaks) were reported for c-Myc, and 32,138 for Mad2. Because the number of bound genomic sequences varied greatly between the two TFs, it would be difficult to perform a comparative analysis directly. The fact that different types of controls were used in the c-Myc and Mad2 ChIP experiments (standard versus no primary antibody) probably contributes to the larger number of peaks reported for Mad2. However, a close examination of the ChIP-seq data also revealed that the p-value cutoffs used for reporting the peaks

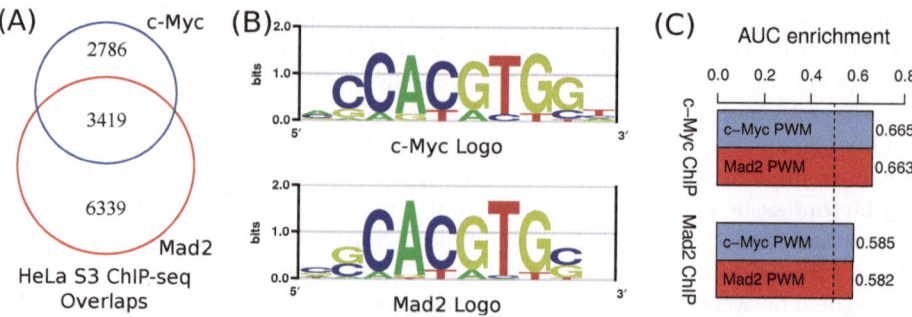

Fig. 1. (A) Overlap between the sets of genomic regions bound by c-Myc and Mad2 in a ChIP-seq experiment [15]. (B) c-Myc and Mad2 DNA binding motifs derived from *in vitro* PBM data. The logos were generated using enoLOGOS [21]. (C) AUC enrichment for the c-Myc and Mad2 DNA binding motifs in the ChIP-seq data for the two TFs in HeLa S3 cells. The dotted line shows the expected AUC for a random motif.

were different: $10^{-8.8}$ for c-Myc and $10^{-2.4}$ for Mad2. To make the two data sets more comparable, we applied a cutoff of 10 for the $-\log_{10}$ of the ChIP-seq p-value. This resulted in more balanced sets of *in vivo* targets for c-Myc and Mad2, with 6205 and 9758 bound regions, respectively. We used these sets of targets for all the analyses described henceforth. As shown in Fig. 1A, although there is a significant overlap between the two sets of targets, c-Myc and Mad2 also bind unique genomic targets in HeLa S3 cells.

Currently, the molecular mechanisms that allow paralogous TFs, such as c-Myc and Mad2, to interact with different sets of DNA sites *in vivo* are not well understood. One hypothesis is that the two TFs exhibit slightly different DNA binding preferences, and this may contribute to their differential *in vivo* binding. To test this hypothesis, it is essential to have high-quality DNA binding site motifs or other types of data that reflect the intrinsic DNA binding preferences of these TFs. Although such data is available for c-Myc [5, 6], none of the Mad factors have been thoroughly characterized, and the only DNA motif available for Mad2 is a general E-box motif of low quality [5]. For this reason, we performed PBM experiments [3] to thoroughly characterize the DNA binding preferences of c-Myc and Mad2. We tested the two TFs either alone or in combination with TF Max. As expected, the c-Myc:c-Myc and Mad2:Mad2 homodimers bound DNA very weakly even when tested at high concentrations, while c-Myc:Max and Mad2:Max bound DNA with high affinity. His-tagged versions of c-Myc, Mad2, and Max were used in the PBM experiments, and they were a kind gift from Richard Young and Peter Rahl (Whitehead Institute). To ensure that the DNA binding signal detected on PBMs corresponds to heterodimers and not the Max:Max homodimer, we used concentrations of c-Myc/Mad2 10 times higher than the concentration of Max. We will henceforth refer to the c-Myc:Max PBM data as c-Myc PBM data, and to the Mad2:Max PBM data as Mad PBM data.

From the universal PBM data for c-Myc and Mad2, we computed several measures of the DNA binding specificity of the two factors: 1) we used the

Seed-and-Wobble algorithm [3] to derive DNA binding site motifs, or position weight matrices (PWMs) [22]; 2) we computed the median fluorescence intensity for each possible 8-mer, as described previously [3], with high median intensities corresponding to 8-mers strongly preferred by the TF; and 3) we computed enrichment scores (E-scores) for each possible 8-mer, as described previously [3]. E-scores range from -0.5 to +0.5, with higher values corresponding to higher sequence preference. Compared to 8-mer median intensities, the E-scores are more robust to changes in experimental conditions (e.g., binding buffers) and protein concentrations. However, 8-mer median intensities can be used to approximate the median intensities for longer k-mers, intensities that are not directly measured on the PBMs (see Supplementary Material online).

2.1   DNA Motifs Cannot Explain Differential *in vivo* DNA Binding by c-Myc versus Mad2

The DNA motifs of c-Myc and Mad2 are very similar, but not identical (Fig. 1B). For example, c-Myc appears to have a slightly higher preference for a C nucleotide immediately upstream of the CACGTG core. To assess whether such differences are significant *in vivo* and potentially explain the differences in *in vivo* DNA binding between the two proteins, we first compared the enrichment of the c-Myc and Mad2 motifs in the ChIP-seq data, using a method based on the area under the receiver operating characteristic curve (AUC) (see [23] and Supplementary Materials online).

Fig. 1C shows AUC enrichments for the c-Myc and Mad2 motifs in the ChIP-seq data. If these motifs could explain, even to a small extent, why c-Myc and Mad2 bind different sets of targets *in vivo*, then we would expect the c-Myc motif to be significantly more enriched than the Mad2 motif in the c-Myc ChIP-seq data, and the Mad2 motif to be significantly more enriched than the c-Myc motif in the Mad2 ChIP-seq data. However, the AUC enrichments of these motifs are almost identical: 0.665 and 0.663 in c-Myc ChIP-seq data, and 0.585 and 0.582 in Mad2 ChIP-seq for the HeLa S3 cell line. In conclusion, we cannot use DNA motifs to differentiate between the c-Myc and Mad2 ChIP-seq data sets.

2.2   *In vitro* Universal PBM Data Cannot Explain Differential *in vivo* Binding by c-Myc versus Mad2

Another way of assessing the enrichment of a DNA motif (PWM) in a ChIP-seq data set is by looking at how many of the ChIP-bound sequences contain a PWM match above a certain cutoff (and possibly use this to compute a hypergeometric p-value). The shortcoming of this method is that it depends greatly on the chosen cutoff, and there is no systematic way of choosing the "best" cutoff for a given PWM [23]. To overcome this problem, we considered a range of cutoffs and, for each cutoff, we computed the fraction of ChIP-bound sequences that contain at least one DNA site with a score above the cutoff. Furthermore, since cutoffs based on PWM scores would not be readily comparable between the c-Myc and Mad2 PWMs, we chose cutoffs based on the number of possible k-mers with scores

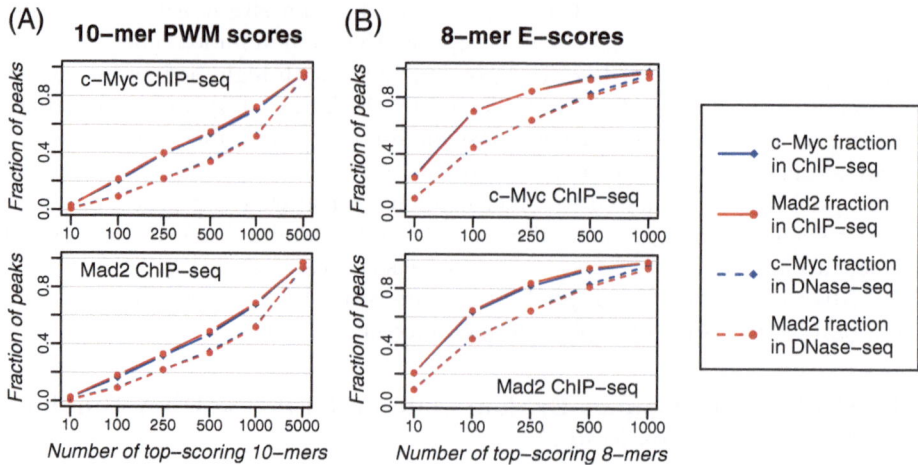

Fig. 2. Fractions of ChIP-seq/DNase-seq peaks that contain DNA sites with (A) PWM scores or (B) 8-mer E-scores above certain cutoffs. The cutoffs represent the number of top-scoring k-mers, ranked by either PWM scores or PBM 8-mer E-scores. The full lines correspond to ChIP-seq data. The dotted lines correspond to DNase-seq data.

above that cutoff, where k is the width of the PWM. The results presented here are for PWMs of size $k = 10$. We obtained similar results for other values of k.

The results of this analysis are illustrated in Fig. 2A. For each ChIP-seq data set and each TF, we counted the number of ChIP-seq peaks that contained at least one 10-mer in the set of 10, 100, 250, 500, 1000, 5000, and 10000 top-scoring 10-mers. We compared the fractions of peaks corresponding to c-Myc versus Mad2 ChIP-seq data. Also, in order to compare those results with the background distribution, we computed similar fractions for DNase-seq [24] peaks. As shown in Fig. 2A, for both ChIP-seq data sets the values corresponding to the c-Myc and Mad2 PWMs are very similar, and we observe the same pattern of slightly higher fractions for Mad2. Thus, we cannot use the PWM scores in this manner to differentiate between c-Myc and Mad2 ChIP-seq targets. We also notice that, as expected, a larger fraction of ChIP-seq peaks contain high-scoring PWM matches compared to DNase-seq peaks (compare the full and dotted lines in Fig. 2, which correspond to ChIP- and DNase-seq data, respectively).

We note that PWMs are in fact summaries of the comprehensive data that we obtain from PBM experiments. Thus, it is possible that differences between the DNA binding specificities of c-Myc and Mad2 do exist, but are not captured by PWMs. To test this hypothesis, we performed an analysis similar to the one described above, but instead of using PWM scores we used 8-mer E-scores derived directly from PBM data. Fig. 2B shows the fractions of peaks containing top-scoring 8-mers. As in the case of PWM scores, the difference between ChIP-seq data and DNase-seq data is significant for both TFs, but the curves for c-Myc and Mad2 are almost identical. Thus, the 8-mer PBM data is still not sufficient to differentiate between the *in vivo* targets of c-Myc compared to Mad2.

3   Binding of Putative Co-factors Can Explain Differences in *in vivo* DNA Binding between c-Myc and Mad2

Given that the intrinsic DNA binding preferences of c-Myc and Mad2 cannot be used to differentiate between the *in vivo* targets of the two TFs, we next focused on the hypothesis that DNA binding of co-factors in the neighborhood of c-Myc or Mad2 binding sites might contribute to the differences we observe between their sets of *in vivo* targets. To test this hypothesis, we built classifiers that can accurately distinguish between sequences bound uniquely by c-Myc versus Mad2 according to the ChIP-seq data, using features derived from either PBM data or PWMs of putative co-factors. We implemented our approach in the **COUGER** (**co**-factors associated with **u**niquely-bound **g**enomic **r**egions) framework. The steps of the framework are summarized in Algorithm 1.

3.1   Classes and Features

Classes. We used the ChIP-seq data to define two classes of sequences: c-Myc-specific sequences (i.e., c-Myc ChIP-seq peaks that do not overlap any of the Mad2 peaks), and Mad2-specific sequences (i.e., Mad2 ChIP-seq peaks that do not overlap any of the c-Myc peaks). In HeLa S3 cells we identified 2786 c-Myc-specific sequences and 6308 Mad2-specific sequences, which account for approximately 45% and 65% of the total ChIP-seq peaks of c-Myc and Mad2, respectively. These percentages are surprisingly high given the similarity between the DNA binding preferences of the two TFs.

After identifying c-Myc- and Mad2-specific sequences, we filtered out some of the Mad2-specific peaks and kept only the top 2786 , sorted according to the Mad2 ChIP-seq p-value. Thus, we obtained two sets containing the same number of DNA sequences, which eliminates a potential classification bias toward one of the two classes. Finally, before computing the features for the selected DNA sequences, we trimmed each sequence to ±100 bp on each side of the ChIP-seq peak summit. This was necessary because many peaks are a few hundred to a few thousand bases long. Given that we are interested in finding co-factors that bind close to c-Myc and Mad2, we should look for DNA sites of these putative co-factors only in close proximity of the c-Myc and Mad2 ChIP-seq peak summits.

Features. We computed features using two types of information on the DNA binding specificities of putative co-factors: PBM 8-mer E-scores and PWM scores. We used 3 different types of features: 1) "**PBM features**" derived from 8-mer E-scores for the mammalian TFs in UniPROBE [4] (420 PBM data sets), plus 9 PBM data sets from our laboratory; 2) "**PWM features**" derived from PWMs computed from the PBM data sets; and 3) "**T (Transfac) features**" derived from the PWMs in Transfac [5] (1226 PWMs). For a given PBM data set and a DNA sequence, we generated: an "**M**" feature that represents the maximum E-score over all the 8-mers in that sequence, and an "**A**" feature that represents the average E-score over the top 3 highest-scoring 8-mers in that sequence (lines 4 and 5 of Algorithm 1). Similarly, we generated **M** and **A** features from PWMs.

Algorithm 1. Classification

Input: D – data set with classification sequences for Myc and Mad; PBM (E-scores from PBM data); PWM (PWMs from PBM data); T (PWMs from TRANSFAC).
Output: Lists of selected features SF, and accuracies A.
1: $D_{\text{train}} \leftarrow \text{random.sample}(D, 2/3 \cdot |D|)$ such that $|\{X \in D_{\text{train}}, \text{class}(X) = \text{Myc}\}| = |\{X \in D_{\text{train}}, \text{class}(X) = \text{Mad}\}|$; $D_{\text{test}} \leftarrow D - D_{\text{train}}$
2: **for** $F \in \{PBM, PWM, T\}$ **do**
3:     **for** $X \in D$ **do**
4:         $F_M(X) = \{\max_{x \in X} f(x) | \forall f \in F\}$
5:         $F_A(X) = \{\text{avg}(\max_{x \in X} f(x), \max_{y \in X - \{x\}} f(y), \max_{z \in X - \{x,y\}} f(z)) | \forall f \in F\}$
6:     $F_{MA}(\text{train}) \leftarrow \{F_M(X), F_A(X) | \forall X \in D_{\text{train}}\}$
7:     $SF_t \leftarrow \text{feature.selection}(D_{\text{train}}, F_t(\text{train}))$
8:     **for** $C \in \{SVM_{\text{lin}}, SVM_{\text{rbf}}, RF_{\text{gi}}, RF_{\text{pi}}\}$ **do**
9:         $bestp(C, SF_t) \leftarrow \arg\max_{p \in \text{params}(C)} \text{accuracy}(\text{train}(C, D_{\text{train}}, SF_t, p))$
10:        $Model(C, SF_t) \leftarrow \text{train}(C, D_{\text{train}}, SF_t, bestp(C, SF_t))$
11:        $A_{\text{test}}(C, SF_t) \leftarrow \text{accuracy}(\text{predict}(C, D_{\text{test}}, Model(C, SF_t)))$
12: $SF \leftarrow \{SPBM_{MA}, SPWM_{MA}, ST_{MA}\}$
13: $A \leftarrow \{A_{\text{test}}(C, F) | \forall C \in \{SVM_{\text{lin}}, SVM_{\text{rbf}}, RF_{\text{gi}}, RF_{\text{pi}}\}, \forall F \in SF\}$
14: **return** SF, A

3.2 Classification Algorithms

We used two state-of-the-art supervised classification algorithms: support vector machine (SVM) and random forest (RF), both available as free software packages (LIBSVM [25], Random Jungle [26]). The SVM is widely used due to its high accuracy on linear and nonlinear classification problems. In addition, the SVM can successfully handle high-dimensional data, which makes it ideal for our classification task. We trained SVMs using linear and radial basis function kernels (**SVM$_{\text{lin}}$** and **SVM$_{\text{rbf}}$**, respectively). The RF classifier is essentially an ensemble of classification trees. RF is comparable in performance with SVM, but one of its distinguishing characteristics is that it explicitly computes a measure of the importance of each variable for the classification task. Random Jungle (RJ) implements 2 variable importance scores: Gini importance (the sum of impurity decreases over all nodes in the forest in which the corresponding variable was selected for splitting), and permutation importance (the average decrease in accuracy when the values of a variable are randomly permuted). We ran RJ with both the Gini importance (**RF$_{\text{gi}}$**) and the permutation importance (**RF$_{\text{pi}}$**).

We split the c-Myc- and Mad2-specific sequences into two sets: 1) a training set containing 2/3 of the sequences (i.e., 3714), randomly chosen from the original set; and 2) a test set containing the remaining 1/3 of the sequences. For each algorithm, we first searched for optimal parameter values using only the training data, and then, using the best model obtained on the training set, we predicted the class for each sequence in the test set. We measured the performance of each algorithm using its accuracy on the test set.

To optimize the parameters we performed grid searches over the parameter space. For **SVM$_{\text{lin}}$** we optimized C, the cost of misclassifying examples.

For $\mathbf{SVM_{rbf}}$ we optimized C and the RBF kernel parameter γ. For $\mathbf{RF_{gi}}$ and $\mathbf{RF_{pi}}$ we optimized *ntree*, the number of trees in the forest, and *mtry*, the number of input variables tried in each split (see Supplementary Material online).

Feature Selection. We performed feature selection on each of the three feature types (**PBM, PWM,** and **T (Transfac)**) using both the maximum score of any DNA site and the average over the top three highest scores. We used RF with a backward elimination technique [26], an iterative process in which a RF is grown at each step and a subset of variables is discarded. The eliminated features are those with the smallest importance. In this instance we used only the unscaled permutation importance, which is recommended for feature selection [27]. We stopped the algorithm when the number of features fell below 100. We performed two variants of selection: **FS1**, with 50% of features dropped at each iteration, and **FS2**, with 33% of features dropped at each iteration.

3.3   Classification Accuracy on the Test Sets

We ran SVM and RF on the HeLa S3 ChIP-seq data using the features types described above (**PBM, PWM,** and **Transfac**). The results for $\mathbf{SVM_{lin}}$ and $\mathbf{RF_{pi}}$ are presented in Table 1 as classification accuracies, and vary between 85.52% and 88.05% depending on the algorithm and the set of features. Results for the other two classifiers ($\mathbf{SVM_{rbf}}$ and $\mathbf{RF_{gi}}$), as well as results on the K562 ChIP-seq data, are available in the Supplementary Material online.

Table 1 shows that our SVM and RF classifiers can accurately distinguish between c-Myc-specific and Mad2-specific genomic targets. This suggests that a potential mechanism by which these TFs achieve differential DNA binding *in vivo* is by interacting with co-factors that bind DNA in the neighborhood of c-Myc or Mad2 DNA binding sites. We will perform follow-up analyses to study the spacing between c-Myc/Mad2 sites and DNA sites of their putative co-factors, to assess the likelihood of direct TF-TF interactions.

We note that $\mathbf{SVM_{lin}}$ with **Transfac** features achieved the best classification accuracy on the HeLa S3 ChIP-seq data: 88.05% when using all 2452 features. However, the accuracy decreased after feature selection and became comparable to the accuracy for **PBM** and **PWM** features.

Table 1. Classification accuracy on the test sets. Table shows the results of SVM and RF on HeLa S3 ChIP-seq data using 3 feature types: PWM, Transfac, and PBM. Bold: best classification accuracy obtained by SVM_{lin} and RF_{pi} for a particular feature type.

Features	PBM			PWM			Transfac		
Feature set	*ALL*	*FS1*	*FS2*	*ALL*	*FS1*	*FS2*	*ALL*	*FS1*	*FS2*
Number of features	*858*	*53*	*74*	*840*	*52*	*73*	*2452*	*76*	*94*
SVM_{lin}	86.87	87.03	**87.08**	**86.60**	85.74	86.33	**88.05**	86.60	86.65
RF_{pi}	86.65	**86.92**	86.71	85.52	**86.01**	85.95	86.60	86.44	**86.71**

3.4   Selected Features. Putative Co-factors

SVM_{lin} with **PBM** features obtained the best classification accuracy with a limited number of features: 87.08% with a total of 74 features (FS2) and 87.03% with a total of 53 features (FS1). Importantly, 52 of the 53 features in FS1 are among the 74 features in FS2. We note that **COUGER** can be used to reduce the number of features even further, although this might lead to a decrease in classification accuracy. For this particular data set, for example, reducing the number of selected features to 10 resulted in an accuracy of 85.47%.

We analyzed the top 53 selected features to identify putative co-factors that might contribute to differential *in vivo* DNA binding by c-Myc versus Mad2. The top 4 putative co-factors (according to RF variable importance score) are: E2F, Sp100, Zfp161, and Sp4, all associated with Mad2-specific sequences. A literature search revealed that at least 3 of these TFs are indeed good candidate co-factors for Mad2: E2F binding site elements are important for autorepression of the c-myc gene [28], Sp100 is a transcriptional repressor (similarly to Mad2) and plays an important role as a tumor suppressor [29], and Zfp161 is a putative c-myc repressor [30]. The fourth TF, Sp4, is not known to act as a repressor but it has been shown to be aberrantly expressed in many cancers [5], which supports a connection with Myc/Mad. For the highest confidence candidate co-factor, E2F, we performed an enrichment analysis similar to the one in Fig. 2. We note that E2F is a family of TFs with highly similar DNA binding specificities. All E2Fs for which PBM data is currently available have been selected by our classifiers, and they are similarly enriched in Mad2-specific targets. In Fig. 3 we show the enrichment of two representative E2F family members: E2F3 and E2F4.

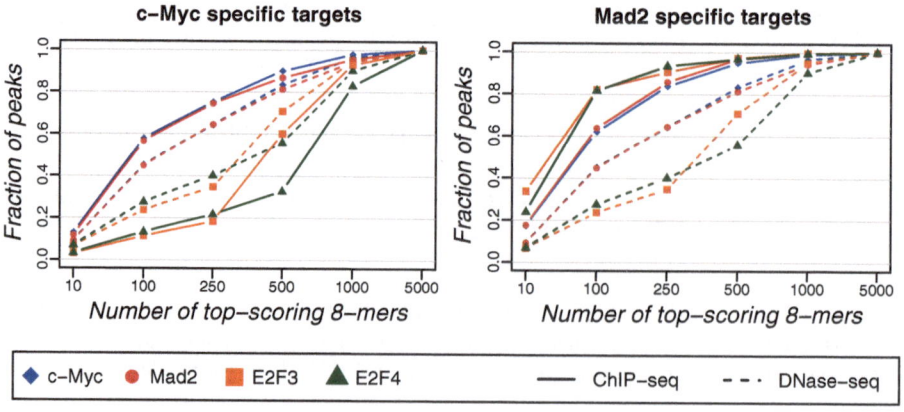

Fig. 3. Enrichment analysis for c-Myc, Mad, and E2F factors in the DNA regions bound uniquely by either c-Myc or Mad2. Right plot shows that E2Fs are more enriched than Mad2 in the Mad2-specific targets, although their enrichment in the DNase-seq peaks is much lower. By contrast, in the c-Myc-specific targets (left plot), E2F sites are depleted, as their enrichment is generally lower than in the DNase-seq peaks. These results are in agreement with our classification analyses, which found that E2F sites are strongly associated with Mad2-specific and not c-Myc-specific targets.

4   Discussion

Identifying the molecular mechanisms that allow paralogous TFs to bind different sets of *in vivo* targets is essential for understanding eukaryotic transcription. Due to recent advances in high-throughput technologies for measuring TF-DNA binding both *in vivo* and *in vitro* (such as ChIP-seq and PBM), it is now possible to quantify the contributions of both intrinsic TF-DNA binding specificity and interacting co-factors to differential *in vivo* DNA binding by related TFs. Here, we focus on paralogous TFs c-Myc and Mad2, and show that differences in their intrinsic sequence preferences cannot account for the large number of targets bound uniquely by each TF. Instead, interactions with putative co-TFs are a likely mechanism used by c-Myc and Mad2 to select their specific genomic sites.

To identify c-Myc and Mad2 co-factors, we designed **COUGER**, a novel framework that uses *in vitro* DNA binding specificity data for putative co-factors to distinguish between the genomic targets of paralogous TFs (here, c-Myc and Mad2). We are not aware of other tools that aim to identify co-factors that interact specifically with paralogous TFs. However, similar classification approaches have been previously used to distinguish between sets of genomic regions. Chen and Zhou [31], for example, use Naïve Bayes to identify co-factors that can distinguish between the regulatory regions of genes upregulated versus downregulated in mouse ES cells. We note that our choice of classification algorithms is very important. When using features derived from TF-DNA binding specificity data (either PWMs or PBM data), it is likely to obtain features that are highly correlated. While Naïve Bayes is not appropriate in this case, both SVM and RF classifiers can be used. Furthermore, as the number of features increases (in our case, as more and more PBM data is being generated), a Naïve Bayes approach may start to overfit the training data, while SVMs and RFs are more robust. Finally, the advantage of using RF for feature selection (as opposed to Wilcoxon rank-sum test) is that RF can easily handle interactions among features, which would not be captured by a statistical test on individual features.

De novo motif discovery tools or methods that search for DNA motifs enriched in particular sets of sequences could also be used, in theory, to identify co-factors [32, 33, 34] . However, these approaches would only search for one co-factor at a time, and would be able to find only DNA motifs that appear in a significant fraction of the DNA sequences of interest (here, the c-Myc- or Mad2-specific sequences). Recent evidence from the ENCODE project shows that the co-association of TFs is highly context specific, i.e., distinct combinations of TFs bind at specific genomic locations [35]. Thus, classification approaches such as **COUGER**, that search for sets of putative co-factors in TF-specific genomic targets are more likely to reveal important molecular mechanisms through which paralogous TFs achieve their regulatory specificity in the cell. Future work will include additional computational analyses to select the best candidate co-factors for c-Myc and Mad2, as well as using **COUGER** to identify co-factors for paralogous TFs from other protein families.

References

[1] Ren, B., Robert, F., Wyrick, J.J., et al.: Genome-wide location and function of DNA binding proteins. Science 290, 2306–2309 (2000)

[2] Johnson, D.S., Mortazavi, A., Myers, R.M., Wold, B.: Genome-wide mapping of in vivo protein-DNA interactions. Science 316, 1497–1502 (2007)

[3] Berger, M.F., Philippakis, A.A., Qureshi, A.M., et al.: Compact, universal DNA microarrays to comprehensively determine transcription-factor binding site specificities. Nat. Biotech. 24, 1429–1435 (2006)

[4] Robasky, K., Bulyk, M.L.: UniPROBE, update 2011: expanded content and search tools in the online database of protein-binding microarray data on protein-DNA interactions. Nucleic Acids Research 39, D124–D128 (2011)

[5] Matys, V., Kel-Margoulis, O.V., Fricke, E., et al.: TRANSFAC and its module TRANSCompel: transcriptional gene regulation in eukaryotes. Nucleic Acids Research 34, D108–D110 (2006)

[6] Portales-Casamar, E., Thongjuea, S., Kwon, A.T., et al.: JASPAR 2010: the greatly expanded open-access database of transcription factor binding profiles. Nucleic Acids Research 38, D105–D110 (2010)

[7] Badis, G., Berger, M.F., Philippakis, A.A., et al.: Diversity and complexity in DNA recognition by transcription factors. Science 324, 1720–1723 (2009)

[8] Wells, J., Graveel, C.R., Bartley, S.M., et al.: The identification of E2F1-specific target genes. Proc. Natl. Acad. Sci. U S A 99, 3890–3895 (2002)

[9] Wu, Z., Zheng, S., Yu, Q.: The E2F family and the role of E2F1 in apoptosis. Int. J. Biochem. Cell Biol. 41, 2389–2397 (2009)

[10] Tao, Y., Kassatly, R., Cress, W., Horowitz, J.: Subunit composition determines E2F DNA-binding site specificity. Mol. Cell Biol. 17, 6994–7007 (1997)

[11] Hollenhorst, P.C., Shah, A.A., Hopkins, C., Graves, B.J.: Genome-wide analyses reveal properties of redundant and specific promoter occupancy within the ETS gene family. Genes Dev. 21, 1882–1894 (2007)

[12] Wei, G.H., Badis, G., Berger, M.F., et al.: Genome-wide analysis of ETS-family DNA-binding in vitro and in vivo. EMBO J. 29, 2147–2160 (2010)

[13] Soleimani, V.D., Punch, V.G., Kawabe, Y.I., et al.: Transcriptional dominance of Pax7 in adult myogenesis is due to high-affinity recognition of homeodomain motifs. Dev. Cell 22, 1208–1220 (2012)

[14] Xu, X., Bieda, M., Jin, V.X., et al.: A comprehensive ChIP-chip analysis of E2F1, E2F4, and E2F6 in normal and tumor cells reveals interchangeable roles of E2F family members. Genome Research 17, 1550–1561 (2007)

[15] ENCODE Project Consortium, Bernstein, B., Birney, E., Dunham, I., Green, E., Gunter, C., Snyder, M.: An integrated encyclopedia of DNA elements in the human genome. Nature 489, 57–74 (2012)

[16] Farnham, P.J.: Insights from genomic profiling of transcription factors. Nat. Rev. Genet. 10, 605–616 (2009)

[17] Grandori, C., Cowley, S.M., James, L.P., Eisenman, R.N.: The Myc/Max/Mad network and the transcriptional control of cell behavior. Annu. Rev. Cell Dev. Biol. 16, 653–699 (2000)

[18] Cortes, C., Vapnik, V.: Support-vector networks. Machine Learning 20, 273–297 (1995)

[19] Breiman, L.: Random forests. Machine Learning 45, 5–32 (2001)

[20] Rosenbloom, K.R., Dreszer, T.R., Long, J.C., et al.: ENCODE whole-genome data in the UCSC Genome Browser: update, Nucleic Acids Research 40, D912–D917 (2012)

[21] Workman, C.T., Yin, Y., Corcoran, D., et al.: enoLOGOS: a versatile web tool for energy normalized sequence logos. Nucl. Acids Res. 33, W389 (2005)

[22] Stormo, G.D.: DNA binding sites: representation and discovery. Bioinformatics 16, 16–23 (2000)

[23] Gordân, R., Hartemink, A., Bulyk, M.: Distinguishing direct versus indirect transcription factor-DNA interactions. Genome Res. 19, 2090–2100 (2009)

[24] Song, L., Crawford, G.E.: DNase-seq: A high-resolution technique for mapping active gene regulatory elements across the genome from mammalian cells. Cold Spring Harbor Protocols 2010, pdb.prot5384 (2010)

[25] Chang, C.C., Lin, C.J.: LIBSVM: A library for support vector machines. ACM Transactions on Intelligent Systems and Technology 2, 1–27 (2011)

[26] Schwarz, D.F., König, I.R., Ziegler, A.: On safari to random jungle: a fast implementation of random forests for high-dimensional data. Bioinformatics 26, 1752–1758 (2010)

[27] Díaz-Uriarte, R., Alvarez de Andrés, S.: Gene selection and classification of microarray data using random forest. BMC Bioinformatics 7, 3 (2006)

[28] Luo, Q., Li, J., Cenkci, B., Kretzner, L.: Autorepression of c-myc requires both initiator and E2F-binding site elements and cooperation with the p107 gene product. Oncogene 23, 1088–1097 (2004)

[29] Negorev, D.G., Vladimirova, O.V., Kossenkov, A.V., et al.: Sp100 as a potent tumor suppressor: accelerated senescence and rapid malignant transformation of human fibroblasts through modulation of an embryonic stem cell program. Cancer Research 70, 9991–10001 (2010)

[30] Sobek-Klocke, I., Disque-Kochem, C., Ronsiek, M., Klocke, R., et al.: The human gene ZFP161 on 18p11.21-pter encodes a putative c-myc repressor and is homologous to murine Zfp161 (Chr 17) and Zfp161-rs1 (X Chr). Genomics 43, 156–164 (1997)

[31] Chen, G., Zhou, Q.: Searching ChIP-seq genomic islands for combinatorial regulatory codes in mouse ES cells. BMC Genomics 12, 515 (2011)

[32] Machanick, P., Bailey, T.L.: MEME-ChIP: motif analysis of large DNA datasets. Bioinformatics 27, 1696–1697 (2011)

[33] Thomas-Chollier, M., Herrmann, C., Defrance, M., et al.: RSAT peak-motifs: motif analysis in full-size ChIP-seq datasets. NAR 40, e31 (2012)

[34] Whitington, T., Frith, M.C., Johnson, J., Bailey, T.L.: Inferring transcription factor complexes from ChIP-seq data. NAR 39, e98 (2011)

[35] Gerstein, M.B., Kundaje, A., Hariharan, M., et al.: Architecture of the human regulatory network derived from ENCODE data. Nature 489, 91–100 (2012)

Assembling Genomes and Mini-metagenomes from Highly Chimeric Reads

Sergey Nurk[1,*], Anton Bankevich[1,*], Dmitry Antipov[1], Alexey Gurevich[1],
Anton Korobeynikov[1,2], Alla Lapidus[1,3], Andrey Prjibelsky[1], Alexey Pyshkin[1],
Alexander Sirotkin[1], Yakov Sirotkin[1], Ramunas Stepanauskas[4],
Jeffrey McLean[5], Roger Lasken[5], Scott R. Clingenpeel[6], Tanja Woyke[6],
Glenn Tesler[7], Max A. Alekseyev[8], and Pavel A. Pevzner[1,9]

[1] Algorithmic Biology Laboratory, St. Petersburg Academic University,
Russian Academy of Sciences, St. Petersburg, Russia
[2] Dept. of Mathematics and Mechanics, St. Petersburg State University,
St. Petersburg, Russia
[3] Theodosius Dobzhansky Center for Genome Bioinformatics,
St. Petersburg State University, St. Petersburg, Russia
[4] Bigelow Laboratory for Ocean Sciences, East Boothbay, ME, USA
[5] J. Craig Venter Institute, La Jolla, California, USA
[6] DOE Joint Genome Institute, Walnut Creek, California, USA
[7] Dept. of Mathematics, University of California, San Diego, La Jolla, CA, USA
[8] Dept. of Computer Science and Engineering, University of South Carolina,
Columbia, SC, USA
[9] Dept. of Computer Science and Engineering, University of California, San Diego,
La Jolla, CA, USA

Abstract. Recent advances in single-cell genomics provide an alternative to gene-centric metagenomics studies, enabling whole genome sequencing of uncultivated bacteria. However, single-cell assembly projects are challenging due to (i) the highly non-uniform read coverage, and (ii) a greatly elevated number of chimeric reads and read pairs. While recently developed single-cell assemblers have addressed the former challenge, methods for assembling highly chimeric reads remain poorly explored. We present algorithms for identifying chimeric edges and resolving complex bulges in de Bruijn graphs, which significantly improve single-cell assemblies. We further describe applications of the single-cell assembler SPADES to a new approach for capturing and sequencing "dark matter of life" that forms small pools of randomly selected single cells (called a *mini-metagenome*) and further sequences all genomes from the mini-metagenome at once. We demonstrate that SPADES enables sequencing mini-metagenomes and benchmark it against various assemblers. On single-cell bacterial datasets, SPADES improves on the recently developed E+V-SC and IDBA-UD assemblers specifically designed for single-cell sequencing. For standard (multicell) datasets, SPADES also improves on A5, ABySS, CLC, EULER-SR, Ray, SOAPdenovo, and Velvet.

* These authors contributed equally.

M. Deng et al. (Eds.): RECOMB 2013, LNBI 7821, pp. 158–170, 2013.
© Springer-Verlag Berlin Heidelberg 2013

1   Introduction

The standard techniques for Next Generation Sequencing (NGS) require at least a million bacterial cells to sequence a genome. Since most bacteria cannot be cultured in the laboratory [1,2] and thus cannot be sequenced, most bacterial diversity remains below the radar of NGS projects. The "dark matter of life" describes microbes and even entire bacterial phyla that have yet to be cultured and sequenced. For example, only a fraction of the 10,000+ bacterial species in the human microbiome have been sequenced [3,4]. Single-cell sequencing [5,6] has recently emerged as a powerful approach to complement largely gene-centric metagenomic data with whole-genome assemblies of uncultivated organisms.

Currently, *Multiple Displacement Amplification (MDA)*, pioneered by Roger Lasken and colleagues [7], is the dominant approach to whole genome amplification prior to single-cell sequencing. However, assembly of reads from MDA-amplified genomes is challenging because of highly non-uniform read coverage, as well as elevated levels of chimeric reads and read pairs. While recent computational advances (Chitsaz et al., 2011 [8]; Peng et al., 2012 [9], Bankevich et al., 2012 [10]) have opened the possibility of sequencing the genome of any bacterial cell, sequencing the vast majority of bacteria in the human microbiome still remains a distant goal. The bottleneck is that it remains unclear how to isolate and capture low-abundance cells from a complex sample. In particular, while there is great interest in investigating the rare bacterial species in the human microbiome, currently there is no technology for comprehensively surveying the diversity of such a complex sample. Indeed, capturing and sequencing even 100,000 randomly chosen single cells from the human microbiome is unlikely to comprehensively sample the bacterial diversity, since many of the 10,000+ species in the human microbiome are underrepresented [11,12]. Since sequencing 100,000 single cells is prohibitively expensive, the question is how to sample bacterial diversity in a more economical way.

McLean et al., 2012 (submitted) recently developed a new approach for analyzing the "dark matter of life" based on forming random pools of single flow sorted cells and sequencing all cells in the resulting *mini-metagenome* at once. These pools only contain a small number of cells as opposed to metagenomics samples, which often contain billions of cells from different species. Since the artificially formed mini-metagenome has lower complexity than the original metagenome, high quality sequencing of mini-metagenomes (as opposed to metagenomes) becomes feasible.

Assembly of mini-metagenome MDA reads is even more challenging than for single-cell MDA reads, and thus requires additional algorithmic developments. Mini-metagenome sequencing can be thought of as sequencing a giant bacterial genome (formed by all genomes within a mini-metagenome) with extremely non-uniform coverage. Moreover, the elevated number of chimeric reads and read pairs (typical for single-cell sequencing) is likely to present an even more difficult challenge in the case of mini-metagenomes, where *intergenomic* chimeric reads (resulting from concatenated fragments from different genomes) can be formed.

This paper addresses computational challenges arising in single-cell and mini-metagenome sequencing. This includes detection of chimeric edges in de Bruijn graphs and analyzing complex bulges. We incorporate these algorithmic developments into the SPADES assembler [10] and demonstrate that it improves on existing single-cell sequencing tools E+V-SC [8] and IDBA-UD [9]. SPADES also performs well on standard (multicell) projects. (We refer to a conventional sequencing project using cultivated strains as *multicell* sequencing.) In particular, we show that it improves on A5 [13], ABySS [14], CLC[1], EULER-SR [15], Ray [16], SOAPdenovo [17]), and Velvet [18] in multicell bacterial assemblies. We also benchmark SPADES on simulated mini-metagenomes obtained by mixing various single-cell read datasets. We demonstrate that SPADES enables mini-metagenome sequencing, and we investigate the computational limits of mini-metagenome sequencing for assessing low-abundance bacterial species.

2   Identifying Chimeric Edges in de Bruijn Graphs

MDA often results in *chimeric reads* (formed by concatenating fragments from different regions of the genome) and *chimeric read-pairs* (formed by two reads sampled from distant regions of the genome). See [8,19,20] for the extent of chimeric reads and read-pairs in single-cell projects. Chimeric reads result in *chimeric edges* in de Bruijn graphs.

Double-Stranded de Bruijn Graphs. Let $DB(\text{GENOME}, k)$ be the de Bruijn graph of a circular genome GENOME and its reverse complement GENOME$'$, where vertices and edges correspond to $(k-1)$-mers and k-mers, respectively. GENOME and GENOME$'$ each traverse a cycle in this graph; these two cycles form the *genome traversal* of the graph. If a genome has multiple chromosomes or linear chromosomes, the genome traversal of $DB(\text{GENOME}, k)$ may consist of multiple paths or cycles. The *genomic multiplicity* of an edge is the number of times the traversal passes through this edge. We often work with *condensed graphs* [10], where each edge is assigned a *length* (in k-mers) and the length of a path is the sum of its edge lengths (rather than the number of edges).

Chimeric Edges in de Bruijn Graphs. Let $DB(\text{READS}, k)$ be the de Bruijn graph constructed from a set READS of reads from GENOME and their reverse complements. In the idealized case with full coverage of GENOME and no read errors, the graphs $DB(\text{READS}, k)$ and $DB(\text{GENOME}, k)$ coincide; however, in reality these graphs differ because of coverage gaps and read errors. Edges in $DB(\text{READS}, k)$ may correspond to genome fragments (*correct* edges) as well as arise either from errors in reads or from chimeric reads (*false* edges). While in $DB(\text{GENOME}, k)$ the genome traversal consists of a pair of cycles, in $DB(\text{READS}, k)$ these cycles may be broken into multiple paths. The genome traversal defines the genomic multiplicities of edges in $DB(\text{READS}, k)$ (or the condensed graph). In particular,

---

[1] CLC Assembly Cell 3.22.55708 (CLC Bio, http://www.clcbio.com).

since false edges are not traversed by the genome traversal in $DB(READS, k)$, they have genomic multiplicity zero.

Assemblers use various algorithms to iteratively remove false edges and transform the de Bruijn graph $DB(READS, k)$ into a smaller *assembly graph*. We use the notation $DB^+(READS, k)$ to denote the current assembly graph at any intermediate stage of assembly, and $DB^*(READS, k)$ to denote the final assembly graph.

While most false edges correspond to easily detectable subgraphs, called *tips* and *bulges*, some form *chimeric edges*, which are hard to identify. Chimeric edges arise from chimeric reads, which are abundant in single-cell datasets. While chimeric edges in the de Bruijn graph represent a major obstacle to constructing long contigs, in standard (multicell) assembly datasets, chimeric edges usually have low coverage and thus are easily identified as false and removed by the conventional assemblers. However, this approach does not work for single-cell datasets, where coverage is non-uniform and the level of chimerism is high [8,19]. For such datasets, low coverage does not characterize false edges since many correct edges also have low coverage.

Our *chimeric edge identification* procedure is based on the following assumptions for bacterial genomes: (i) since chimeric edges in the condensed de Bruijn graphs are typically short,[2] we assume that edges longer than $n = 250$ have genomic multiplicity at least 1, and (ii) since edges longer than $N = 1500$ (referred to as *long* edges) in the condensed de Bruijn graph tend to have genomic multiplicity 1, we assume that all long edges have genomic multiplicity 1.[3]

Since genomic multiplicities of edges in $DB^+(READS, k)$ are unknown, we attempt to bound them. An edge e with genomic multiplicity bounded by $c_{LOWER}(e)$ from below and by $c_{UPPER}(e)$ from above has *capacity* $(c_{LOWER}(e), c_{UPPER}(e))$. We assign capacities to all edges in the condensed graph of $DB^+(READS, k)$ as follows, where the second and third categories are dictated by assumptions (i) and (ii) above:

$$(c_{LOWER}(e), c_{UPPER}(e)) = \begin{cases} (0, \infty) & \text{if } LENGTH(e) \leq n; \\ (1, \infty) & \text{if } n < LENGTH(e) \leq N; \\ (1, 1) & \text{if } N < LENGTH(e). \end{cases}$$

To simplify this presentation, we assume that an assembly algorithm successfully removes all (or the vast majority of) bulges and tips, resulting in an intermediate assembly graph $DB^+(READS, k)$, but fails to remove chimeric edges. Thus, the search for chimeric edges amounts to finding edges of genomic multiplicity zero.

---

[2] Out of 117 chimeric edges in the graph $DB^+(READS, 55)$ constructed for the single-cell *E. coli* dataset ECOLI-SC (described in Results), 115 have length $\leq n = 250$. Here and in the further statistics $DB^+(READS, 55)$ is a graph that we have after doing initial simplifications including removing condensed edges with average coverage below 10 that satisfy some additional length and topology conditions.

[3] This holds for 97% of long edges in $DB(GENOME, 55)$ for the *E. coli* reference genome.

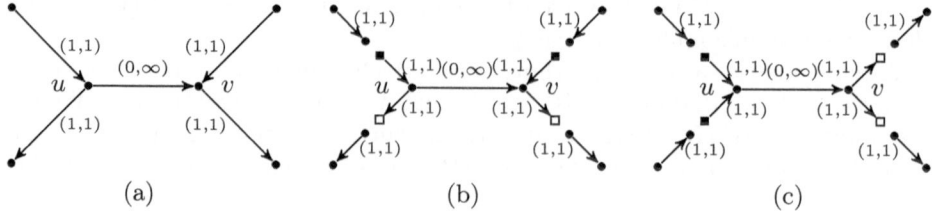

(a)   (b)   (c)

Fig. 1. Example of breaking long edges in an assembly graph. (a) Subgraph of assembly graph where the four diagonal edges are long edges, while the horizontal edge in the center is not long. (b) Result of breaking the four long edges contains a connected component (in the center) with 2 sources (solid square vertices) and 2 sinks (hollow square vertices). The capacities of the edges starting (ending) at the newly formed sources (sinks) are inherited from the capacities of the broken edges. (c) Result of breaking long edges in a subgraph similar to the subgraph in (a) but with different directions on some edges.

Chimeric Edges and Circulations in Networks. A graph with capacity constraints on the edges is referred to as a *network*. Given a vertex v and a function f on edges of a network G, we define $\text{INFLUX}_f(v) = \sum_e f(e)$, where the sum is taken over all incoming edges e of the vertex v. We define $\text{OUTFLUX}_f(v)$ similarly. A function f is called a *circulation* in the network G if $\text{INFLUX}_f(v) = \text{OUTFLUX}_f(v)$ for each vertex v in G, and $c_{\text{LOWER}}(e) \le f(e) \le c_{\text{UPPER}}(e)$ for each edge e in G.

The *Circulation Problem* is to find a circulation in a network [21]. Genomic multiplicities define a circulation in the network $\text{DB}^+(\text{READS}, k)$ with capacity constraints. There are usually multiple circulations in this network and we do not know which of them corresponds to the actual genomic multiplicities. However, if an edge e has $f(e) = 0$ in all circulations, then it must be a false edge and in most cases represents a chimeric edge. A polynomial-time algorithm for finding all such edges in the network will be described elsewhere.

We remark that this strategy is based on the assumption that the genome corresponds to a cycle in the graph, which often fails for real data. Since in $\text{DB}^+(\text{READS}, k)$ the genome traversal may be broken into multiple subpaths, a circulation in it may not even exist. To address this complication, we break the network into smaller subnetworks and analyze their subcirculations.

The operation of *breaking an edge* (v, w) in a graph G removes (v, w) from G; adds two new vertices v^* and w^* (called the *sink* and the *source*, respectively); and adds two new edges (v, v^*) and (w^*, w) with the same capacity as the edge (v, w). Given a weighted graph G and a positive integer t, we define G_t as the graph obtained from G by breaking all edges longer than t. To break the de Bruijn graph into subnetworks, we break all long edges (Fig. 1). After this transformation, the graph $\text{DB}^+(\text{READS}, k)$ is typically decomposed into many connected components. For the ECOLI-SC dataset described in the Results section, the graph is decomposed into 114 non-trivial connected components (containing more than one vertex).

A circulation (genome traversal) in $DB^+(\text{GENOME}, k)$ defines a *flow* [21] between sources and sinks of every connected component of $DB_N^+(\text{GENOME}, k)$ satisfying the capacity constraints. Similarly to $DB(\text{GENOME}, k)$, for many components in $DB^+(\text{READS}, k)$, the genome traversal also defines a flow satisfying the capacity constrains. Thus, the search for chimeric edges in $DB^+(\text{READS}, k)$, can be performed independently in each component.

Many components have a particularly simple structure with two sources and two sinks (Fig. 1b and Fig. 1c). The only flow that satisfies the capacity constraints in Fig. 1b (resp., Fig. 1c assigns flow 0 (resp., flow 2) to the edge (u, v). Thus (u, v) is classified as chimeric in Fig. 1b and as correct in Fig. 1c.

Unfortunately, the above procedure fails for some connected components (e.g., when the outgoing edge from vertex v in Fig. 1a is missing). Furthermore, such components tend to be large. For example, after assembling reads from a single *E. coli* cell, the procedure fails only for 10% of all components, but these components contain most (52%) vertices of the graph. Below we describe an approach to identify chimeric edges in such components.

Chimeric Edges and Critical Cut-sets. Given a subset U of vertices in the graph, the *cut-set*, denoted $\text{CUT}(U)$, is the set of all edges (u, v) in the graph such that $u \in U$ and $v \in \overline{U}$ (where $\overline{U}$ denotes the set of vertices of the graph that do not belong to U). We define $c_{\text{LOWER}}(U)$ (resp., $c_{\text{UPPER}}(U)$) as the sum of lower (resp., upper) capacities of all edges from $\text{CUT}(U)$. A cut-set $\text{CUT}(U)$ is *balanced* if $c_{\text{LOWER}}(U) \le c_{\text{UPPER}}(\overline{U})$ and *unbalanced* otherwise. A cut-set $\text{CUT}(U)$ is *critical* if $c_{\text{LOWER}}(U) = c_{\text{UPPER}}(\overline{U})$. According to Hoffman's Circulation Theorem [21], a circulation exists if and only if every cut-set in the network is balanced.

It is easy to see that for a critical cut-set $\text{CUT}(U)$, all edges $(u, v) \in \text{CUT}(U)$ must have genomic multiplicity equal to $c_{\text{LOWER}}(u, v)$, while all edges $(v, u) \in \text{CUT}(\overline{U})$ must have genomic multiplicity equal to $c_{\text{UPPER}}(v, u)$. Indeed, if the lower capacity of any edge $(u, v) \in \text{CUT}(U)$ is increased by 1, the cut-set would become unbalanced (as $c_{\text{LOWER}}(U) + 1 > c_{\text{UPPER}}(\overline{U})$), implying that no circulation exists. Similarly, the upper capacity of any edge $(v, u) \in \text{CUT}(\overline{U})$ cannot be decreased, implying that the genomic multiplicity of (v, u) must be equal to $c_{\text{UPPER}}(v, u)$. In particular, for a critical cut-set $\text{CUT}(U)$, all *crossing* edges $(u, v) \in \text{CUT}(U)$ with $c_{\text{LOWER}}(u, v) = 0$ must be chimeric.

SPADES analyzes only certain types of critical cut-sets that are common in de Bruijn graphs of reads (details are to be described elsewhere).

3 Removing Complex Bulges

Bulges and Bulge Corremoval. Errors in reads often result in two short paths between the same two vertices in the de Bruijn graph, where the two paths have roughly the same length and represent similar sequences. Such pairs of paths may aggregate into larger subgraphs called *bulges*. Assemblers use various *bulge removal* algorithms (and additional steps) to transform the de Bruijn graph $DB(\text{READS}, k)$ into a smaller *assembly graph* $DB^*(\text{READS}, k)$. While they remove the vast majority of bulges, they fail to remove some *complex bulges*.

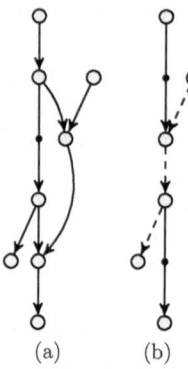 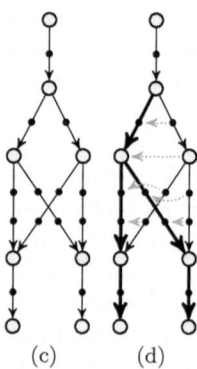

(a)   (b)   (c)   (d)

Fig. 2. Illustration of bulge removal algorithm. The vertices of the condensed graph are shown in white. Dotted arrows indicate projection operations (not graph edges). (a–b): Merging paths instead of projecting paths. Merging two paths in (a) results in a graph (b) with an artificial (dashed) path violating condition (ii). (c–d): Blob corremoval. Complex bulge (c) is not removed by the bulge corremoval procedure from [10]. Applying the new "blob corremoval procedure" to blob (c) simplifies it via the projections shown in (d). Thick edges denote the tree to which we project the blob.

In this section, we describe an algorithm for removal of complex bulges that evade the "bulge corremoval" algorithm from [10]. One approach to removing bulges is to map the de Bruijn graph onto a smaller graph. SPADES tries to find a mapping that satisfies the following conditions:

(i) Every path in the de Bruijn graph maps to a path in the assembly graph.
(ii) For every path ρ in the assembly graph, there exists a path in the de Bruijn graph that maps onto ρ.[4]

Some bulge removal algorithms either do not explicitly map the de Bruijn graph onto the assembly graph or use mappings that may violate conditions (i) and/or (ii). For example, they may find a bulge formed by two paths in the de Bruijn graph and either remove one of the paths or merge these paths into a single one, without considering the impact on other edges incident to these paths. Removing one of the paths may lead to deterioration of assemblies, since important information (along with some correct paths) may be lost. Merging the paths may introduce artificial paths into the assembly graph, violating condition (ii) (see Fig. 2a,b).

SPADES [10] introduced the *bulge corremoval* procedure, which satisfies conditions (i) and (ii). For each edge (u, v) (with length below a threshold) in the condensed de Bruijn graph, SPADES searches for a path from u to v of length approximately equal to the length of (u, v). If such an *alternative path* exists, the two paths P and (u, v) form a *simple bulge*. To remove a simple bulge, the edge

---

[4] In fact, SPADES creates the assembly graph as a subgraph of the de Bruijn graph so that paths in the assembly graph also represent paths in the de Bruijn graph.

(u, v) is *projected* onto this path and is removed afterwards. Applied iteratively, the bulge corremoval strategy eliminates the vast majority of bulges. However, for some bulges, no edge in the bulge has an alternative path, implying that the algorithm from [10] will not be able to remove such bulges (Fig. 2c). Below we describe an algorithm satisfying conditions (i) and (ii) for removing the majority of the remaining complex bulges.

Blob Corremoval. Let G be a directed acyclic graph (DAG) with vertex set V and edge set E. For vertices v and w in G, we define $v \prec w$ if there exists a directed path from v to w in G. A mapping $f : V \to V, E \to E$ is called a *projection* if (1) for every vertex $v \in V$, we have $f(f(v)) = f(v)$, (2) for every pair of vertices v and w, if $v \prec w$ then $f(v) \prec f(w)$, and (3) for every edge $e = (u, v)$, we have $f(e) = (f(u), f(v))$.

A projection f defines the induced DAG G_f on the vertex set $V_f = f(V)$. We limit our attention to projections of DAGs onto *directed trees* (i.e., projections f such that G_f is a directed tree) and additionally require that every path and its projection have similar lengths. Fig. 2d shows the directed tree and projection of DAG shown in Fig. 2c.

Breaking edges longer than t in the assembly graph $DB^+(\text{READS}, k)$ results in a graph $DB_t^+(\text{READS}, k)$ that typically consists of many connected components. A *blob* is a component of $DB_t^+(\text{READS}, k)$ that is a DAG with a single source and one or more sinks. SPADES analyzes blobs in $DB^+(\text{READS}, k)$ and for each blob, attempts to find a directed tree (with root at the source and leaves at the sinks of the blob) such that there exists a projection of the blob onto this tree.

This leads to a *blob corremoval* procedure, which generalizes the bulge corremoval procedure from [10] and satisfies conditions (i), (ii). A generalized notion of blob and efficient algorithm to search for trees and projections will be described elsewhere.

4   Results

Metrics. The N50 (resp., NG50) metric is the maximum contig size such that using blocks of that size or larger gives at least 50% of the assembly length (resp., reference genome length). We use metrics NA50 and NGA50, introduced and justified in [22], instead of the standard N50 or NG50 metrics. To count NA50, contigs are alligned to reference genome. If a contig has a misassembly or has nonaligning sequence such as large gaps or indels, the contig is broken into blocks that do align. Then we compute N50 using these aligned blocks instead of using the original contigs. Similarly, NGA50 is computed as NG50 applied to these adjusted blocks.

In some of our experiments, the fraction of the genome assembled is below 50%, so NGA50 would be 0 for all assemblers, and thus, we use NA50.

Benchmarking. We compared a number of single-cell and conventional assemblers on two *E. coli* paired-end Illumina libraries described in [8]: a single-cell

Table 1. Comparison of assemblers on ECOLI-SC, a single-cell *E. coli* dataset, using QUAST [22]. In each column, the best assembler by that criteria is indicated in bold. Only contigs of length ≥ 500 bp were used. For single cell projects, the total assembly size often exceeds the genome length due to contaminants and other reasons (see [23]). The "GF (%)" (Genome fraction) column filters out these issues. MA: number of misassemblies. Misassemblies are locations on an assembled contig where the left flanking sequence aligns over 1 kb away from the right flanking sequence on the reference. MM: Mismatch (substitution) error rate per 100 kb. IND: number of indels per 100 kb. MM and IND are measured in aligned regions of the contigs.

Assembler	NGA50	# contigs	Longest contig	Total length	MA	MM	IND	GF (%)	# genes
Conventional (multicell) assemblers									
A5	14399	745	101584	4441145	8	11.97	0.19	90.141	3453
ABySS	68534	**179**	178720	4345617	5	2.71	2.66	88.268	3704
CLC	32506	503	113285	4656964	3	4.76	2.87	92.378	3768
EULER-SR	26662	429	140518	4248713	18	9.37	218.72	85.005	3419
Ray	55395	296	210612	4649552	13	2.34	0.87	91.864	3838
SOAPdenovo	18468	569	87533	4098032	7	114.38	11.08	79.861	3038
Velvet	22648	261	132865	3501984	**2**	2.07	1.23	74.254	3098
Single-cell assemblers									
E+V-SC	32051	344	132865	4540286	**2**	1.85	0.70	92.162	3793
IDBA-UD	96947	250	224018	4791744	10	**1.61**	**0.16**	95.661	4046
SPAdes 2.3	**110539**	276	**268756**	4875378	**2**	4.24	0.70	**95.737**	**4057**

library (ECOLI-SC) and a multicell library (ECOLI-MC). They consist of 100 bp paired-end reads with average insert sizes 266 bp for ECOLI-SC and 215 bp for ECOLI-MC. Both *E. coli* datasets have 600× coverage. The *E. coli* K-12 MG1655 reference length is 4639675 bp with 4324 annotated genes.

Tables 1 and 2 present the benchmarking results for various assemblers.[5] Table 1 illustrates that single-cell assemblers significantly improve on the conventional assemblers in *single-cell* projects. Table 2 shows that recently developed single-cell assemblers IDBA-UD and SPAdes also improve on the conventional assemblers in *standard (multicell)* projects by most metrics.

From Genomes to Mini-metagenomes. Below we investigate the performance of SPAdes on artificially simulated mini-metagenomes and demonstrate that it is capable of assembling a significant portion of each genome in a mini-metagenome. In addition to simulations, we also applied this assembly algorithm to a real mini-metagenome dataset; details are in McLean et al., 2012 (submitted).

---

[5] ABySS 1.3.4, EULER-SR 2.0.1, Ray 2.0.0, Velvet, Velvet-SC, and E+V-SC were run with vertex size 55. A5 and CLC 3.22.55708 were run with default parameters. SOAPdenovo 1.0.4 was run with vertex sizes 27–31. IDBA-UD 1.0.9 was run in its default iterative mode.

Table 2. Comparison of assemblers on ECOLI-MC, a multicell *E. coli* dataset. See the caption of Table 1 for further details.

Assembler	NGA50	# contigs	Longest contig	Total length	MA	MM	IND	GF (%)	# genes
Conventional (multicell) assemblers									
A5	43651	176	181690	4551797	**0**	**0.40**	**0.13**	98.476	4178
ABySS	106155	**96**	221861	4619631	2	2.45	0.52	99.202	4242
CLC	69146	122	221533	4547925	2	0.73	0.15	98.547	4233
EULER-SR	110153	100	221409	4574240	8	2.98	47.15	98.438	4206
RAY	83128	113	221942	4563341	2	2.10	0.20	98.162	4194
SOAPDENOVO	62512	141	172567	4519621	1	26.56	5.58	97.405	4134
VELVET	82776	120	242032	4554702	3	0.70	0.20	98.824	4211
Single-cell assemblers									
E+V-SC	54856	171	166115	4539639	**0**	1.21	0.15	98.329	4149
IDBA-UD	111789	107	236470	4562955	1	0.53	0.15	98.834	4215
SPADES 2.3	**119880**	104	**265405**	**4634928**	2	1.86	0.63	**99.420**	**4250**

We applied SPADES to a simulated mini-metagenome that consists of four bacterial species with known genomes. We mixed together reads (in various proportions), from four different MDA-amplified single-cell bacterial projects at Joint Genome Institute and Bigelow Laboratory. The genomes of these bacteria vary in GC content and genome length: *Prochlorococcus marinus* (31% GC, 1.7 Mb genome length) [24], *Pedobacter heparinus* [25] (42% GC, 5.0 Mb genome length), *Escherichia coli* [26] (51% GC, 4.6 Mb genome length), and *Meiothermus ruber* [27] (63% GC, 3.0 Mb genome length). Note that as simulated datasets, these are highly idealized and do not exactly match what would be found in the environment, but they are useful for modeling the ability of the assembler to deal with different mixtures.

In the first simulation, we randomly selected a fixed fraction of reads from each genome, mixed them together, and assembled the resulting dataset with SPADES. This simulation was repeated 10 times, varying the fraction as $1/2^m$ with $m = 0, 1, \ldots, 9$ (the same fraction $1/2^m$ applies to all genomes). The assembled contigs were aligned against individual genomes to compute the assembly statistics (in Table 3 we present statistics for *M. ruber* and *P. heparinus*). In particular, even with a relatively small fraction $1/64$ of selected reads, SPADES assembled 1779 out of 4339 genes for *P. heparinus*, 1366 out of 4324 genes for *E. coli*, and 710 out of 3105 genes for *M. ruber*. This is significantly larger than the number of complete genes captured in a typical metagenomics project. However, for *P. marinus*, only 55 out of 1732 genes were assembled.

In the second simulation, we formed a mini-metagenome using all reads from three species and varied the coverage for the fourth. For the fourth species, we selected either a genome with high GC content (*M. ruber*) or low GC content (*P. heparinus*). Table 4 illustrates that SPADES recovers a substantial fraction of an underrepresented genome within a mini-metagenome. Even with a

Table 3. SPADES assembly of a simulated mini-metagenome with equal fractions of each genome

	$1/1$	$1/2$	$1/4$	$1/8$	$1/16$	$1/32$	$1/64$	$1/128$	$1/256$	$1/512$
M. ruber:										
Misassemblies	14	12	5	15	13	10	14	8	8	12
NA50 (kb)	44	33	39	24	20	14	9	10	8	3
Longest contig (kb)	113	133	119	114	108	109	93	61	50	39
Genome fraction (%)	76.3	69.5	62.4	56.7	49.1	39.7	29.9	22.2	15.7	10.9
# genes	2160	1950	1777	1533	1219	939	710	521	357	214
P. heparinus:										
Misassemblies	1	2	6	11	11	13	26	27	17	12
NA50 (kb)	185	207	165	96	70	27	12	4	2	1
Longest contig (kb)	946	410	426	307	337	379	225	102	34	32
Genome fraction (%)	97.8	96.6	94.3	88.4	82.5	71.3	57.6	40.4	24.9	12.1
# genes	4148	4038	3855	3524	3133	2401	1779	1125	548	189

Table 4. SPADES assembly of a mini-metagenome with variable fraction of *M. ruber* and *P. heparinus*

	$1/2$	$1/4$	$1/8$	$1/16$	$1/32$	$1/64$	$1/128$	$1/256$	$1/512$
M. ruber:									
Misassemblies	14	6	8	11	13	14	14	17	11
NA50 (kb)	47	45	27	15	11	10	7	4	2
Longest contig (kb)	137	120	118	106	114	114	62	45	49
Genome fraction (%)	70.0	63.7	58.3	52.4	43.0	34.2	27.1	21.5	16.7
# genes	1973	1797	1585	1327	1022	822	618	464	330
P. heparinus:									
Misassemblies	2	5	8	12	19	23	14	19	15
NA50 (kb)	163	165	131	87	33	11	3	2	1
Longest contig (kb)	426	439	396	339	333	227	89	40	33
Genome fraction (%)	96.6	94.0	88.7	81.6	71.2	56.7	40.6	24.5	12.9
# genes	4043	3815	3527	3116	2488	1752	1113	534	228

small fraction of reads in the underrepresented genome (e.g., $1/256$), we recovered a significantly larger number of genes (more than 450 genes for *M. ruber* and *P. heparinus*) as compared to a typical metagenomics project. Tables 3 and 4 demonstrate that the assembly quality of an individual genome depends mainly on the coverage of this genome, rather than on what fraction of the mini-metagenome this genome represents.

5   Discussion

Since 2008, when the first NGS assemblers were released, many excellent assemblers have become available. Since most of them use a de Bruijn graph approach, they often generate rather similar assemblies, at least for bacterial projects. Recent developments in single-cell genomics tested the limits of conventional assemblers and demonstrated that they all have room for improvement. Our benchmarking illustrates that single-cell assemblers not only enable single-cell sequencing but also improve on conventional assemblers on their own turf.

References

1. Rappe, M.S., Giovannoni, S.J.: The uncultured microbial majority. Annu. Rev. Microbiol. 57, 369–394 (2003)
2. Tringe, S.G., Rubin, E.M.: Metagenomics: DNA sequencing of environmental samples. Nat. Rev. Genet. 6(11), 805–814 (2005)
3. Nelson, K.E., Weinstock, G.M., Highlander, S.K., Worley, K.C., Creasy, H.H., et al.: A catalog of reference genomes from the human microbiome. Science 328(5981), 994–999 (2010)
4. Wylie, K.M., Truty, R.M., Sharpton, T.J., Mihindukulasuriya, K.A., Zhou, Y., et al.: Novel bacterial taxa in the human microbiome. PLoS ONE 7(6), e35294 (2012)
5. Stepanauskas, R.: Single cell genomics: an individual look at microbes. Current Opinion in Microbiology 15(5), 613–620 (2012)
6. Lasken, R.S.: Genomic sequencing of uncultured microorganisms from single cells. Nat. Rev. Microbiol. 10(9), 631–640 (2012)
7. Lasken, R.S.: Single-cell genomic sequencing using Multiple Displacement Amplification. Curr. Opin. Microbiol. 10(5), 510–516 (2007)
8. Chitsaz, H., Yee-Greenbaum, J., Tesler, G., Lombardo, M., Dupont, C., et al.: Efficient de novo assembly of single-cell bacterial genomes from short-read data sets. Nat. Biotechnol. 29(10), 915–921 (2011)
9. Peng, Y., Leung, H.C.M., Yiu, S.M., Chin, F.Y.L.: IDBA-UD: a de novo assembler for single-cell and metagenomic sequencing data with highly uneven depth. Bioinformatics 28(11), 1420–1428 (2012)
10. Bankevich, A., Nurk, S., Antipov, D., Gurevich, A.A., Dvorkin, M., et al.: SPAdes: A New Genome Assembly Algorithm and Its Applications to Single-Cell Sequencing. Journal of Computational Biology 19(5), 455–477 (2012)
11. Huttenhower, C., Gevers, D., et al.: Structure, function and diversity of the healthy human microbiome. Nature 486(7402), 207–214 (2012)
12. Li, K., Bihan, M., Yooseph, S., Methe, B.A.: Analyses of the microbial diversity across the human microbiome. PLoS ONE 7(6), e32118 (2012)
13. Tritt, A., Eisen, J.A., Facciotti, M.T., Darling, A.E.: An integrated pipeline for de novo assembly of microbial genomes. PLoS ONE 7(9), e42304 (2012)
14. Simpson, J., et al.: ABySS: a parallel assembler for short read sequence data. Genome Res. 19(6), 1117–1123 (2009)
15. Chaisson, M., Brinza, D., Pevzner, P.: De novo fragment assembly with short mate-paired reads: Does the read length matter? Genome Res. 19(2), 336–346 (2009)
16. Boisvert, S., Laviolette, F., Corbeil, J.: Ray: Simultaneous assembly of reads from a mix of high-throughput sequencing technologies. Journal of Computational Biology 17(11), 1519–1533 (2010)
17. Li, R., Zhu, H., Ruan, J., Qian, W., Fang, X., et al.: De novo assembly of human genomes with massively parallel short read sequencing. Genome Res. 20(2), 265–272 (2010)
18. Zerbino, D., Birney, E.: Velvet: algorithms for de novo short read assembly using de Bruijn graphs. Genome Res. 18(5), 821–829 (2008)

19. Lasken, R.S., Stockwell, T.B.: Mechanism of chimera formation during the multiple displacement amplification reaction. BMC Biotechnol. 7, 19 (2007)
20. Woyke, T., Xie, G., Copeland, A., González, J.M., Han, C., Kiss, H., Saw, J.H., Senin, P., Yang, C., Chatterji, S., Cheng, J.F., Eisen, J.A., Sieracki, M.E., Stepanauskas, R.: Assembling the marine metagenome, one cell at a time. PLoS ONE 4(4), e5299 (2009)
21. Ford, L.R., Fulkerson, D.R.: Flows in Networks. Princeton University Press (1962)
22. Gurevich, A., Saveliev, V., Vyahhi, N., Tesler, G.: QUAST: Quality Assessment for Genome Assemblies (2012) (submitted)
23. Woyke, T., Sczyrba, A., Lee, J., Rinke, C., Tighe, D., et al.: Decontamination of MDA reagents for single cell whole genome amplification. PLoS ONE 6(10), e26161 (2011)
24. Dufresne, A., Salanoubat, M., Partensky, F., Artiguenave, F., Axmann, I.M., et al.: Genome sequence of the cyanobacterium Prochlorococcus marinus SS120, a nearly minimal oxyphototrophic genome. Proceedings of the National Academy of Sciences 100(17), 10020–10025 (2003)
25. Han, C., et al.: Complete genome sequence of Pedobacter heparinus type strain (HIM 762-3 T). Standards in Genomic Sciences 1(1) (2009)
26. Blattner, F.R., Plunkett, G., Bloch, C.A., Perna, N.T., Burland, V., et al.: The complete genome sequence of Escherichia coli K-12. Science 277(5331), 1453–1462 (1997)
27. Tindall, B., Sikorski, J., Lucas, S., Goltsman, E., Copeland, A., et al.: Complete genome sequence of Meiothermus ruber type strain (21 T). Standards in Genomic Sciences 3(1) (2010)

Inferring Intra-tumor Heterogeneity from High-Throughput DNA Sequencing Data

Layla Oesper, Ahmad Mahmoody, and Benjamin J. Raphael

Department of Computer Science, Brown University, Providence, RI
Center for Computational Molecular Biology, Brown University, Providence, RI
{layla,ahmad,braphael}@cs.brown.edu

Background

Cancer is a disease driven in part by somatic mutations that accumulate during the lifetime of an individual. The clonal theory [1] posits that the cancerous cells in a tumor are descended from a single founder cell and that descendants of this cell acquired multiple mutations beneficial for tumor growth through rounds of selection and clonal expansion. A tumor is thus a heterogeneous population of cells, with different subpopulations of cells containing both *clonal* mutations from the founder cell or early rounds of clonal expansion, and *subclonal* mutations that occurred after the most recent clonal expansion. Most cancer sequencing projects sequence a mixture of cells from a tumor sample including admixture by normal (non-cancerous) cells and different subpopulations of cancerous cells. In addition most solid tumors exhibit extensive aneuploidy and copy number aberrations. Intra-tumor heterogeneity and aneuploidy conspire to complicate analysis of somatic mutations in sequenced tumor samples.

Methods

We describe an algorithm – $\underline{T}$umor $\underline{H}$eterogeneity $\underline{A}$nalysis (THetA) – to infer tumor purity (and more generally, the fraction of each subpopulation in the sample) and clonal/subclonal tumor subpopulations directly from high-throughput DNA sequencing data. We focus on copy number aberrations to estimate tumor purity and distinct subpopulations. Suppose that a mixture $\mathcal{T}$ of cells is sequenced, with each cell differing from the normal (reference) genome by some number of copy number aberrations. We assume that these cells are from a small number of subpopulations. The genome of each subpopulation is represented by the number of copies of each genomic interval, or its *copy number profile*. Therefore, the mixture $\mathcal{T}$ is determined by: (1) a copy number profile $\mathbf{c}_i$ for each subpopulation i; (2) the *mixture fraction* μ_i for each subpopulation i. By aligning DNA sequence reads from $\mathcal{T}$ to the reference (human) genome, we observe a read depth vector $\mathbf{r} = (r_1, ..., r_m)$ where r_j is the number reads that align within the jth genomic interval. We formulate the **Maximum Likelihood Mixture Decomposition Problem** of finding the copy number profiles $\mathbf{c}_i$ and mixture fractions μ_i whose mixture best explains the observed sequencing data. We solve

M. Deng et al. (Eds.): RECOMB 2013, LNBI 7821, pp. 171–172, 2013.
© Springer-Verlag Berlin Heidelberg 2013

an instance of the problem using techniques from convex optimization, where the likelihood of our observed data **r** is generated from a multinomial distribution. We determine the number of subpopulations using the Bayesian Information Criterion (BIC). In contrast to existing methods, THetA optimizes an explicit probabilistic model for the generation of the observed tumor sequencing data from a mixture of a normal genome and one *or more* cancer genomes.

Results

Our THetA algorithm compares favorably to earlier methods for inferring tumor purity from SNP array data [2], or DNA sequencing data [3]. We applied THetA to 3 of the breast cancer samples analyzed in [4]. Our analysis of sample PD4120a (sequenced at ~188X coverage) is similar to that reported in [4] but with several notable differences including a clonal deletion of 16q and a clonal vs. subclonal distinction for aberrations on chromosomes 1 and 22 (Fig. 1). Further investigation of these differences – employing sequencing data not used by our algorithm – support our findings. We also find evidence for multiple subclonal populations in another sample sequenced at ~40X coverage.

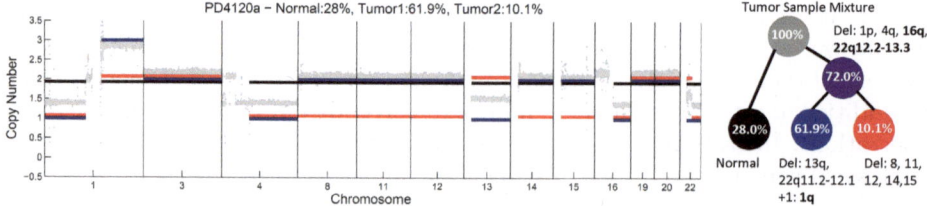

Fig. 1. Analysis of the 188X coverage breast tumor PD4120a. **(Left)** Read depth ratios (gray) and the copy number aberrations inferred by our THetA algorithm, including the normal population (black), dominant (i.e. clonal) tumor population (blue), and subclonal tumor population (red). **(Right)** A reconstruction of the tumor mixture with the inferred aberrations and estimated fraction of cells in each subpopulation. Differences between our reconstruction and [4] are in bold.

References

1. Nowell, P.C.: The clonal evolution of tumor cell populations. Science 194(4260), 23–28 (1976)
2. Van Loo, P., et al.: Allele-specific copy number analysis of tumors. Proc. Natl. Acad. Sci. U.S.A. 107(39), 16910–16915 (2010)
3. Gusnanto, A., et al.: Correcting for cancer genome size and tumour cell content enables better estimation of copy number alterations from next-generation sequence data. Bioinformatics 28(1), 40–47 (2012)
4. Nik-Zainal, S., et al.: The life history of 21 breast cancers. Cell 149(5), 994–1007 (2012)

NP-MuScL: Unsupervised Global Prediction of Interaction Networks from Multiple Data Sources

Kriti Puniyani and Eric P. Xing

School of Computer Science, Carnegie Mellon University
{kpuniyan,epxing}@cs.cmu.edu

Abstract. Inference of gene interaction networks from expression data usually focuses on either supervised or unsupervised edge prediction from a single data source. However, in many real world applications, multiple data sources, such as microarray and ISH measurements of mRNA abundances, are available to offer multi-view information about the same set of genes. We propose NP-MuScL (nonparanormal multi-source learning) to estimate a gene interaction network that is consistent with such multiple data sources, which are expected to reflect the same underlying relationships between the genes. NP-MuScL casts the network estimation problem as estimating the structure of a sparse undirected graphical model. We use the semiparametric Gaussian copula to model the distribution of the different data sources, with the different copulas sharing the same precision (i.e., inverse covariance) matrix, and we present an efficient algorithm to estimate such a model in the high dimensional scenario. Results are reported on synthetic data, where NP-MuScL outperforms baseline algorithms significantly, even in the presence of noisy data sources. Experiments are also run on two real-world scenarios: two yeast microarray data sets, and three Drosophila embryonic gene expression data sets, where NP-MuScL predicts a higher number of known gene interactions than existing techniques.

Keywords: interaction networks, gene expression, multi-source learning, sparsity, Gaussian graphical models, nonparanormal, copula.

1 Introduction

With the prevalence of high throughput technologies such as microarray and RNA-seq for measuring gene expressions, computational inference of gene regulatory or interaction networks from large-scale gene expression datasets has emerged as a popular technique to improve our understanding of cellular systems [1,2,3]. In numerous studies, gene interactions reverse engineered from analysis of such high-throughput data have been experimentally validated [4,5], demonstrating the credibility of such data-driven algorithmic approaches.

There have been two popular approaches to reverse engineering gene networks. The first approach is to build a generative model of the data, and learn a graphical model that captures the conditional independencies in the data. Learning

M. Deng et al. (Eds.): RECOMB 2013, LNBI 7821, pp. 173–185, 2013.

the structure of a graphical model under a multivariate Gaussian assumption of the data has received wide attention in recent years [6,7,8]; various algorithms have been proposed [6,7,8], many with theoretical analysis offering asymptotic guarantee of consistent estimation of the interactions between genes in the network. Empirically, these algorithms are computationally efficient and the results obtained have been encouraging.

However, a limitation of this class of network inference approach is that, it assumes data are identically and independently distributed (i.e., *iid*), which implicitly means that they are from a single experimental source. In reality, many real world biological problems sit on multiple sources of information that can be used to predict interactions between genes. For example, there can be multiple microarray data sets from different laboratories available for the same organism, sometimes measured at the same conditions where the main differences lie in the data sampling strategy or measurement technologies. Biologically, it is often plausible to assume that multiple experimental means resulting in the different datasets may have captured the same information from different viewpoints, e.g., both microarray and *in-situ* hybridization can capture gene expression information, even though the technology used to measure mRNA abundances is different. It remains unclear how to integrate such multiple sources of data in a statistically valid and computational efficient way to infer the underlying network. One may imagine inferring independently a network from each data source, and then averaging across multiple resultant networks, but such an *ad hoc* method is not only un-robust (e.g., each view may have only a small amount of samples), but also lacks statistically justification and consistence guarantee (e.g., on the "average" operator). In this paper, we address the question of inferring a network by analyzing multiple sources of information simultaneously.

An alternative approach to tackle this problem is via supervised learning methods, where a classifier (e.g., SVM) is trained by using examples of known gene interactions (edges in the network) as training data to learn the importance of each data source in predicting unknown interactions between other genepairs [9]. This approach suffers from some intrinsic limitations which prevent it from being widely applicable. First, while such an approach works well for problems where there are sufficient examples of known edges in the network, e.g., in the form of a *reference network* or *reference interactions* obtained from reliable sources, it fails for problems where few or no examples of known edges are available. Gene networks for humans or yeast may be learned by supervised methods where reference interactions are available from extensive prior studies; but for organisms where prior research is limited, this approach cannot be used. Furthermore, one can argue that predicting gene networks is of high importance for such organisms with few known edges, to help biologists who are starting research for regulatory mechanisms of these organisms.

Secondly, using a classifier to predict edges implicitly utilizes the notion of *marginal independence* between nodes. To classify an edge as "positive", i.e., to predict an edge between a given pair of nodes, the correlation between the data for these nodes must be high. Gene networks usually have pathways in which genes interact with each other in a sequential order, which results in high marginal correlation between all pairs of genes in the same pathway. Predicting each edge locally and independently of all other edges will often result in an non-stringent prediction of a clique for all genes in the same pathway, leading to high false positive rates. To reduce such false positives and increase accuracy, we wish to analyze *conditional independence* between the genes instead, which must be done by building a global graphical model that captures simultaneously all the conditional independencies among genes. Each edge resultant from such an estimator enjoys *global* statistical interpretability and consistency guarantee, and such an estimator does not require supervised training, although prior knowledge of interactions on the "reference gene pairs" can still be utilized via introducing a prior over the model, if desired. Thus, it is desirable to develop an *unsupervised* and global inference method which can incorporate multiple data sources to predict a consensus graphical model that explains all the data sources, without using any examples of known edges for training the model.

This paper proposes NP-MuScL (NonParanormal Multi-Source Learning), a machine learning technique for estimating the structure of a sparse undirected graphical model that is consistent with multiple sources of data. The multiple data sources are all defined over the same feature space, and it is assumed that they share the same underlying relationships between the genes (nodes). We use the semiparametric Gaussian copula to model the distribution of the different data sources, where the copula for each data source has its own mean and transformation functions, but all data sources share the same precision matrix (i.e., the inverse covariance matrix, which captures the topological structure of the network). We propose an efficient algorithm to estimate such a model in the high dimensional scenario. The likelihood-related objective function used in NP-MuScL is convex, and results in a globally optimal estimator. Furthermore, the implementation of our algorithm is simple and efficient, computing a network over 2000 nodes using 3 data sources in a matter of minutes. Results are reported on synthetic data, where NP-MuScL outperforms baseline algorithms significantly, even in the presence of noisy data sources. We also use NP-MuScL to estimate a gene network for yeast using two microarray data sets: one over time series expression, and the other over knockout mutants. Finally, we run NP-MuScL on three data sets of Drosophila embryonic gene expression using ISH images and microarray. In both yeast and Drosophila, we find that NP-MuScL predicts a higher number of gene interactions that are known to interact in the literature, than existing techniques.

1.1 Related Work

Previous work on analyzing multiple data sources for network prediction has either specifically taken time into account[10,11], or has different source and target organisms via transfer learning [12]. Katenka et. al.[13] propose a strategy to learn a network from multi-attribute data, where aligned vector observations are made for each node. The NP-MuScL algorithm on the other hand works for data sources which are not aligned, hence each data source may have a different number of observations. Honorio et. al. [14] proposed techniques for multi-task structure learning of Gaussian Graphical Models, to share knowledge across multiple problems, using multi-task learning. However, their method estimates a separate graphical model for each data source, unlike our problem which requires a consensus network common to all data sources. To the best of our knowedge, the NP-MuScL algorithm is the first work that builds a consensus graphical model to explain the relationship between genes by combining information from multiple data sources without explicitly constraining the data to be time-series, or about different organisms.

2   Nonparanormal Multi-Source Learning (NP-MuScL)

Let the k input data sources be defined as $\mathbf{X}^{(1)} \in \mathbb{R}^{n_1 \times d}$, $\mathbf{X}^{(2)} \in \mathbb{R}^{n_2 \times d}$, ..., $\mathbf{X}^{(k)} \in \mathbb{R}^{n_k \times d}$ with total number of data samples $n = \sum_{i=1}^{k} n_i$. Each data source i may have a different number of measurements or samples n_i, but they all measure information about the same feature space of d genes. The goal of NP-MuScL is to learn the structure of a graphical model over the feature space, such that the graphical model will encapsulate global conditional independencies between the genes.

2.1   Glasso

Given a single source of data $\mathbf{X} \in \mathbb{R}^{n \times d}$ drawn from a Gaussian distribution $\mathcal{N}(\mathbf{0}, \boldsymbol{\Sigma})$, a Gaussian graphical model (GGM) may be estimated by computing the inverse covariance matrix $\boldsymbol{\Sigma}^{-1}$ of the Gaussian. Zeros in the inverse covariance matrix imply conditional independence between the features, and thus the absence of an edge between them in the corresponding GGM. Given the empirical covariance matrix $\mathbf{S}$ of the data, the inverse covariance matrix may be computed by maximizing the log likelihood of the data, with an L_1 regularizer to encourage sparsity.

$$\hat{\boldsymbol{\Sigma}}^{-1} = \arg\max_{\boldsymbol{\Theta} \succ 0} \left\{ \log \det \boldsymbol{\Theta} - \text{tr}(\mathbf{S}\boldsymbol{\Theta}) - \lambda||\boldsymbol{\Theta}||_1 \right\} \tag{1}$$

where λ is a tuning parameter that controls the sparsity of the solution; as λ increases, fewer edges are predicted in the GGM. Rothman et. al. [15] showed the consistence of such estimators in Frobenius and Operator norms in high dimensions when $d >> n$; Friedman et. al.[8] proposed a block coordinate descent algorithm for this objective - they named their technique glasso. The glasso algorithm uses a series of L_1 penalized regressions, called Lasso regressions [16], that can be solved in time $O(d^3)$.

2.2   Joint Estimation of the GGM

Given k data sources $\mathbf{X}^{(1)}, \mathbf{X}^{(2)}, \cdots \mathbf{X}^{(k)}$ with corresponding sample covariances $\mathbf{S}^{(1)}, \mathbf{S}^{(2)}, \cdots, \mathbf{S}^{(k)}$, a joint estimator of the underlying GGM may be computed as

$$\hat{\boldsymbol{\Sigma}}^{-1} = \arg\max_{\boldsymbol{\Theta} \succ 0} \sum_{i=1}^{k} w_i \left\{ \log \det \boldsymbol{\Theta} - \text{tr}(\mathbf{S}^{(i)}\boldsymbol{\Theta}) \right\} - \lambda ||\boldsymbol{\Theta}||_1 \tag{2}$$

where w_i defines the relative importance of each data source, and must be defined by the user such that $\sum_{i=1}^{k} w_i = 1$. Assuming the data in each data source is drawn i.i.d., an appropriate choice for the weights may be $w_i = \frac{n_i}{n}$. It can be seen that if each data source is assumed to have mean 0, then for this choice of w_i

$$\hat{\boldsymbol{\Sigma}}^{-1} = \arg\max_{\boldsymbol{\Theta} \succ 0} \log \det \boldsymbol{\Theta} - \sum_{i=1}^{k} \frac{n_i}{n} \text{tr}\left(\mathbf{S}^{(i)}\boldsymbol{\Theta}\right) - \lambda ||\boldsymbol{\Theta}||_1$$

$$= \arg\max_{\boldsymbol{\Theta} \succ 0} \log \det \boldsymbol{\Theta} - \text{tr}\left(\frac{1}{n} \sum_{i=1}^{k} \sum_{l=1}^{n_i} \mathbf{X}^{(i)}(l,\cdot)^T \mathbf{X}^{(i)}(l,\cdot) \, \boldsymbol{\Theta}\right) - \lambda ||\boldsymbol{\Theta}||_1 \tag{3}$$

Thus, our objective function is equivalent to calling glasso with covariance matrix $\frac{1}{n} \sum_{i=1}^{k} \sum_{l=1}^{n_i} \mathbf{X}^{(i)}(l,\cdot)^T \mathbf{X}^{(i)}(l,\cdot)$. We call this method "glasso-bag of data". With an appropriate choice of weights, this model concatenates the data from all data sources into a single matrix, and uses the second moment of the data to estimate the inverse covariance matrix.

Such a procedure highlights the underlying assumption of Gaussianity of the data. If we assume that all data is being drawn from the same Gaussian distribution, then it is reasonable to construct a single sample covariance matrix from the data to estimate the network. However, real data is not always Gaussian; and such an assumption can be limiting, especially when analyzing multiple data sources simultaneously, since non-Gaussianity in a single data source will result in the non-Gaussianity of the combined data. A lot of previous work has been done to drop the Gaussianity assumption in the solution to classic problems like sparse regression [17], estimating GGMs [18], sparse CCA [19] etc., and propose non-parametric solutions to the same. We will also drop the assumption that the data is drawn from the same Gaussian distribution in the next section.

However, if the data is not drawn from the same Gaussian distribution, then how can we characterize the underlying network that generated the data? We propose a generative model where we assume that each data source is drawn from a semi-parametric Gaussian copula, where the copulas for the different data sources share the same covariance matrix, but have different functional transformations. To justify this model, we assume that for each data source, the data is sampled from a multi-variate Gaussian, but this sample is not directly observed. Instead, due to non-linearities introduced during data measurement, a transformed version of the data is measured. Each data source will have its own transformation, hence, the observed distribution of each data source will be different. The key idea of NP-MuScL is then to estimate the non-linear transformation, so that all data can be assumed Gaussian, and the network can be estimated using Equation 2.

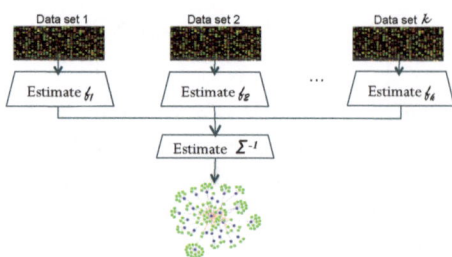

Algorithm 1. Data generation model for NP-MuScL

Input: True covariance matrix Σ with $\sigma_{jj} = 1 \quad \forall\, j \in \{1, \cdots, d\}$
Input: Transformation function g_{ij}, mean μ_{ij} and variance ρ_{ij} for each feature j for each data source i.
for $i = 1$ to k **do**
  for $l = 1$ to n_i **do**
    $y \sim N(0, \Sigma)$
    for $j = 1$ to d **do**
      $\mathbf{X}^{(i)}(l, j) = \mu_{ij} + \rho_{ij}g_{ij}(y(j))$
    end for
  end for
end for
return Observed data $\mathbf{X}^{(i)}$ from k data sources.

Fig. 1. The overall algorithm for NP-MuScL. Each data source is transformed into a Gaussian, using a nonparanormal, and the Gaussian data is then used to jointly estimate a inverse covariance matrix, giving the structure of the Gaussian Graphical Model, underlying the data.

2.3   Dropping the Gaussianity Assumption

We model that each data source is drawn from an underlying Gaussian distribution with mean 0, and covariance matrix Σ, where the variance of each feature $\sigma_{jj} = 1, \forall\, j \in \{1, \cdots, d\}$. However, the observed data may be some unknown transformation of the Gaussian data; thus, if $y \sim \mathcal{N}(0, \Sigma)$, then the observed data is $X^{(i)}(j) = \mu_{ij} + \rho_{ij}g_{ij}(y(j))$ where μ_{ij} and ρ_{ij} is the mean and standard deviation respectively of feature j in data source i.

The function g_{ij} is some (unknown) transformation that depends on the data source, our task is to estimate $f_{ij} = g_{ij}^{-1}$ from the data, so that $f_{ij}(X_j^{(i)})$ is Gaussian. The data generation process is then described in Algorithm 1.

2.4   NP-MuScL Algorithm

A random vector X has a nonparanormal distribution $NPN(\mu, \Sigma, f)$ if there exists a function $f(X) = (f_1(X_1), f_2(X_2), \cdots, f_d(X_d))$ such that $f(X)$ has a multi-variate Gaussian distribution $\mathcal{N}(\mu, \Sigma)$ [18]. To preserve identifiability, we constrain each f_j to have mean 0 and standard deviation 1. The nonparanormal distribution is a Gaussian copula when the fs are monotone and differentiable. For our model, we assume that each data source $X^{(i)} \sim NPN(\mathbf{0}, \Sigma, f_i)$, that is, while each data source has its own functional transformation, they all share the same underlying relationship between the nodes, represented by Σ. The mean of each copula is zero, since we constrain the estimated functions f_j to have zero means. Then, for nonparanormal data, it can be shown that conditional independence in the corresponding graph is equivalent to zeros in the inverse covariance matrix Σ^{-1} [18].

This suggests the following two step algorithm. For each data source i and each feature j, we first estimate the sample mean μ_{ij} and sample variance ρ_{ij}.

$$\hat{\mu}_{ij} = \frac{1}{n_i} \sum_{l=1}^{n_i} \mathbf{X}^{(i)}(l, j); \qquad \hat{\rho}_{ij}^2 = \frac{1}{n_i} \sum_{l=1}^{n_i} \left(\mathbf{X}^{(i)}(l, j) - \hat{\mu}_{ij} \right)^2 \qquad (4)$$

The data in each data source is normalized by the appropriate μ and ρ to have mean 0 and standard deviation 1. Non-parametric functions f_{ij} are estimated for each data source i and feature j, so that $f_{ij} \sim \mathcal{N}(0, 1)$. The details of estimating f are discussed in Sec. 2.6.

In the second step, the inverse covariance matrix is estimated jointly from the transformed f_is. We can define $\mathbf{Y}^{(i)} \in \mathbb{R}^{n_i \times d}$ as

$$\mathbf{Y}^{(i)}(\cdot, j) = \hat{f}_{ij} \left(\mathbf{X}^{(i)}(\cdot, j) \right) \forall j \in \{1, 2, \cdots d\} \qquad (5)$$

The distribution of $\mathbf{Y}^{(i)}$ is then Gaussian with covariance matrix $\mathbf{\Sigma}$. The graphical model corresponding to all data sources can be jointly estimated as

$$\hat{\mathbf{\Sigma}}^{-1} = \arg\max_{\Theta \succeq 0} \sum_{i=1}^{k} w_i \left\{ \log \det \Theta - \mathrm{tr}(\Theta \, \hat{\mathbf{S}}_{\mathbf{f}}^{(i)}) \right\} - \lambda ||\Theta||_1 \qquad (6)$$

where

$$\hat{\mathbf{S}}_{\mathbf{f}}^{(i)} = \frac{1}{n_i} \sum_{l=1}^{n_i} \mathbf{Y}^{(i)}(l, \cdot)^T \mathbf{Y}^{(i)}(l, \cdot) \qquad (7)$$

Setting the weights $w_i = \frac{n_i}{n}$ is equivalent to the data in each data source being drawn i.i.d. from the corresponding Gaussian copula; while setting different weights suggests that the effective sample size of a data source is not the observed sample size.

2.5 Optimization

The objective function in Equation 6 can be rewritten as

$$\hat{\mathbf{\Sigma}}^{-1} = \arg\max_{\Theta \succeq 0} \log \det \Theta - \mathrm{tr}(\Theta \sum_{i=1}^{k} w_i \hat{\mathbf{S}}_{\mathbf{f}}^{(i)}) - \lambda ||\Theta||_1 \qquad (8)$$

Thus, by using $\sum_{i=1}^{k} w_i \hat{\mathbf{S}}_{\mathbf{f}}^{(i)}$ as the covariance matrix, we can optimize the above objective by using efficient, known algorithms like glasso. The overall NP-MuScL algorithm is summarized in Figure 1.

2.6 Estimating $\hat{f}$

For each feature j in data source i, we can compute the empirical distribution function as (where $\mathbb{I}$ is the indicator function)

$$\hat{F}_{ij}(t) = \frac{1}{n_i} \sum_{l=1}^{n_i} \mathbb{I}(X^{(i)}(l, j) \leq t) \qquad (9)$$

The variance of such an estimate may be very large, when computed in the high dimensional scenario $d >> n$. Liu et. al.[18] propose using a Windsorized estimator, for the same, where very small and large values of $\hat{F}_{ij}(t)$ are bounded away from 0 and 1 respectively. Thus,

$$
\tilde{F}_{ij}(t) = \begin{cases} \delta_n & \hat{F}_{ij}(t) < \delta_n \\ \hat{F}_{ij}(t) & \delta_n \leq \hat{F}_{ij}(t) \leq 1 - \delta_n \\ 1 - \delta_n & \hat{F}_{ij}(t) \geq 1 - \delta_n \end{cases} \tag{10}
$$

where δ_n is a truncation parameter. A value of δ_n chosen to be $\delta_n = \frac{1}{4n^{1/4}\sqrt{\pi \log n_i}}$ is found to give good convergence properties for estimating the network for a single data source [18]; and we use the same estimate for NP-MuScL.

Now, for any continuous pdf f, the distribution of the cdf $F(x) = P(X \leq x)$ is uniform. Then, the distribution of $\Phi^{-1}(F(x))$ is Gaussian with mean zero, and standard deviation one, as required (where Φ is the cdf of the standard Gaussian). Thus, we can estimate the required function by using the marginal empirical distribution function defined above: $\hat{f}_{ij}(x) = \Phi^{-1}(\tilde{F}_{ij}(x))$.

3   Results

We first demonstrate that when multiple data sources have different distributions, NP-MuScL can extract the underlying network more accurately than other methods. Next, we show that NP-MuScL can identify the correct network, even when one of the data sources is noise. To analyze NP-MuScL on real data, we run NP-MuScL on two microarray yeast data sets, and find that the network obtained by NP-MuScL predicts more known edges of the yeast interaction network than other methods. Finally, we analyze NP-MuScL on Drosophila embryonic gene expression data from 3 data sets of ISH images and microarray.

3.1   Multiple Data Sources with Different Distributions

Data generation. The details of generating the data for different experiments is described in detail in the supplementary material. In brief, we construct an inverse covariance matrix with an equivalent random sparse Gaussian graphical model. Data is sampled from the Gaussian, and then transformed into non-Gaussian distribution using different transformations. For $d = 50$ with $k = 2$ data sources, we use the Gaussian cdf ($\mu_0 = 0.05, \sigma_0 = 0.4$) and power transform ($\alpha = 3$) for the two data sources respectively (see supp. material for details). The task then is to jointly use the data from the two sources to extract the network. For $k = 3$ data sources, we use the identity transform for the third data source, so that the data sampled from the third source is truly Gaussian. For $k = 4$ data sources, the fourth data source is Gaussian noise, to test the performance of the algorithms in the presence of noise. We generate the same amount of data in each source (n), and run the experiment as n varies. Each result is reported as the average of 10 randomized runs of the experiment.

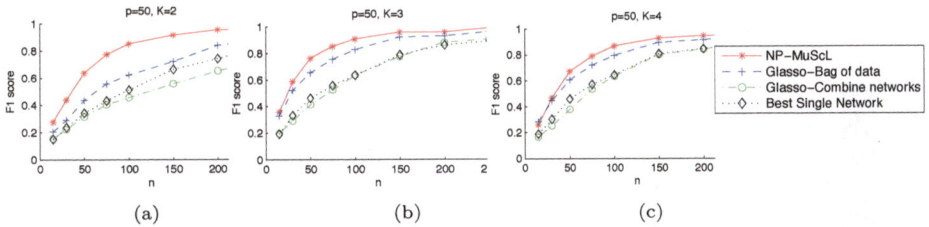

Fig. 2. $F1$ score for predicting edges in simulated data, as n is varied, for (a) $k = 2$, (b) $k = 3$, and (c) $k = 4$ data sources. The standard deviation in the results is small and almost constant across the different experiments; it ranges from (0.01-0.03), and is hence not displayed on the plot.

Metrics. We report the $F1$ measure, which is the harmonic mean of precision and recall, as a measure of the accuracy of predicting the edges in the network.

Baselines. We report three baselines. The first baseline is to report the best accuracy found by a single data source (Best Single Network). We assume that an oracle tells us which data source is most predictive. In our data experiments, we found that it was not possible to predict the most informative data source without using an oracle. Even when $k = 3$, the identity transformed source was not always the most informative. The second baseline is the glasso-bag of data, described in Section 2.2. The third baseline is to compute a separate network for each data source using glasso, and combine the networks to predict a single network (glasso-combine networks). An edge in the final network is present if it is present in m out of the k networks from the k data sources. We assume an oracle defines the best value of m for a given data set, the best value of m varied with different data sets.

As can be seen in Figure 2, NP-MuScL outperforms all three baselines significantly in all three scenarios. Interestingly, using the best single source outperforms estimating separate networks, and combining them in a second step. Note that an oracle is used for identifying the best source, as well as the optimal m used to combine networks. Hence, in a real world scenario, we may expect combining different data sources to perform as well as using only the best single data source for network prediction. When $k = 4$ (Figure 2(c)), one of the data sources is Gaussian noise, however, the use of the oracle in the "Glasso-combine networks" and the "single best source" baselines allows these baselines to ignore the noise source completely. However, NP-MuScL is still able to identify more correct edges in the network. Using a paired t-test, we found that the difference in $F1$ scores between NP-MuScL and "glasso-bag of data" is significant in all conditions, with P-value $p = 10^{-4}$.

3.2   Yeast Data

In this experiment, we look at two different yeast microarray data sets, and make joint predictions via NP-MuScL. Data source 1 is a set of 18 expression profiles from Cho et. al. [20], where each expression corresponds to a different stage in the

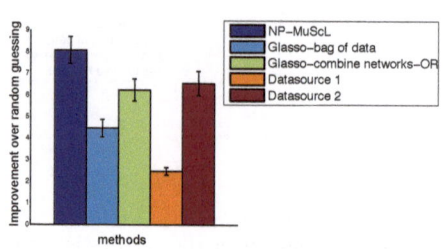

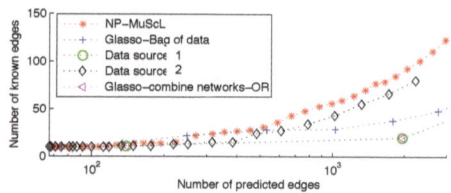

Fig. 4. Effect of varying tuning parameter on different methods. For a fixed number of predicted edges, the NP-MuScL method predicts more known edges than the other methods.

Fig. 3. Performance of different methods on predicting edges in the yeast network

cell cycle of the the yeast. Data source 2 is a set of 300 expression profiles from Hughes et. al.[21], where each expression corresponds to a different knockout mutant of the yeast. Both data sets are processed using standard microarray processing algorithms [22].

We use a list of known interactions from BioGrid [23] to test how well do the different algorithms predict the known edges. Note that since the known gene interactions is an incomplete set, predicted gene interactions may be interactions that have not been observed yet, and thus, have not been added to the BioGrid data base. Hence, measuring recall is no longer appropriate, and we report the improvement in accuracy over random prediction of edges, as suggested by Liben-Nowell & Kleinberg [24].

The total data is over 6120 genes, we sample 1000 genes at a time, and run the algorithms for them. Results are reported for 10 random sub-samples of the genes. Figure 3 shows the improvement over random prediction for edges predicted by each method. Due to the amount of data available, the knockout mutant expression profiles capture more information (and hence more known edges) than the time series expression. Surprisingly, both methods of combining information without taking non-Gaussianity into account, perform worse than using only data source 2. NP-MuScL is the only method where using both data sets into account increases the number of correctly predicted edges. The same results were found to hold true when the network is predicted over the entire set of 6120 genes - NP-MuScL did significantly better than all other methods, and both glasso bag-of-data and glasso-OR did worse than using only data set 2.

To test the effect of varying tuning parameter λ, Figure 4 plots the number of known edges predicted by each method, versus the total number of edges predicted, as λ is varied. For very large values of λ when few edges are predicted, NP-MuScL and "glasso-Bag of data" perform equally well, however, as the amount of predictions increase, NP-MuScL outperforms other methods significantly.

Figure 5 shows the transformations learned for the two data sets by NP-MuScL for 4 random genes. A straight line corresponds to Gaussian data,

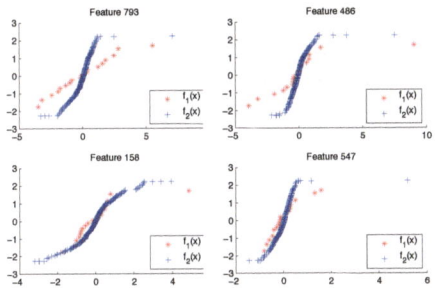

Fig. 5. Examples of the transformations made for data in source 1 (red) and source 2 (blue) for different features

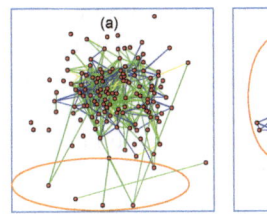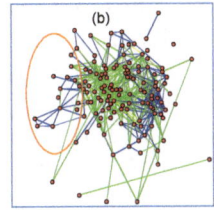

Fig. 6. Difference between the NP-MuScL network and (a) the 13-16 ISH network alone and (b) microarray network alone. Green edges are only predicted in the NP-MuScL network. Blue edges are only present in the (a) 13-16 ISH network and (b) microarray network.

Table 1. Improvement in prediction over random guessing for predicting gene interactions using Drosophila embryonic data

NP-MuScL	Glasso Bag-of-data	Glasso OR	ISH 13-16	ISH 9-10	Microarray
7.29	4.88	4.06	5.98	2.35	3.66

non-linearities are clearly detected by the NP-MuScL algorithm. The transformations also seem to be damping extremely large values observed in the features.

3.3 Drosophila Embryonic Data

We study three data sets of Drosophila embryonic gene expression for 146 genes [25]. The first data set measures spatial gene expression in embryonic stage 9-10 of Drosophila development via in-situ hybridization (ISH) images (4.3 to 5.3 hours after fertilization), when germ band elongation of the embryo is observed. The second data set also studies ISH images measuring spatial gene expression in the 13-16 stage of embryonic development (9.3 to 15 hours after fertilization), when segmentation has already been established. The last data set is of microarray expression at 12 time points spaced evenly in embryonic development.

The ISH images were processed to extract 311 data points for each data set, as described in Puniyani & Xing [26]. The microarray data was processed using standard microarray processing algorithms. Since the number of data points extracted from the ISH data is dependent on the image processing algorithm used, using weights proportional to the number of data points is no longer suitable. We expect the microarray data to be as informative as the ISH data, hence we use $w_i = 0.25$ for each of the two ISH data sources, and $w_i = 0.5$ for the microarray data. The results in Table 1 show that NP-MuScL outperforms using the data separately, and glasso bag-of-data and glasso-combine networks (m=1, called glasso-OR).

We visualized the differences in edge prediction between the NP-MuScL network and the networks predicted by analyzing only one single data source at a time. The orange ellipse in Figure 6(a) highlights gene interactions predicted by NP-MuScL by analyzing all 3 data sources, which were not predicted by any single data source. Figure 6(b) highlights interactions predicted by the microarray data that were not predicted either by the ISH data or the NP-MuScL network. The 9-10 ISH network is similar to the 13-16 ISH network, and hence, is not shown. A detailed analysis of the specific differences in the gene interactions predicted by the different methods is ongoing.

4   Conclusions

We proposed NP-MuScL, an algorithm that predicts gene interaction networks in a global, unsupervised fashion by jointly analyzing multiple data sources to capture the conditional independencies observed in the data. NP-MuScL models each data source as a non-parametric Gaussian copula, with all data sources having different mean and transformation functions, but sharing the covariance matrix across the underlying copulas. The network can then be efficiently estimated in a two step process, of transforming each data source into Gaussian, and then estimating the inverse covariance matrix of the Gaussian using all data sources jointly. We found that NP-MuScL significantly outperforms baseline methods in both synthetic data, and two experiments predicting a gene interaction network from two yeast microarray data sets, and three Drosophila ISH images and microarray data sets.

One limitation of NP-MuScL is that the weights giving the importance of each data source must be assigned by the user. While a good estimate of the weights may be obtained if all data sources are truly drawn i.i.d. from their nonparanormal distributions, and have similar noise levels; in practice, some data sources may be known to be noisier than others, or known to not be i.i.d. (eg. microarray experiments over time are not truly independent draws from the distribution). The question of automatically learning the weights from data remains an open challenge.

References

1. Segal, E., Koller, D., Friedman, N.: Module networks: identifying regulatory modules and their condition-specific regulators from gene expression data. Nature Genetics 34, 166–176 (2003)
2. Basso, K., Magolin, A., Califano, A.: Reverse engineering of regulatory networks in human b cells. Nature Genetics 37, 382–390 (2005)
3. Morrissey, E.R., Juárez, M.A., Denby, K.J., Burroughs, N.J.: On reverse engineering of gene interaction networks using time course data with repeated measurements. Bioinformatics 26(18), 2305–2312 (2010)
4. Carro, M.S., Califano, A., Iavarone, A.: The transcriptional network for mesenchymal transformation of brain tumours. Nature 463, 318–325 (2010)
5. Wang, K., Saito, M., Califano, A.: Genome-wide identification of post-translational modulators of transcription factor activity in human b-cells. Nature Biotechnology 27(9), 829–839 (2009)

6. Meinshausen, N., Bühlmann, P.: High-dimensional graphs and variable selection with the lasso. Annals of Statistics (2006)
7. Banerjee, O., Ghaoui, L.E., d'Aspremont, A., Natsoulis, G.: Convex optimization techniques for fitting sparse gaussian graphical models. In: ICML (2006)
8. Friedman, J., Hastie, T., Tibshirani, R.: Sparse inverse covariance estimation with the graphical lasso. Biostatistics (2007)
9. Ben-Hur, A., Noble, W.S.: Kernel methods for predicting protein–protein interactions. In: ISMB, vol. 21, pp. i38–i46 (2005)
10. Wang, Y., Joshi, T., Zhang, X.S., Xu, D., Chen, L.: Inferring gene regulatory networks from multiple microarray datasets. Bioinformatics 22(19), 2413–2420 (2006)
11. Ahmed, A., Xing, E.P.: Tesla: Recovering time-varying networks of dependencies in social and biological studies. Proc. Natl. Acad. Sci. 106, 11878–11883 (2009)
12. Xu, Q., Hu, D.H., Yang, Q., Xue, H.: Simpletrppi: A simple method for transferring knowledge between interaction networks for ppi prediction. In: Bioinformatics and Biomedicine Workshops (2012)
13. Katenka, N., Kolaczyk, E.D.: Inference and characterization of multi-attribute networks with application to computational biology. Arxiv (2012)
14. Honorio, J., Samaras, D.: Multi-task learning of gaussian graphical models. In: ICML (2011)
15. Rothman, A.J., Bickel, P.J., Levina, E., Zhu, J.: Sparse permutation invariant covariance estimation. Electronic Journal of Statistics 2 (2008)
16. Tibshirani, R.: Regression shrinkage and selection via the lasso. J. R. Statist. Soc. B 58(1), 267–288 (1996)
17. Ravikumar, P., Liu, H., Lafferty, J., Wasserman, L.: Spam: Sparse additive models. In: NIPS (2007)
18. Liu, H., Lafferty, J., Wasserman, L.: The nonparanormal: Semiparametric estimation of high dimensional undirected graphs. Journal of Machine Learning Research 10, 2295–2328 (2009)
19. Balakrishnan, S., Puniyani, K., Lafferty, J.: Sparse additive functional and kernel cca. In: ICML (2012)
20. Cho, R., Campbell, M., Winzeler, E., Davis, R.: A genome-wide transcriptional analysis of the mitotic cell cycle. Mol. Cell 2(1), 65–73 (1998)
21. Hughes, T., Marton, M., Jones, A., Roberts, C., Friend, S.: Functional discovery via a compendium of expression profiles. Cell 102(1) (2000)
22. Hibbs, M., Hess, D., Myers, C., Troyanskaya, O.: Exploring the functional landscape of gene expression: directed search of large microarray compendia. Bioinformatics (2007)
23. Stark, C., Breitkreutz, B., Chatr-Aryamontri, A., Boucher, L., Tyers, M.: The biogrid interaction database: update. Nucleic Acids Res. 39(D), 698–704 (2011)
24. Liben-Nowell, D., Kleinberg, J.: The link prediction problem for social networks. In: CIKM (2003)
25. Tomancak, P., Beaton, A., Weiszmann, R., Kwan, E., Shu, S., Lewis, S., Richards, S., Celniker, S., Rubin, G.: Systematic determination of patterns of gene expression during drosophila embryogenesis. Genome Biol. 3(2), 14 (2002)
26. Puniyani, K., Xing, E.P.: Inferring Gene Interaction Networks from ISH Images via Kernelized Graphical Models. In: Fitzgibbon, A., Lazebnik, S., Perona, P., Sato, Y., Schmid, C. (eds.) ECCV 2012, Part VI. LNCS, vol. 7577, pp. 72–85. Springer, Heidelberg (2012)

High Resolution Modeling
of Chromatin Interactions

Christopher Reeder and David Gifford

Massachusetts Institute of Technology
{reeder,gifford}@mit.edu

Abstract. SPROUT is a novel generative model for ChIA-PET data that characterizes physical chromatin interactions and points of contact at high spatial resolution. SPROUT improves upon other methods by learning empirical distributions for pairs of reads that reflect ligation events between genomic locations that are bound by a protein of interest. Using these learned empirical distributions Sprout is able to accurately position interaction anchors, infer whether read pairs were created by self-ligation or inter-ligation, and accurately assign read pairs to anchors which allows for the identification of high confidence interactions. When SPROUT is run on CTCF ChIA-PET data it identifies more interaction anchors that are supported by CTCF motif matches than other approaches with competitive positional accuracy. SPROUT rejects interaction events that are not supported by pairs of reads that fit the empirical model for inter-ligation read pairs, producing a set of interactions that are more consistent across CTCF biological replicates than established methods.

Keywords: Chromatin Interactions, ChIA-PET, CTCF.

1 Introduction

Chromatin interactions are a key component of gene regulation as looping induced interactions bring distal genomic regulatory sequences spatially proximal to their regulatory targets [8]. Identifying the connections between regulatory elements and the genes they regulate is required for understanding transcriptional regulation. Thus, the precise characterization of looping based interactions would help refine our understanding of how genes are controlled. Other forms of looping can implement other kinds of transcriptional regulation, such as isolating regions of the genome from transcriptional activity [4, 13–15, 17, 20].

Recently developed molecular approaches [19] identify chromatin interactions by producing single DNA molecules that combine pieces of DNA from both ends of an interaction event under appropriate ligation conditions. The base sequences at the ends of these DNA molecules are evidence in support of chromatin interactions at the genomic coordinates where the observed sequences originated. ChIA-PET is one such approach that measures chromatin interactions between genomic sites bound by a particular protein [5]. In ChIA-PET the dilute ligation step is preceded by fixation by formaldehyde, fragmentation by sonication, and

M. Deng et al. (Eds.): RECOMB 2013, LNBI 7821, pp. 186–198, 2013.
© Springer-Verlag Berlin Heidelberg 2013

chromatin immunoprecipitation using an antibody designed to target the protein of interest. By using an antibody against a protein that is known to play a role in maintaining genome structure [16], subsequent analysis can focus specifically on chromatin contacts that involve that protein. However, ChIA-PET experimental data are polluted by pairs of reads whose ends do not correspond to binding events for the protein of interest. Such pairs of reads are much like the background reads observed in ChIP-Seq data [18]. This combined with the noisy positioning of reads around binding events presents two challenges for accurately analyzing ChIA-PET data. The first is to accurately identify the positions of the binding events that serve as potential anchors for interactions. The second is to accurately assign read pairs to chromatin interaction anchors or to a background noise model. Focusing on the set of chromatin interactions that are mediated by a specific regulatory protein or complex permits sequencing resources to be focused on the corresponding events. However, sophisticated computational methods are still required to accurately discover interactions from ChIA-PET data.

SPROUT is a novel computational approach for analyzing ChIA-PET data that integrates chromatin interaction discovery with the identification of interaction anchor points. SPROUT accomplishes this by modeling the empirical distribution of read positions around interaction anchors, allowing it to determine the positions of anchors and assign pairs of reads to anchors accurately. Previous approaches to analyzing ChIA-PET data [10] have separated anchor and interaction discovery eliminating the statistical strength that is gained from combining the two procedures. We note that SPROUT is theoretically applicable to datasets generated using related technologies such as Hi-C [11] when sufficient read coverage is available.

In the remainder of the paper we introduce the SPROUT model, discuss our results on CTCF ChIA-PET data, and conclude with observations about SPROUT's applicability.

2   Methods

SPROUT is a hierarchical generative model for ChIA-PET data that discovers interaction anchors, and a set of binary interactions between anchors. There are two types of pairs of reads that are present in ChIA-PET data. Self-ligation pairs arise from the ligation of a DNA molecule to itself. These pairs do not provide direct information about interactions between anchors and can be thought of as providing the same information as paired-end ChIP-Seq data. Inter-ligation pairs arise from the ligation of two distinct DNA molecules from the same chromosome or different chromosomes and thus provide information about a potential interaction.

SPROUT models read-pair data with a mixture over distributions describing the generation of self-ligation pairs and inter-ligation pairs. The components of the model describing these two types of read pairs are themselves mixtures of distributions corresponding to the way pairs of reads are expected to be distributed around anchors. We assume that the paired-end sequencing data generated by a ChIA-PET experiment have been processed appropriately resulting in a set $\mathbf{R} = \{r_1, \ldots, r_N\}$ such that each $r_i = \langle r_i^{(1)}, r_i^{(2)} \rangle$ is a pair of genomic coordinates corresponding to the aligned positions of a pair of reads. Such processing includes removing linker tags from the reads, filtering out pairs that are identified as chimeric because of their heterogeneous linker tags, and aligning the reads to the genome. The following is the likelihood of $\mathbf{R}$

$$\Pr(\mathbf{R}, \pi, \psi, \rho, l) = \prod_{i=1}^{N} \left[\rho \left[\sum_{j=1}^{M} \pi_j \Pr(r_i|l_j) \right] + (1-\rho) \left[\sum_{j=1}^{M} \sum_{k=1}^{M} \psi_{j,k} \Pr(r_i|l_j, l_k) \right] \right] \quad (1)$$

Where $0 \le \rho \le 1$, $\sum_{i=1}^{N} \pi_i = 1$, $\sum_{i=1}^{N} \sum_{j=1}^{N} \psi_{i,j} = 1$

SPROUT identifies a set $l = \{l_1, \ldots, l_M\}$ that specifies the locations of sites that are bound by the protein of interest and are potential anchors for interactions. ρ is the probability that a pair of reads was generated by self-ligation. Self-ligation pairs reflect the ligation of a DNA fragment to itself to form a circular fragment. Such pairs are associated with one anchor and the self-ligation component of the model is a mixture of distributions each taking a single parameter to specify the location of the anchor position. These distributions take the form $\Pr(r_i|l_j)$ (Fig. 1a). A relative weight π_j is associated with each anchor j. These distributions describe the length and arrangement of fragments around an anchor which are induced by the fragmentation step of the ChIA-PET protocol.

Inter-ligation pairs can be associated with either the same anchor or two different anchors that were in close proximity in the nucleus. The inter-ligation component of the model is a mixture of distributions each taking two parameters that specify the locations of the anchor(s) that the fragments were associated with. A relative weight $\psi_{j,k}$ is associated with each pair of anchors j and k. The distributions $\Pr(r_i|l_j, l_k)$ take different forms because if $j = k$ (Fig. 1b) then there are constraints on the ends of the fragments involved in the ligation. For example, the fragments cannot have been overlapping since they were part of the same chromosome prior to fragmentation. If $j \ne k$ (Fig. 1c) it is assumed that the ends were generated independently by two one-dimensional distributions centered around the two anchors $\Pr(r_i|l_j, l_k) = \Pr(r_i^{(1)}|l_j)\Pr(r_i^{(2)}|l_k)$. We also assume that r_i implicitly carries information about the strandedness of the reads because in both the case where $j = k$ and $j \ne k$ the distributions depend on the strandedness of the reads.

ChIA-PET data are noisy, and we observe reads that do not correspond to anchors. To account for these reads, we introduce a noise component with dummy variable l_B ($B \notin \{1, \ldots, M\}$). In this work we consider uniform $\Pr(r_i|l_B)$, however knowledge about the propensity for genomic regions to generate background noise could be incorporated into a more refined noise distribution. We assume that $\Pr(r_i|l_j, l_k)$ where $j = B$ or $k = B$ is defined in the same way as the case in which j and k specify two different anchors: $\Pr(r_i|l_j, l_k) = \Pr(r_i^{(1)}|l_j)\Pr(r_i^{(2)}|l_k)$ and $\Pr(r_i^{(\cdot)}|l_j)$ is uniform when $j = B$.

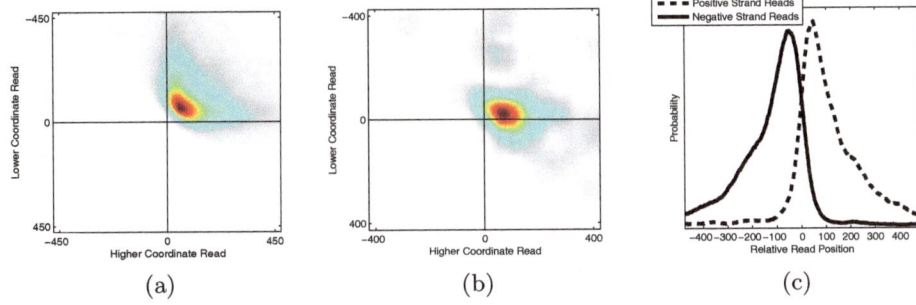

Fig. 1. These are examples of read distributions learned from CTCF ChIA-PET data. Sprout is initially run with "generic" distributions and then the distributions are re-estimated using the strongest events and Sprout is re-run with the empirically learned distributions to discover more accurate predictions. (a) The positions of the ends of self-ligation pairs are modeled using a two dimensional distribution. (b) The positions of the ends of inter-ligation pairs where both ends are assigned to the same anchor are also modeled using two dimensional distributions. Each of the four possible strand combinations has its own constraints in terms of where the ends are likely to be positioned relative to each other and to the anchor. This figure demonstrates the distribution associated with inter-ligation pairs where both ends map to the positive strand. (c) The positions of the ends of inter-ligation pairs are modeled separately using one dimensional distributions.

To avoid overfitting, we wish to find a minimal number of anchors that explain the data well while allowing the noise distribution to account for reads that are not accounted for by anchors. Additionally, we assume that among all possible pairs of anchors most pairs are not interacting. Thus, we wish to find a minimal number of interacting pairs of anchors that explain the observed data. To achieve both of these types of sparsity we introduce negative Dirichlet priors [3] on π and ψ as specified by Eq. 2 and Eq. 3.

$$\Pr(\pi|\alpha) \propto \prod_{j=1}^{M} \pi_j^{-\alpha} \tag{2}$$

$$\Pr(\psi|\beta) \propto \prod_{j=1}^{M} \prod_{k=1}^{M} \psi_{j,k}^{-\beta} \tag{3}$$

As will become apparent when the inference procedure is described, the α and β parameters have the effect of specifying the minimum number of pairs of reads that must be associated with an anchor or an interaction, respectively, in order to avoid being eliminated from the model.

We also introduce priors on l and ρ. For l we introduce a Bernoulli prior which reflects our prior belief that an anchor exists at a particular genomic coordinate and that at most one anchor exists at any genomic coordinate. Given L possible genomic coordinates,

$$\Pr(l|k) = \prod_{i=1}^{L} k_i^{\mathbf{1}(i \in l)}(1 - k_i)^{\mathbf{1}(i \notin l)} \tag{4}$$

$$= \prod_{i=1}^{L}(1 - k_i) \prod_{j=1}^{M} \frac{k_{l_j}}{1 - k_{l_j}} \tag{5}$$

$$\propto \prod_{j=1}^{M} \frac{k_{l_j}}{1 - k_{l_j}} \tag{6}$$

In this work we consider uniform k, but k could be made non-uniform to reflect any prior belief about where anchors should be located. For ρ we introduce a Beta prior

$$\Pr(\rho|a, b) \propto \rho^{a-1}(1 - \rho)^{b-1} \tag{7}$$

In this work we let $a = 1$ and $b = 1$ which is a uniform prior on ρ.

Each pair of reads is either a result of a self-ligation event or an inter-ligation event and is associated with one or two anchors. We introduce latent variables $\mathbf{Z} = \{z_1, \ldots, z_N\}$ such that each $z_i = \langle z_i^{(1)}, z_i^{(2)} \rangle$ is a pair of anchor indices $1 \ldots M$ or special index B reflecting the noise distribution. Another special index is used to indicate that a pair of reads was generated by self-ligation i.e. $z_i = \langle j, - \rangle$.

The complete data likelihood is

$$\Pr(\mathbf{R}, \mathbf{Z}|\pi, \psi, \rho, l) = \Pr(\mathbf{R}|\mathbf{Z}, l) \Pr(\mathbf{Z}|\pi, \psi, \rho) \tag{8}$$

$$= \prod_{i=1}^{N} \left[\prod_{j=1}^{M} [\rho \pi_j \Pr(r_i|l_j)]^{\mathbf{1}(z_i = \langle j, - \rangle)} \prod_{k=1}^{M} [(1 - \rho)\psi_{j,k} \Pr(r_i|l_j, l_k)]^{\mathbf{1}(z_i = \langle j, k \rangle)} \right] \tag{9}$$

We are interested in inferring likely values for π, ψ, ρ, and l. To accomplish this we employ a variant of the EM algorithm [2] to maximize the complete data log posterior

$$\log \Pr(l, \pi, \psi, \rho|\mathbf{R}, \mathbf{Z}, k, \alpha, \beta, a, b) = \sum_{i=1}^{N} \left[\sum_{j=1}^{M} \left[\mathbf{1}(z_i = \langle j, - \rangle) (\log \rho + \log \pi_j + \log \Pr(r_i|l_j)) \right. \right.$$

$$\left. \left. + \sum_{k=1}^{M} \mathbf{1}(z_i = \langle j, k \rangle) (\log(1 - \rho) + \log \psi_{j,k} + \log \Pr(r_i|l_j, l_k)) \right] \right] \tag{10}$$

$$- \alpha \sum_{j=1}^{M} \log \pi_j - \beta \sum_{j=1}^{M} \sum_{k=1}^{M} \log \psi_{j,k} + \sum_{j=1}^{M} \log \frac{k_{l_j}}{1 - k_{l_j}} + (a - 1)\log \rho + (b - 1)(1 - \rho) + C$$

E Step:

$$\gamma(z_i) = \frac{\prod_{j=1}^{M} \left[[\rho \pi_j \Pr(r_i|l_j)]^{\mathbf{1}(z_i = \langle j, - \rangle)} \prod_{k=1}^{M} [(1 - \rho)\psi_{j,k} \Pr(r_i|l_j, l_k)]^{\mathbf{1}(z_i = \langle j, k \rangle)} \right]}{\sum_{j=1}^{M} \left[[\rho \pi_j \Pr(r_i|l_j)] + \sum_{k=1}^{M} [(1 - \rho)\psi_{j,k} \Pr(r_i|l_j, l_k)] \right]} \tag{11}$$

M Step:

$$\hat{l}_j = \underset{x}{\operatorname{argmax}} \left\{ \sum_{i=1}^{N} [\gamma(z_i = \langle j, -\rangle) \log \Pr(r_i|x) \right. \tag{12}$$

$$\left. + \sum_{k=1}^{M} [\gamma(z_i = \langle j, k\rangle) \log \Pr(r_i|x, l_k)] \right] + \log \frac{k_x}{1-k_x} \right\}$$

$$\hat{\pi}_j = \frac{\max(N_j - \alpha, 0)}{N_\pi} \tag{13}$$

$$N_\pi = \sum_{j=1}^{M} \max(N_j - \alpha, 0) \tag{14}$$

$$N_j = \sum_{i=1}^{N} \gamma(z_i = \langle j, -\rangle) \tag{15}$$

$$\hat{\psi}_{j,k} = \frac{\max(N_{j,k} - \beta, 0)}{N_\psi} \tag{16}$$

$$N_\psi = \sum_{j=1}^{M} \sum_{k=1}^{M} \max(N_{j,k} - \beta, 0) \tag{17}$$

$$N_{j,k} = \sum_{i=1}^{N} \gamma(z_i = \langle j, k\rangle) \tag{18}$$

$$\hat{\rho} = \frac{N_\pi + a}{N + a + b} \tag{19}$$

The E and M steps are repeated until the posterior approximately converges. The components of l that correspond to non-zero components of π are the estimated anchor locations. Non-zero components of ψ indicate pairs of anchors that are candidates for significance testing as interactions.

The algorithm is initialized with uniform π and l set at regular intervals throughout the genome. Components of π that do not assign probability to any pairs of reads are set to 0 and effectively eliminated from the model. Components with $N_j < \alpha$ are eliminated shortly thereafter. In the estimation of $\hat{l}_j$ during each M step the components of l other than the jth component are held fixed making this algorithm an instance of the expectation-conditional maximization algorithm [12]. Thus, the posterior is not necessarily maximized at each iteration but convergence to a local maximum is still guaranteed. The estimation of $\hat{l}_j$ is tractable, despite the lack of a closed form solution, because for the set of pairs of reads such that $\gamma(z_i = \langle j, \cdot\rangle) > 0$, $Pr(r_i|x) > 0$ for any pair of reads in the set for x in only a small neighborhood around the previous value of l_i. Only x in that neighborhood need be considered which reduces the search space for the optimal x considerably.

To test the significance of a component $\psi_{j,k}$, the posterior is recomputed with that component removed. The greater the ratio of the posterior with the component to the posterior without the component, the greater the significance of the corresponding interaction. Making the conservative assumption that all components with $N_{j,k} \leq 2$ are false positives, we set a threshold for the posterior ratio to be the value such that 5% of the components deemed significant have $N_{j,k} \leq 2$.

3   Results

We compared the performance of SPROUT to other methods by analyzing CTCF ChIA-PET data published by Handoko et al. [7]. Reads were processed using the LinkerRemover component of the ChIA-PET tool [10]. The pairs of reads that were positively identified as chimeric were discarded and the rest of the reads were aligned to the mouse genome as unpaired reads using BOWTIE [9]. Only pairs with both ends that map uniquely were considered for further analysis. In cases where more than one pair of reads aligns to the same location at both ends, only one pair is retained because such positional duplicates are likely to be PCR artifacts.

For comparison, we downloaded the significant intra-chromosomal interactions and CTCF binding events published by Handoko et al. SPROUT discovers both inter- and intra-chromosomal interactions, but for this analysis we limited our comparison to intra-chromosomal interactions only. SPROUT does not impose a lower bound on the distance between pairs of anchors that it will consider for identifying interactions. However, linearly proximal anchors are expected to be spatially proximal due to random polymeric movement of the chromosome in the space of the nucleus. Therefore, linearly proximal anchors are expected to be called interacting by SPROUT. To investigate the distance at which this effect diminishes, we looked at the frequency at which pairs of anchors detected by SPROUT interact as a function of distance between the anchors (Fig. 2). By 4000 bp, detected interactions become very infrequent suggesting that interactions of this distance or greater are unlikely to be due to the linear proximity effect. The shortest range interaction published by Handoko et al. is 5928 bp, so for comparison we only consider interactions discovered by SPROUT that span at least this distance. But, we note that Fig. 2 suggests that functional interactions may be discoverable by SPROUT at distances as low as 4000 bp.

By comparing the positions of the anchors discovered by SPROUT to matches to the CTCF motif, we discovered that SPROUT positions anchors with high accuracy, and is very sensitive compared to other methods of discovering CTCF binding events while maintaining a high degree of specificity. For comparison, we examined the CTCF binding event calls published by Handoko et al. as well as binding events identified by the GEM peak calling algorithm [6] which was run on an independent ChIP-Seq dataset [1]. It is worth noting that Handoko et al. based their binding event predictions on the ChIA-PET data but that their method for identifying interactions is independent of their binding event predictions. Overall there were more motif supported events in the set of events identified by SPROUT than the other two sets. The maximum height of each

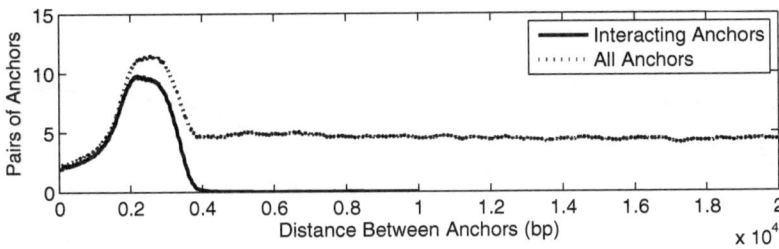

Fig. 2. Smoothed plots of the frequency at which interacting anchors identified by SPROUT exist at distances up to 20000 bp. Beyond 4000 bp anchors are very infrequently interacting relative to the number of possible interactions at a given distance and individual. This suggests that interactions that are detected by SPROUT that span more than 4000 bp are not explained by the linear proximity of the anchors.

curve in Fig. 3b indicates the total number of motif supported events discovered by each method. Furthermore, the weight assigned to events by SPROUT is a better classifier of motif supported events than the weights assigned by Handoko et al. to the ChIA-PET events or the weights assigned to ChIP-Seq events by GEM. The fact that the SPROUT curve in Fig. 3b is always greater than the other curves indicates that SPROUT achieves greater specificity.

The anchor regions identified by Handoko et al. tend to be relatively broad with an average width of 1997.7 bp (Fig. 4). By identifying binding events within the anchor regions, it may be possible to recover the true anchors for the interactions as a post-processing step. However, as an example of the difficulty in interpreting such broad interaction anchor regions, 63 of the 4077 interacting anchors identified by Handoko et al. contain more than one motif supported binding event. One of the strengths of SPROUT is that interactions called by SPROUT are directly associated with binding events, thereby reducing ambiguity in interpreting the results.

Upon comparing the significant interactions identified by SPROUT and Handoko et al., we noticed that certain significant interactions were missed by Handoko et al. Of the 420 significant interactions that span more than 5928 bp identified by SPROUT, 87 interactions lack a corresponding interaction identified by Handoko et al. with both anchors within 4 kb of the SPROUT identified anchors. Of these interactions, 64 have binding events identified by Handoko et al. within 250 bp of the SPROUT identified anchors. The fact that Handoko et al. failed to identify several interactions between binding events that they identify with their own method for detecting binding events indicates one of the benefits of SPROUT's approach of integrating interaction detection with anchor detection.

Handoko et al. identified 2241 significant interactions, however many of these interactions do not fit the model of an interaction between two distinct anchors as defined by SPROUT (Fig. 5). 200 of the Handoko et al. interactions do subsume SPROUT identified interactions and an additional 11 Handoko et al. interactions have SPROUT identified interactions with anchors within 4 kb of their anchors. 1181 of the Handoko et al. interactions that do not subsume SPROUT identified

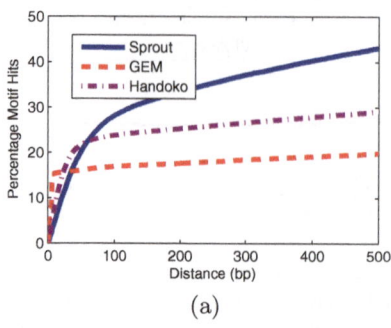
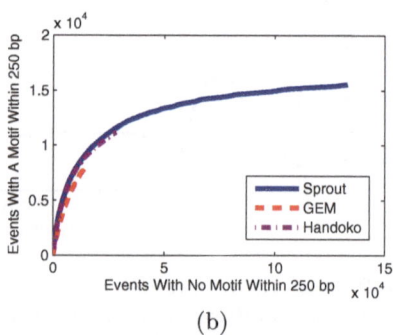

(a)   (b)

Fig. 3. Evaluation of the accuracy of CTCF binding events predicted by SPROUT and Handoko et al. from the ChIA-PET data as well as by GEM from an independent ChIP-Seq dataset. (a) The percentage of CTCF motif matches in the genome that have a binding event identified within distances up to 500 bp. (b) We used the presence of a CTCF motif match within 250 bp of an event as an approximate indicator of true positive anchor calls. As thresholds for significance are varied for each method, the number of true positive and false positive calls are plotted. This results in a receiver operating characteristic curve for each method.

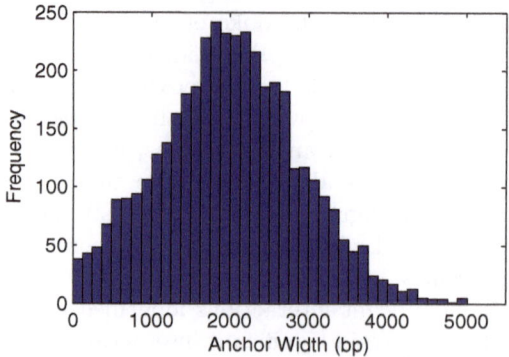

Fig. 4. A histogram of the widths of anchors identified by Handoko et al. illustrating the breadth of many of the anchor regions.

interactions do not contain a CTCF binding event (by their own definition) at one or both anchors. This clearly indicates that these are unlikely to reflect true interactions between CTCF-bound anchors. Of the remaining 860 Handoko et al. interactions, 52 involve 0 pairs of reads and 123 involve 1 pair of reads according to our alignment of the data. Handoko et al. used a rescue procedure in which reads that align to multiple locations are in some cases assigned to one location. We did not use this procedure when we aligned the reads which may explain why Handoko et al. assign significance to interactions that do not seem to be supported by enough read pairs without the rescue procedure. This leaves 685 Handoko et al. interactions that are supported by at least 2 pairs of reads that

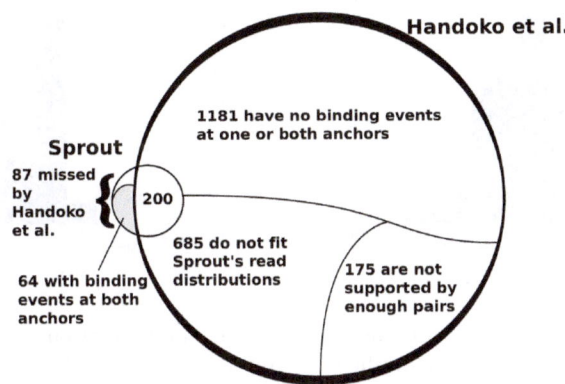

Fig. 5. Most of the interactions identified by Handoko et al. are not supported by pairs of reads with ends that fit SPROUT's read distribution.

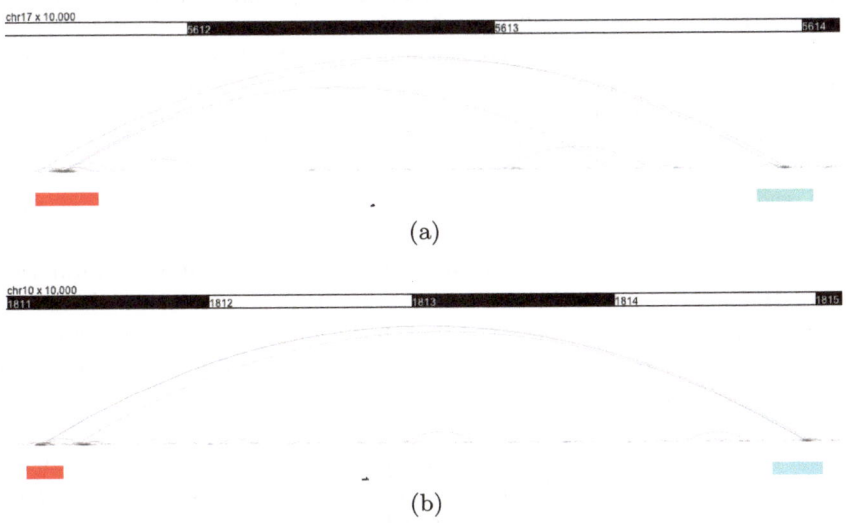

Fig. 6. Two interactions that are identified by Handoko et al. The boxes indicate the anchor regions that they identify. (a) This interaction is not called significant by SPROUT because the pairs of reads that connect the anchor regions do not fit SPROUT's model. (b) SPROUT does call a significant interaction between the anchors that fall within the Handoko et al. anchor regions because the pairs of reads that connect the regions were likely to have been generated by the anchors within the regions according to SPROUT's model. Note that there is a second potential anchor on the left side that falls outside of the Handoko et al. identified region. This anchor is identified by both SPROUT and Handoko et al. and is identified by SPROUT but not by Handoko et al. as an independent interaction with the anchor on the right.

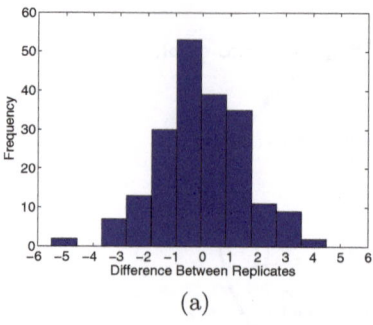
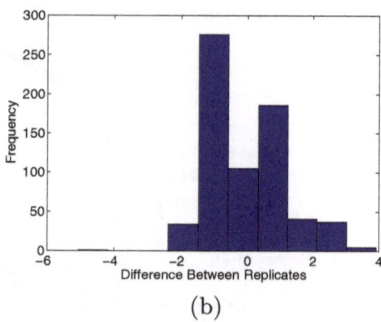

(a)   (b)

Fig. 7. Evaluation of biological replicate consistency in interactions discovered by both methods and in interactions identified by Handoko et al. that do not fit SPROUT's read distributions. (a) A histogram of the difference in the number of pairs of reads from each biological replicate that connect anchors identified by Handoko et al. that subsume interactions called by SPROUT. To account for the overall difference in signal strength, the values were subtracted by the mean per interaction difference. There are interactions that differ in support between the biological replicates. However, the normalized difference in pairs between the biological replicates is most frequently close to 0. (b) A histogram of the difference in the number of pairs of reads from each biological replicate that connect anchors identified by Handoko et al. that are supported by a plausible number of pairs of reads but do not fit SPROUT's read distributions. As in (a), the differences are subtracted by the mean difference. The biological replicates differ by one pair of reads much more frequently than they agree. This difference is significant given that 491 out of 685 interactions in this set are only supported by 2 pairs of reads total.

do not subsume SPROUT identified interactions. However, upon examination of many of these interactions (Fig. 6), the broadness of the interaction anchors allow pairs of reads to be considered together even though the positions of the reads do not fit SPROUT's model of how reads should be distributed around anchors.

Interactions supported by pairs of reads that fit SPROUT's read distributions are more consistent across biological replicates and therefore are more likely to represent true interactions. To demonstrate this we consider two sets of interactions. One set, which we call the good fit set, includes the 200 interactions identified by Handoko et al. that subsume SPROUT identified interactions. The other set, which we call the bad fit set, includes the 685 Handoko et al. interactions that contain binding events at both anchors and are connected by at least 2 pairs of reads but do not subsume interactions discovered by SPROUT. The first thing we noticed is that the interactions in the good fit set tend to be supported by more pairs of reads. The average number of pairs per interaction in the good fit set is 4.15 while for the bad fit set the average number of pairs is 2.73. We then identified which of the biological replicates each pair of reads came from. As can be seen in Fig 7, the biological replicates assign pairs of reads to the interactions in the good fit set more consistently than interactions in the bad fit set.

4   Conclusion

SPROUT uses all pairs of reads to estimate anchor positions and learns empirical interaction read distributions to more accurately assign pairs of reads to anchors. SPROUT interaction calls are more consistent across biological replicates than the method proposed by Handoko et al. Identifying high confidence interactions between accurately positioned anchors is a task that is increasing in importance as more genome structure data are produced. Utilizing data from various types of high throughput sequencing based approaches, several successful approaches to identifying regulatory elements have been developed. However, it is impossible to fully understand how these regulatory elements function without putting them in their spatial context in the nucleus. The interaction results produced by SPROUT from ChIA-PET data allow for a more accurate understanding of this spatial context.

References

1. Chen, X., Xu, H., Yuan, P., Fang, F., Huss, M., Vega, V.B., Wong, E., Orlov, Y.L., Zhang, W., Jiang, J., Loh, Y., Yeo, H.C., Yeo, Z.X., Narang, V., Govindarajan, K.R., Leong, B., Shahab, A., Ruan, Y., Bourque, G., Sung, W., Clarke, N.D., Wei, C., Ng, H.: Integration of External Signaling Pathways with the Core Transcriptional Network in Embryonic Stem Cells. Cell 133, 1106–1117 (2008)
2. Dempster, A.P., Laird, N.M., Rubin, D.B.: Maximum Likelihood Estimation from Incomplete Data via the EM Algorithm. J. R. Stat. Soc. B 39, 1–38 (1977)
3. Figueiredo, M.A., Jain, A.K.: Unsupervised Learning of Finite Mixture Models. IEEE T. Pattern Anal. 4, 381–396 (2002)
4. Francastel, C., Schübeler, D., Martin, D.I.K., Groudine, M.: Nuclear Compartmentalization and Gene Activity. Nat. Rev. Mol. Cell Biol. 1, 137–143 (2000)
5. Fullwood, M.J., Liu, M.H., Pan, Y.F., Liu, J., Xu, H., Mohamed, Y.B., Orlov, Y.L., Velkov, S., Ho, A., Mei, P.H., Chew, E.G.Y., Huang, P.Y.H., Welboren, W., Han, Y., Ooi, H.S., Ariyaratne, P.N., Vega, V.B., Luo, Y., Tan, P.Y., Choy, P.Y., Wansa, K.D.S.A., Zhao, B., Lim, K.S., Leow, S.C., Yow, J.S., Joseph, R., Li, H., Desai, K.V., Thomsen, J.S., Lee, Y.K., Karuturi, R.K.M., Herve, T., Bourque, G., Stunnenberg, H.G., Ruan, X., Cacheux-Rataboul, V., Sung, W., Liu, E.T., Wei, C., Cheung, E., Ruan, Y.: An Oestrogen-Receptor-α-Bound Human Chromatin Interactome. Nature 462, 58–64 (2009)
6. Guo, Y., Mahony, S., Gifford, D.K.: High Resolution Genome Wide Binding Event Finding and Motif Discovery Reveals Transcription Factor Spatial Binding Constraints. P.L.O.S. Comput. Biol. 8, e1002638 (2012)
7. Handoko, L., Xu, H., Li, G., Ngan, C.Y., Chew, E., Schnapp, M., Lee, C.W.H., Ye, C., Ping, J.L.H., Mulawadi, F., Wong, E., Sheng, J., Zhang, Y., Poh, T., Chan, C.S., Kunarso, G., Shahab, A., Bourque, G., Cacheux-Rataboul, V., Sung, W., Ruan, Y., Wei, C.: CTCF-Mediated Functional Chromatin Interactome in Pluripotent Cells. Nat. Genet. 43, 630–638 (2011)
8. Hatzis, P., Talianidis, I.: Dynamics of Enhancer-Promoter Communication During Differentiation-Induced Gene Activation. Mol. Cell 10, 1467–1477 (2002)
9. Langmead, B., Trapnell, C., Pop, M., Salzber, S.L.: Ultrafast and Memory-Efficient Alignment of Short DNA Sequences to the Human Genome. Genome Biol. 10, R25 (2009)

10. Li, G., Fullwood, M.J., Xu, H., Mulawadi, F.H., Velkov, S., Vega, V., Ariyaratne, P.N., Mohamed, Y.B., Ooi, H., Tennakoon, C., Wei, C., Ruan, Y., Sung, W.: ChIA-PET Tool for Comprehensive Chromatin Interaction Analysis with Paired-End Tag Sequencing. Genome Biol. 11, R22 (2010)
11. Lieberman-Aiden, E., van Berkum, N.L., Williams, L., Imakaev, M., Ragoczy, T., Telling, A., Amit, I., Lajoie, B.R., Sabo, P.J., Dorschner, M.O., Sandstrom, R., Bernstein, B., Bender, M.A., Groudine, M., Gnirke, A., Stamatoyannopoulos, J., Mirny, L.A., Lander, E.S., Dekker, J.: Comprehensive Mapping of Long-Range Interactions Reveals Folding Principles of the Human Genome. Science 326, 289–293 (2009)
12. Meng, X., Rubin, D.B.: Maximum Likelihood Estimation via the ECM Algorithm: A General Framework. Biometrika 80, 267–278 (1993)
13. Meshorer, E., Misteli, T.: Chromatin in Pluripotent Embryonic Stem Cells and Differentiation. Nat. Rev. Mol. Cell Biol. 7, 540–546 (2006)
14. Misteli, T.: Beyond the Sequence: Cellular Organization of Genome Function. Cell 128, 787–800 (2007)
15. Misteli, T., Soutoglou, E.: The Emerging Role of Nuclear Architecture in DNA Repair and Genome Maintenance. Nat. Rev. Mol. Cell Biol. 10, 243–254 (2009)
16. Philips, J.E., Corces, V.G.: CTCF: Master Weaver of the Genome. Cell 137, 1194–1211 (2009)
17. Pombo, A., Branco, M.R.: Functional Organisation of the Genome During Interphase. Curr. Opin. Genet. Dev. 17, 451–455 (2007)
18. Rozowski, J., Euskirchen, G., Auerbach, R.K., Zhang, Z.D., Gibson, T., Bjornson, R., Carriero, N., Snyder, M., Gerstein, M.B.: PeakSeq enables systematic scoring of ChIP-seq Experiments Relative to Controls. Nat. Biotechnol. 27, 66–75 (2009)
19. van Steensel, B., Dekker, J.: Genomics Tools for Unraveling Chromosome Architecture. Nat. Biotechnol. 28, 1089–1095 (2010)
20. Zhao, R., Bodnar, M.S., Spector, D.L.: Nuclear Neighborhoods and Gene Expression. Curr. Opin. Genet. Dev. 19, 172–179 (2009)

A Linear Inside-Outside Algorithm for Correcting Sequencing Errors in Structured RNAs

Vladimir Reinharz[1], Yann Ponty[2,*], and Jérôme Waldispühl[1,*]

[1] School of Computer Science, McGill University, Montreal, Canada
[2] Laboratoire d'informatique, École Polytechnique, Palaiseau, France
jeromew@cs.mcgill.ca, yann.ponty@lix.polytechnique.fr

Abstract. Analysis of the sequence-structure relationship in RNA molecules are essential to evolutionary studies but also to concrete applications such as error-correction methodologies in sequencing technologies. The prohibitive sizes of the mutational and conformational landscapes combined with the volume of data to proceed require efficient algorithms to compute sequence-structure properties. More specifically, here we aim to calculate which mutations increase the most the likelihood of a sequence to a given structure and RNA family.

In this paper, we introduce `RNApyro`, an efficient linear-time and space inside-outside algorithm that computes exact mutational probabilities under secondary structure and evolutionary constraints given as a multiple sequence alignment with a consensus structure. We develop a scoring scheme combining classical stacking base pair energies to novel isostericity scales, and apply our techniques to correct point-wise errors in 5s rRNA sequences. Our results suggest that `RNApyro` is a promising algorithm to complement existing tools in the NGS error-correction pipeline.

Keywords: RNA, mutations, secondary structure.

1 Introduction

Ribonucleic acids (RNAs) are found in every living organism, and exhibit a broad range of functions, ranging from catalyzing chemical reactions, as the RNase P or the group II introns, hybridizing messenger RNA to regulate gene expressions, to ribosomal RNA (rRNA) synthesizing proteins. Those functions require specific structures, encoded in their nucleotide sequence. Although the functions, and thus the structures, need to be preserved through various organisms, the sequences can greatly differ from one organism to another. This sequence diversity coupled with the structural conservation is a fundamental asset for evolutionary studies. To this end, algorithms to analyze the relationship between RNA mutants and structures are required.

For half a century, biological molecules have been studied as a proxy to understand evolution [1], and due to their fundamental functions and remarkably

---

[*] Corresponding authors.

M. Deng et al. (Eds.): RECOMB 2013, LNBI 7821, pp. 199–211, 2013.
© Springer-Verlag Berlin Heidelberg 2013

conserved structures, rRNAs have always been a prime candidate for phylogenetic studies [2, 3]. In recent years, studies as the *Human Microbiome Project* [4] benefited of new technologies such as the NGS techniques to sequence as many new organisms as possible and extract an unprecedented flow of new information. Nonetheless, these high-throughput techniques typically have high error rates that make their applications to metagenomics (a.k.a. environmental genomics) studies challenging. For instance, pyrosequencing as implemented by Roche's 454 produces may have an error rate raising up to 10%. Because there is no cloning step, resequencing to increase accuracy is not possible and it is therefore vital to disentangle noise from true sequence diversity in this type of data [5]. Errors can be significantly reduced when large multiple sequence alignments with close homologs are available, but in studies of new or not well known organisms, such information is rather sparse. In particular, it is common that there is not enough similarity to differentiate between the sequencing errors and the natural polymorphisms that we want to observe, often leading to artificially inflated diversity estimates [6]. A few techniques have been developed to remedy to this problem [7, 8] but they do not take into account all the available information. It is therefore essential to develop methods that can exploit any type of signal available to correct errors.

In this paper, we introduce `RNApyro`, a novel algorithm that enables us to calculate precisely mutational probabilities in RNA sequences with a conserved consensus secondary structure. We show how our techniques can exploit the structural information embedded in physics-based energy models, covariance models and isostericity scales to identify and correct point-wise errors in RNA molecules with conserved secondary structure. In particular, we hypothesize that conserved consensus secondary structures combined with sequence profiles provide an information that allow us to identify and fix sequencing errors.

Here, we expand the range of algorithmic techniques previously introduced with `RNAmutants` [9]. Instead of exploring the full conformational landscape and sample mutants, we develop an inside-outside algorithm that enables us to explore the complete mutational landscape with a *fixed* secondary structure and to calculate exactly mutational probability values. In addition to a gain into the numerical precision, this strategy allows us to drastically reduce the computational complexity ($\mathcal{O}(n^3 \cdot m^2)$ for the original version of `RNAmutants` to $\mathcal{O}(n \cdot m^2)$ for `RNApyro`, where n is the size of the sequence and m the number of mutations).

We design a new scoring scheme combining nearest-neighbor models [10] to isostericity metrics [11]. Classical approaches use a Boltzmann distribution whose weights are estimated using a nearest-neighbour energy model [10]. However, the latter only accounts for canonical and wobble, base pairs. As was shown by Leontis and Westhof [12], the diversity of base pairs observed in tertiary structures is much larger, albeit their energetic contribution remains unknown. To quantify geometrical discrepancies, an isostericity distance has been designed [11], increasing as two base pairs geometrically differ from each other in space. Therefore, we incorporate these scores in the Boltzmann weights used by `RNApyro`.

We illustrate and benchmark our techniques for point-wise error corrections on the 5S ribosomal RNA. We choose the latter since it has been extensively used for phylogenetic reconstructions [13] and its sequence has been recovered for over 712 species (in the Rfam seed alignment with id RF00001). Using a leave one out strategy, we perform random distributed mutations on a sequence. While our methodology is restricted to the correction of point-wise error in structured regions (i.e. with base pairs), we show that RNApyro can successfully extract a signal that can be used to reconstruct the original sequence with an excellent accuracy. This suggests that RNApyro is a promising algorithm to complement existing tools in the NGS error-correction pipeline.

The algorithm and the scoring scheme are presented in Sec. 2. Details of the implementation and benchmarks are in Sec. 3. Finally, we discuss future developments and applications in Sec. 4.

2   Methods

We introduce a probabilistic model, which aims at capturing both the stability of the folded RNA and its ability to adopt a predefined 3D conformation. To that purpose, a Boltzmann weighted distribution is used, based on a pseudo-energy function $E(\cdot)$ which includes contributions for both the free-energy and its putative isostericity towards a multiple sequence alignment. In this model, the probability that the nucleotide at a given position needs to be mutated (i.e. corresponds to a sequencing error) can be computed using a variant of the *Inside-Outside algorithm* [14].

2.1   Probabilistic Model

Let Ω be an gap-free RNA alignment sequence, S its associated secondary structure, then any sequence s has probability proportional to its Boltzmann factor

$$\mathcal{B}(s) = e^{\frac{-E(s)}{RT}}, \qquad \text{with} \qquad E(s) := \alpha \cdot \text{ES}(s, S) + (1 - \alpha) \cdot \text{EI}(s, S, \Omega),$$

where R is the Boltzmann constant, T the temperature in Kelvin, $\text{ES}(s)$ and $\text{EI}(s, S, \Omega)$ are the free-energy and isostericity contributions respectively (further described below), and $\alpha \in [0, 1]$ is an arbitrary parameter that sets the relative weight for both contributions.

Energy Contribution. The free-energy contribution in our pseudo-energy model corresponds to an additive stacking-pairs model, taking values from the Turner 2004 model retrieved from the NNDB [10]. Given a candidate sequence s for a secondary structure S, the free-energy of S on s is given by

$$\text{ES}(s, S) = \sum_{\substack{(i,j) \to (i',j') \in S \\ \text{stacking pairs}}} \text{ES}^{\beta}_{s_i s_j \to s_{i'} s_{j'}}$$

where $ES^{\beta}_{ab \to a'b'}$ is set to 0 if $ab = \varnothing$ (no base-pair to stack onto), the tabulated free-energy of stacking pairs $(ab)/(a'b')$ in the Turner model if available, or $\beta \in [0, \infty]$ for non-Watson-Crick/Wobble entries (i.e. neither GU, UG, CG, GC, AU nor UA). This latter parameter allows to choose whether to simply penalize invalid base pairs, or forbid them altogether ($\beta = \infty$). The loss of precision due to this simplification of the Turner model remains reasonable since the targeted secondary structure is fixed (e.g. multiloops do not account for base-specific contributions). Furthermore, it greatly eases the design and implementation of dynamic-programming equations.

Isostericity Contribution. The concept of isostericity score [11] is based on the geometric discrepancy (superimposability) of two base-pairs, using individual additive contributions computed by Stombaugh *et al* [11]. Let s be a candidate sequence for a secondary structure S, given in the context of a gap-free RNA alignment Ω, we define the isostericity contribution to the pseudo-energy as

$$ES(s, S, \Omega) = \sum_{\substack{(i,j) \in S \\ \text{pairs}}} EI^{\Omega}_{(i,j),s_i s_j}, \quad \text{where} \quad EI^{\Omega}_{(i,j),ab} := \frac{\sum_{s' \in \Omega} ISO((s'_i, s'_j), (a, b))}{|\Omega|}$$

is the average isostericity of a base-pair in the candidate sequence, compared with the reference alignment. The ISO function uses the Watson-Crick/Watson-Crick cis isostericity matrix computed by Stombaugh *et al* [11]. Isostericity scores range between 0 and 9.7, 0 corresponding to a perfect isostericity, and a penalty of 10 is used for missing entries. The isostericity contribution will favor exponentially sequences that are likely to adopt a similar local conformation as the sequences contained in the alignment.

2.2   Mutational Profile of Sequences

Let s be an RNA sequence, S a reference structure, and $m \geq 0$ a desired number of mutations. We are interested in the probability that a given position contains a specific nucleotide, over all sequences having at most m mutations from s (formally $\mathbb{P}(s_i = x \mid s, \Omega, S, m)$). We define a variant of the *Inside-Outside algorithm* [14], allowing us to compute these probability.

The former, defined in Equations (2) and (3), is analogous to the *inside* algorithm. It is the partition function, i.e. the sum of Boltzmann factors, over all sequences within $[i, j]$, knowing that position $i - 1$ is composed of nucleotide a (resp. $j + 1$ is b), within m mutations of s. The latter, defined by Equations (4) and (5), computes the *outside* algorithm, i.e. the partition function over sequences within m mutations of s, restricted to two intervals $[0, i] \cup [j, n - 1]$, and knowing that position $i + 1$ is composed of a (resp. $j - 1$ is b). A suitable combination of these terms, given in Equation (7), gives the total weight, and in turn the probability, of seeing a specific base at a given position.

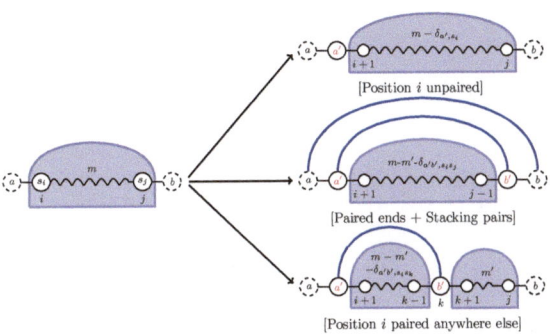

Fig. 1. Principle of the inside computation (partition function). Any sequence with mutations can be decomposed as a sequence preceded by a, possibly mutated, base (Unpaired case), a sequence surrounded by some base-pair (Stacking-pair case), or as two sequences segregated by some base-pair (General base-pairing case), and mutations must be distributed between sub-sequences and instantiated bases.

Definitions. Let $B := \{A, C, G, U\}$ be the set of nucleotides. Given $s \in B^n$ an RNA sequence, let s_i be the nucleotide at position i. Let Ω be a set of un-gapped RNA sequences of length n, and S a secondary structure without pseudoknots. Formally, if (i, j) and (k, l) are base pairs in S, there is no overlapping extremities $\{i, j\} \cap \{k, l\} = \varnothing$ and either the intersection is empty ($[i, j] \cap [k, l] = \varnothing$) or one is included in the other ($[k, l] \subset [i, j]$ or $[i, j] \subset [k, l]$).

Let us then remind the Hamming distance function $\delta : B^* \times B^* \to \mathbb{N}^+$, which takes two sequences s' and s'' as input, $|s'| = |s''|$, and returns the number of differing positions. Finally, let us denote by $E^{\Omega,\beta}_{(i,j),ab \to a'b'}$ the local contribution of a base-pair (i, j) to the pseudo-energy, such that

$$E^{\Omega,\beta}_{(i,j),ab \to a'b'} = \alpha \cdot \mathrm{ES}^{\beta}_{ab \to a'b'} + (1 - \alpha) \cdot \mathrm{EI}^{\Omega}_{(i,j),a'b'}. \tag{1}$$

Inside computation. The *Inside* function $\mathcal{Z}^m_{\substack{(i,j) \\ [a,b]}}$ is the partition function, i.e. the sum of Boltzmann factors over all sequences in the interval $[i, j]$, at distance m of $s_{[i,j]}$, and having flanking nucleotides a and b (at positions $i - 1$ and $j + 1$ respectively). Such terms can be defined by recurrence, for which the following initial conditions holds:

$$\forall i \in [0, n-1] : \mathcal{Z}^m_{\substack{(i+1,i) \\ [a,b]}} = \begin{cases} 1 \text{ If } m = 0 \\ 0 \text{ Otherwise.} \end{cases} \tag{2}$$

In other words, the set of sequences at distance m of the empty sequence is either empty if $m > 0$, or restricted to the empty sequence, having energy 0, if $m = 0$. Since the energetic terms only depend on base pairs, they are not involved in the initial conditions. The main recursion itself is composed of four terms:

$$
\mathcal{Z}^m_{\substack{(i,j) \\ [a,b]}} := \begin{cases} \displaystyle\sum_{\substack{a' \in \mathcal{B}, \\ \delta_{a',s_i} \leq m}} \mathcal{Z}^{m-\delta_{a',s_i}}_{\substack{(i+1,j) \\ [a',b]}} & \text{If } S_i = -1 \\[2em] \displaystyle\sum_{\substack{a',b' \in \mathcal{B}^2, \\ \delta_{a'b',s_i s_j} \leq m}} e^{\frac{-E^{\Omega,\beta}_{(i,j),ab \to a'b'}}{RT}} \cdot \mathcal{Z}^{m-\delta_{a'b',s_i s_j}}_{\substack{(i+1,j-1) \\ [a',b']}} & \text{Elif } S_i = j \wedge S_{i-1} = j+1 \\[2em] \displaystyle\sum_{\substack{a',b' \in \mathcal{B}^2, \\ \delta_{a'b',s_i s_k} \leq m}} \sum_{m'=0}^{m-\delta_{a'b',s_i s_k}} e^{\frac{-E^{\Omega,\beta}_{(i,k),\varnothing \to a'b'}}{RT}} \cdot \mathcal{Z}^{m-\delta_{a'b',s_i s_k}-m'}_{\substack{(i+1,k-1) \\ [a',b']}} \cdot \mathcal{Z}^{m'}_{\substack{(k+1,j) \\ [b',b]}} & \text{Elif } S_i = k \wedge i < k \leq j \\[2em] 0 & \text{Otherwise} \end{cases} \tag{3}
$$

The cases can be broken down as follows:

$S_i = -1$: If nucleotide at position i is unpaired, then any sequence is a concatenation of a, possibly mutated, nucleotide a' at position i, followed by a sequence over $[i+1,j]$ having $m - \delta_{a',s_i}$ mutations (accounting for a possible mutation at position i), and having flanking nucleotides a' and b.

$S_i = j$ and $S_{i-1} = j+1$: Any sequence generated in $[i,j]$ consists of two, possibly mutated, nucleotides a' and b', flanking a sequence over $[i+1,j-1]$ having distance $m - \delta_{a'b',s_i s_j}$. Since positions i and $i-1$ are paired with j and $j+1$ respectively, then a stacking energy contribution is added.

$S_i = k$ and $i < k \leq j$: If position i is paired and not involved in a stacking, then the only term contributing directly to the energy is the isostericity of the base pair (i,k). Any sequence on $[i,j]$ consists of two nucleotide a' and b' at positions i and k respectively, flanking a sequence over interval $[i+1,k-1]$ and preceding a (possibly empty) sequence interval $[k+1,j]$. Since the number of mutations sum to m over the whole sequence must , then a parameter m' is introduced to distribute the remaining mutations between the two sequences.

Else: In any other case, we are in a derivation of the SCFG that does not correspond to the secondary structure S, and we return 0.

Outside Computation. The *Outside* function, $\mathcal{Y}$, is the partition function considering only the contributions of subsequences $[0,i] \cup [j, n-1]$ over the mutants of s having exactly m mutations between $[0,i] \cup [j, n-1]$ and whose nucleotide at position $i+1$ is a (resp. in position $j-1$ it is b). The resulting terms $\mathcal{Y}^m_{\substack{(i,j) \\ [a,b]}}$ can be computed by recurrence, using as initial conditions:

$$
\mathcal{Y}^m_{\substack{(-1,j) \\ [X,X]}} := \mathcal{Z}^m_{\substack{(j,n-1) \\ [X,X]}}. \tag{4}
$$

The recurrence below extends the interval $[i,j]$, by including $i-1$ when position i not base paired, or extended in both directions if i is paired with a

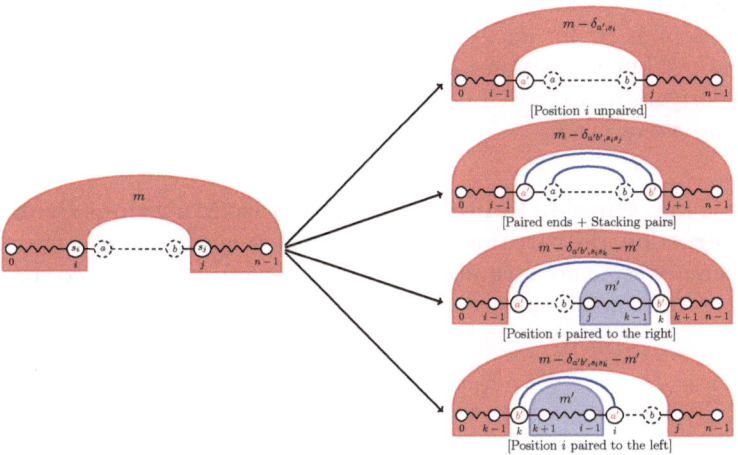

Fig. 2. Principle of the outside computation. Note that the outside algorithm uses intermediate results from the inside algorithm, therefore its efficient implementation requires a precomputation of the inside contributions.

position $k > j$. The recursion itself unfolds as follows:

$$
\mathcal{Y}^m_{\substack{(i,j)\\ [a,b]}} = \begin{cases}
\displaystyle\sum_{\substack{a'\in\mathcal{B},\\ \delta_{a',s_i}\leq m}} \mathcal{Y}^{m-\delta_{a',s_i}}_{\substack{(i-1,j)\\ [a',b]}} & \text{Elif } S_i = -1 \\[3em]
\displaystyle\sum_{\substack{a'b'\in\mathcal{B}^2,\\ \delta_{a'b',s_is_j}\leq m}} e^{\frac{-E^{\Omega,\beta}_{(i,j),ab\to a'b'}}{RT}}\cdot \mathcal{Y}^{m-\delta_{a'b',s_is_j}}_{\substack{(i-1,j+1)\\ [a',b']}} & \text{Elif } S_i = j \wedge S_{i+1} = j-1 \\[3em]
\displaystyle\sum_{\substack{a'b'\in\mathcal{B}^2,\\ \delta_{a'b',s_is_k}\leq m}} \sum_{m'=0}^{m-\delta_{a'b',s_is_k}} e^{\frac{-E^{\Omega,\beta}_{(i,k),\varnothing\to a'b'}}{RT}}\cdot \mathcal{Y}^{m-\delta_{a'b',s_is_k}-m'}_{\substack{(i-1,k+1)\\ [a',b']}}\cdot \mathcal{Z}^{m'}_{\substack{(j,k-1)\\ [b,b']}} & \text{Elif } S_i = k \geq j \\[3em]
\displaystyle\sum_{\substack{a'b'\in\mathcal{B}^2,\\ \delta_{a'b',s_ks_i}\leq m}} \sum_{m'=0}^{m-\delta_{a'b',s_ks_i}} e^{\frac{-E^{\Omega,\beta}_{(k,i),\varnothing\to a'b'}}{RT}}\cdot \mathcal{Y}^{m-\delta_{a'b',s_ks_i}-m'}_{\substack{(k-1,j)\\ [a',b]}}\cdot \mathcal{Z}^{m'}_{\substack{(k+1,i-1)\\ [a',b']}} & \text{Elif } -1 < S_i = k < i \\[3em]
0 & \text{Otherwise}
\end{cases}
\tag{5}
$$

The five cases can be broked down as follows.

$S_i = -1$: If the nucleotide at position i is not paired, then the value is the same as if we decrease the lower interval bound by 1 (i.e. $i-1$), and consider all possible nucleotides a' at position i, correcting the number of mutants in function of δ_{a',s_i}.

$S_i = j$ **and** $S_{i+1} = j-1$: If nucleotide i is paired with j and nucleotide $i+1$ is paired with $j-11$, we are in the only case were stacked base pairs can occur. We thus add the energy of the stacking and of the isostericity of the base pair (i,j). What is left to compute is the *outside* value for the interval $[i-1, j+1]$ over all possible nucleotides $a', b' \in B^2$ at positions i and j respectively.

$S_i = k \geq j$: If nucleotide i is paired with position $k \geq j$, and is not stacked inside, the only term contributing directly to the energy is the isostericity of base pair (i,k). Therefore, we consider the outside interval $[i-1, k+1]$, multiplying

it by the *inside* value of the newly included interval (i.e. $[j, k-1]$), for all possible values $a', b' \in B^2$ for nucleotides at positions i and k respectively.
$-1 < S_i < i$: As above but if position i is paired with a lower value.
Else: Any other derivation of the SCFG does not correspond to the secondary structure S, and we return 0.

Combining Inside and Outside Values into Point-Wise Mutations Probabilities. By construction, the partition function over all sequences at exactly m mutations of a reference sequence s can be either described from the *inside* contribution $\mathcal{Z}^m_{\substack{(0,n-1) \\ [X,X]}}$ of the whole sequence, $\forall X \in B$, or from *outside* terms as:

$$\mathcal{Z}^m_{\substack{(0,n-1) \\ [X,X]}} \equiv \sum_{\substack{a \in B, \\ \delta_{a,s[k]} \leq m}} \mathcal{Y}^{m-\delta_{a,s[k]}}_{\substack{(k-1,k+1) \\ [a,a]}}, \forall k \text{ unpaired.}$$

We are now left to compute the probability that a given position is a given nucleotide. We leverage the *Inside-Outside* construction to immediately obtain the following 3 cases. Given $i \in [0, n-1], x \in B$, and $M \geq 0$ a bound on the number of allowed mutations, one defines

$$\mathbb{P}(s_i = x \mid M) := \frac{\mathcal{W}^M_{i,[x]}}{\sum_{m=0}^{M} \mathcal{Z}^m_{\substack{(0,n-1) \\ [X,X]}}} \tag{6}$$

where $\mathcal{W}^*_*$ is defined by:

$$\mathcal{W}^M_{i,[x]} = \begin{cases} \sum_{m=0}^{M} \mathcal{Y}^{m-\delta_{x,s_i}}_{\substack{(i-1,i+1) \\ [x,x]}} & \text{If } S_i = -1 \\[2em] \sum_{m=0}^{M} \sum_{\substack{b \in B \\ \delta_{xb,s_i s_k} \leq m}} \sum_{m'=0}^{m-\delta_{xb,s_i s_k}} e^{\frac{-E^{\Omega,\beta}_{(i,k),\varnothing \to xb}}{RT}} \cdot \mathcal{Y}^{m-\delta_{xb,s_i s_k}-m'}_{\substack{(i-1,k+1) \\ [x,b]}} \cdot \mathcal{Z}^{m'}_{\substack{(i+1,k-1) \\ [x,b]}} & \text{If } S_i = k > i \\[2em] \sum_{m=0}^{M} \sum_{\substack{b \in B \\ \delta_{bx,s_k s_i} \leq m}} \sum_{m'=0}^{m-\delta_{bx,s_k s_i}} e^{\frac{-E^{\Omega,\beta}_{(k,i),\varnothing \to bx}}{RT}} \cdot \mathcal{Y}^{m-\delta_{bx,s_k s_i}-m'}_{\substack{(k-1,i+1) \\ [b,x]}} \cdot \mathcal{Z}^{m'}_{\substack{(k+1,i-1) \\ [b,x]}} & \text{If } S_i = k < i \end{cases}$$

$$\tag{7}$$

In every case, the denominator is the sum of the partitions function of exactly m mutations, for m smaller or equal to our target M. The numerators are divided in the following three cases.

$S_i = -1$: If the nucleotide at position i is not paired, we are concerned by the weights over all sequences which have at position i nucleotide x, which is exactly the sum of the values of $\mathcal{Y}^{m-\delta_{x,s_i}}_{\substack{(i-1,i+1) \\ [x,x]}}$, for all m between 0 and M.

$S_i = k > i$: Since we need to respect the derivation of the secondary structure S, if position i is paired, we must consider the two partition functions.

The *outside* of the base pair, and the *inside*, for all possible values for the nucleotide at position k, and all possible distribution of the mutant positions between the inside and outside of the base pair. We also add the term of isostericity for this specific base pair.

$S_i = k < i$: Same as above, but with position i pairing with a lower position.

2.3 Complexity Considerations

Equations (3) and (5) can be computed using dynamic programming. Namely, the $\mathcal{Z}_*^*$ and $\mathcal{Y}_*^*$ terms are computed starting from smaller values of m and interval lengths, memorizing the results as they become available to ensure constant-time access during later stages of the computation. Furthermore, energy terms $E(\cdot)$ can be accessed in constant time thanks to a simple precomputation (not described) of the isostericity contributions in $\Theta(n \cdot |\Omega|)$. Computing any given term therefore requires $\Theta(m)$ operations.

In principle, $\Theta(m \cdot n^2)$ terms, identified by (m, i, j) triplets, should be computed. However, a close inspection of the recurrences reveals that the computation can be safely restricted to a subset of intervals (i, j). For instance, the inside algorithm only requires computing intervals $[i, j]$ that do not break any base-pair, and whose next position $j + 1$ is either past the end of the sequence, or is base-paired prior to i. Similar constraints hold for the outside computation, resulting in a drastic limitation of the combinatorics of required computations, dropping from $\Theta(n^2)$ to $\Theta(n)$ the number of terms that need to be computed and stored. Consequently the overall complexity of the algorithm is $\Theta(n \cdot (|\Omega| + m^2))$ arithmetic operations and $\Theta(n \cdot (|\Omega| + m))$ memory.

3 Results

3.1 Implementation

The software was implemented in Python2.7 using the *mpmath* [15] library for arbitrary floating point precision. The source code is freely available at:

$$\texttt{https://github.com/McGill-CSB/RNApyro}$$

The time benchmarks were performed on a MacMini 2010, 2.3GHz dual-core Intel Core i5, 8GB of RAM. Since typical use-cases of `RNApyro` require efficiency and scalability, we present in Table 1 typical runtimes required to compute the probabilities for every nucleotide at every positions for a vast set of parameters. For those tests, both the sequences and the target secondary structure were generated at random.

3.2 Error Correction in 5s rRNA

To illustrate the potential of our algorithm, we applied our techniques to identify and correct point-wise errors in RNA sequences with conserved secondary structures. More precisely, we used `RNApyro` to reconstruct 5s rRNA sequences with

Length	#mutations		
	6	12	24
100	35s	238s	1023s
300	135s	594s	2460s
	25		50
500	5400s		21003s

Table 1. Time required by the computation of probabilities. First column indicates the length and the column indexes indicate the number of mutations. α is set at 0.5, β to 15 and $|\Omega| = 44$.

randomly distributed mutations. This experiment has been designed to suggest further applications to error-corrections in pyrosequencing data.

We built our data set from the 5S rRNA multiple sequence alignment (MSA) available in the Rfam Database 11.0 (Rfam id: RF00001). Since our software does not currently implement gaps (mainly because scoring indels is a challenging issue that cannot be fully addressed in this work), we clustered together the sequences with identical gap locations. From the 54 MSAs without gap produced, we selected the biggest MSA which contains 130 sequences (out of 712 in the original Rfam MSA). Then, in order to avoid overfitting, we used cd-hit [16] to remove sequences with more than 80% of sequence similarity. This operation resulted in a data set of 45 sequences.

We designed our benchmark using a leave-one-out strategy. We randomly picked a single sequence from our data set and performed 12 random mutations, corresponding to an error-rate of 10%. We repeated this operation 10 times. The value of β was set to 15 (larger values gave similar results). To estimate the impact on the distribution of the relative weights of energy and isostericity, we used 4 different values of $\alpha = 0, 0.5, 0.8, 1.0$. Similarly, we also investigated the impact of an under- and over- estimate of the number of errors, by setting the presumed number of errors to 50% (6 mutations) and 200% (24 mutations) of their exact number (i.e. 12).

To evaluate our method, we computed a ROC curve representing the performance of a classifier based on the mutational probabilities computed by RNApyro. More specifically, we fixed a threshold $\lambda \in [0, 1]$, and predicted an error at position i in sequence ω if and only if the probability $P(i, x)$ of a nucleotide $x \in \{A, C, G, U\}$ exceeds this threshold. To correct the errors we used the set of nucleotides having probability greated than λ, that is $C(i) = \{x \mid x \in \{A, C, G, U\}$ and $P(i, x) > \lambda$ and $n \neq \omega[i]\}$, where $\omega[i]$ is the nucleotide at position i in the input sequence. We note that, for lower thresholds, multiple nucleotides may be available in $C(i)$ to correct the sequence. Here, we remind that our aim is to estimate the potential of error-correction of RNApyro, and not to develop a full-fledged error-correction pipe-line, which shall be the subject of further studies. Finally, we progressively varied λ between 0 and 1 to calculate the ROC curve and the area under the curve (AUC). Our results are reported in Figure 3.

Our data demonstrates that our algorithm exhibits interesting potential for error-correction applications. First, the AUC values (up to 0.86) indicate that a signal has been successfully extracted. This result has been achieved with errors in loop regions (i.e. without base pairing information) and thus suggests that

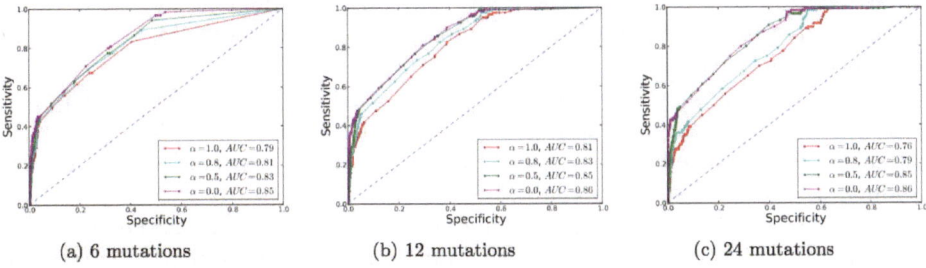

Fig. 3. Performance of error-correction. Subfigures show accuracy with under-estimated error rates (6 mutations), exact estimates (12 mutations) and over estimates (24 mutations). We also analyze the impact of the parameter α distributing the weights of stacking pair energies vs isostericity scores and use values ranging of $\alpha = \{0, 0.5, 0.8, 1.0\}$. The AUC is indicated in the legend of the figures. Each individual ROC curve represent the average performance over the 10 experiments.

correction rates in structured regions (i.e. base paired regions) could be even higher. Next, the optimal values of α tend to be close to 0.0. This finding suggests that, at this point, the information issued from the consideration of stacking energies is currently modest. However, specific examples showed improved performance using this energy term. Further studies must be conducted to understand how to make the best use of it. Finally, our algorithm seems robust to the number of presumed mutations. Indeed, good AUC values are achieved even with conservative estimates for the number of errors (c.f. 50% of the errors, leading to Fig. 3(a)), as well as with large values (cf 200% of the errors in Fig. 3(c)). It is worth noting that scoring schemes giving a larger weight on the isostericity scores (i.e. for low α values) seem more robust to under- and over-estimating the number of errors.

4   Conclusion

In this article we presented a new and efficient way of exploring the mutational landscape of an RNA under structural constraints, and apply our techniques to identify and fix sequencing errors. In addition, we introduce a new scoring scheme combining the nearest-neighbour energy model to new isostericity matrices in order to account for geometrical discrepancies occurring during base pair replacements. The algorithm runs in $\Theta(n \cdot (|\Omega| + m^2))$ time and $\Theta(n \cdot (|\Omega| + m))$ memory, where n is the length of the RNA, m the number of mutations and Ω the size of the multiple sequence alignment.

By combining into `RNApyro` these two approaches, the mutational landscape exploration and the pseudo energy model, we created a tool predicting the positions yielding point-wise sequencing error and correcting them. We validated our model with the 5s rRNA, as presented in Sec. 3. We observed that the models with larger weights on the isostericity seems to hold a higher accuracy on the estimation of errors. This indicates that an exploitable signal is captured by the isostericity. Importantly, the implementation is fast enough for practical applications.

We must recall that our approach is restricted to the correction of pointwise error in structured regions (i.e. base paired nucleotides). Nonetheless it should supplement well existing tools, by using previously discarded information holding, as shown, a strong signal.

Further research, given the potential of error-correction of RNApyro, will evaluate its impact over large datasets with different existing NGS error-correction pipe-line.

Acknowledgments. The authors would like to thank Rob Knight for his suggestions and comments. This work was funded by the French Agence Nationale de la Recherche (ANR) through the MAGNUM ANR 2010 BLAN 0204 project (to YP), the FQRNT team grant 232983 (to VR and JW) and NSERC Discovery grant 219671 (to JW).

References

[1] Zuckerkandl, E., Pauling, L.: Molecules as documents of evolutionary history. Journal of Theoretical Biology 8(2), 357–366 (1965)
[2] Olsen, G.J., Lane, D.J., Giovannoni, S.J., Pace, N.R., Stahl, D.A.: Microbial ecology and evolution: a ribosomal RNA approach. Annual Review of Microbiology 40, 337–365 (1986)
[3] Olsen, G., Woese, C.: Ribosomal RNA: a key to phylogeny. The FASEB Journal 7(1), 113–123 (1993)
[4] Turnbaugh, P.J., Ley, R.E., Hamady, M., Fraser-Liggett, C.M., Knight, R., Gordon, J.I.: The Human Microbiome Project. Nature 449(7164), 804–810 (2007)
[5] Quince, C., Lanzén, A., Curtis, T.P., Davenport, R.J., Hall, N., Head, I.M., Read, L.F., Sloan, W.T.: Accurate determination of microbial diversity from 454 pyrosequencing data. Nat. Methods 6(9), 639–641 (2009)
[6] Kunin, V., Engelbrektson, A., Ochman, H., Hugenholtz, P.: Wrinkles in the rare biosphere: pyrosequencing errors can lead to artificial inflation of diversity estimates. Environmental Microbiology 12(1), 118–123 (2010)
[7] Quinlan, A., Stewart, D., Strömberg, M., Marth, G.: Pyrobayes: an improved base caller for SNP discovery in pyrosequences. Nature Methods 5(2), 179–181 (2008)
[8] Medvedev, P., Scott, E., Kakaradov, B., Pevzner, P.: Error correction of high-throughput sequencing datasets with non-uniform coverage. Bioinformatics 27(13), i137–i141 (2011)
[9] Waldispühl, J., Devadas, S., Berger, B., Clote, P.: Efficient Algorithms for Probing the RNA Mutation Landscape. PLoS Computational Biology 4(8), e1000124 (2008)
[10] Turner, D.H., Mathews, D.H.: NNDB: the nearest neighbor parameter database for predicting stability of nucleic acid secondary structure. Nucleic Acids Research 38(Database issue), D280–D282 (2010)
[11] Stombaugh, J., Zirbel, C.L., Westhof, E., Leontis, N.B.: Frequency and isostericity of RNA base pairs. Nucleic Acids Research 37(7), 2294–2312 (2009)

[12] Leontis, N.B., Westhof, E.: Geometric nomenclature and classification of RNA base pairs. RNA 7(4), 499–512 (2001)

[13] Hori, H., Osawa, S.: Origin and Evolution of Organisms as Deduced from 5s Ribosomal RNA Sequences. Molecular Biology and Evolution 4(5), 445–472 (1987)

[14] Lari, K., Young, S.: The estimation of stochastic context-free grammars using the Inside-Outside algorithm. Computer Speech & Language 4(1), 35–56 (1990)

[15] Johansson, F., et al.: mpmath: a Python library for arbitrary-precision floating-point arithmetic (version 0.14) (February 2010),
http://code.google.com/p/mpmath/

[16] Li, W., Godzik, A.: Cd-hit: a fast program for clustering and comparing large sets of protein or nucleotide sequences. Bioinformatics 22(13), 1658–1659 (2006)

An Accurate Method for Inferring Relatedness in Large Datasets of Unphased Genotypes via an Embedded Likelihood-Ratio Test

Jesse M. Rodriguez[*,1,2], Serafim Batzoglou[1], and Sivan Bercovici[1]

[1] Department of Computer Science, Stanford University
[2] Biomedical Informatics Program, Stanford University
jesserod@cs.stanford.edu

Abstract. Studies that map disease genes rely on accurate annotations that indicate whether individuals in the studied cohorts are related to each other or not. For example, in genome-wide association studies, the cohort members are assumed to be unrelated to one another. Investigators can correct for individuals in a cohort with previously-unknown shared familial descent by detecting genomic segments that are shared between them, which are considered to be identical by descent (IBD). Alternatively, elevated frequencies of IBD segments near a particular locus among affected individuals can be indicative of a disease-associated gene. As genotyping studies grow to use increasingly large sample sizes and meta-analyses begin to include many data sets, accurate and efficient detection of hidden relatedness becomes a challenge. To enable disease-mapping studies of increasingly large cohorts, a fast and accurate method to detect IBD segments is required.

We present PARENTE, a novel method for detecting related pairs of individuals and shared haplotypic segments within these pairs. PARENTE is a computationally-efficient method based on an embedded likelihood ratio test. As demonstrated by the results of our simulations, our method exhibits better accuracy than the current state of the art, and can be used for the analysis of large genotyped cohorts. PARENTE's higher accuracy becomes even more significant in more challenging scenarios, such as detecting shorter IBD segments or when an extremely low false-positive rate is required. PARENTE is publicly and freely available at *http://parente.stanford.edu/*.

Keywords: Population genetics, IBD, relatedness.

1 Introduction

Genomic sequence variants such as single-nucleotide variants, insertions, and deletions, are being constantly introduced to populations with each generation. As mutation rates are considered to be relatively low, [10] and as genetic drift drives allele frequencies to become fixed, it is reasonable to assume that two

---

[*] Corresponding author.

M. Deng et al. (Eds.): RECOMB 2013, LNBI 7821, pp. 212–229, 2013.
© Springer-Verlag Berlin Heidelberg 2013

individuals carrying the same allele have actually inherited it from a common ancestor; in such a case, the alleles can be said to be identical-by-descent (IBD). This strict definition of IBD holds for the majority of evident human germline mutations, and with high probability. Many biological applications, however, are driven by the study of longer shared stretches that cover multiple mutations. Using knowledge of such longer shared segments, inferences can be made regarding ancestry [27], population demographics [15, 19, 23], and perhaps more important, the location of disease susceptibility genes [2, 22, 4]. For such applications, the alleles of two individuals that were inherited from a *recent* common ancestor are called IBD, whereas the alleles that simply have the same allelic state but did not originate from a recent common ancestor are called identical-in-state (IIS). Note that alleles that are IBD are also IIS, but multiple independent mutation events can cause two alleles to be IIS but not IBD. It follows that in the case of a recent common ancestor, IBD alleles are harbored within longer segments containing additional IBD alleles; the more recent the common ancestor, the fewer meiosis occurred, and the longer the shared segment. In this work, we describe two individuals as being *related* to one another if they share an IBD segment from a recent common ancestor.

Identity-by-descent (IBD) inference is defined as the process of detecting genomic segments that were inherited from recent common ancestors in a given set of genotyped individuals. In the problem's simplest form, a pedigree describing the connection between sampled individuals is provided with the genotypes in order to identify the segments. Given the pedigree, a model can be derived to explicitly capture these relationships when the genotypes are examined. The most common model used is based on a factorial hidden Markov model (factorial-HMM) [26, 12] with a hidden state space defined by selector variables that determine the inheritance pattern in the pedigree [18, 1, 13, 16, 20]. More recently, such methods were extended to model linkage disequilibrium (LD) between neighboring markers, enabling the detection of shorter IBD segments [4]. The main use of these models is in the application of genetic linkage analysis. When a hereditary disease is studied in a family of healthy and affected individuals, linkage analysis is applied to identify loci that are associated with the hereditary disease; these loci may contain genes or regulatory elements that increase the probability of having the disease. The premise of linkage analysis is that affected individuals will share an IBD segment around the disease locus, and that this segment is not shared (or less likely to be shared) by healthy individuals [11, 18, 9, 24].

In the large majority of hereditary disease studies, however, the relationship between sampled individuals is unknown. In genome-wide association studies (GWAS), sampled individuals are assumed to be *unrelated*. However, it is common to have hidden relationships (also known as cryptic relationships) within large sampled cohorts [5, 15, 17].

The accurate detection of IBD segments within these samples enables the correction for the cryptic relationships, for example, by removing related individuals from analysis. Conversely, instead of discarding related individuals, IBD mapping [7, 22, 4] can be applied, directly associating the levels of IBD with phenotype in the process of mapping disease susceptibility genes.

Extensive previous work has focused on developing methods for the accurate detection of IBD segments without using pedigree information. Most commonly, an HMM or a factorial-HMM is applied to infer the IBD segments. Purcell et al. presented PLINK [25], which uses a simple three-state model, counting the occurrences of IBD per position given the observed genotypes of two individuals. In BEAGLE, by Browning and Browning [8], a factorial HMM was developed to phase and simultaneously detect the specific haplotypes that are shared between examined individuals. To improve accuracy, the BEAGLE model captured complex linkage-disequilibrium patterns by extending the state space to accommodate the haplotypic structure found in the data and measuring the patterns' frequencies. In the work by Bercovici et al. the inheritance vector capturing the relationship between two individuals was explicitly modeled, and LD was incorporated via a first-order Markov model at the level of the founders [4]. The explicit modeling of both relationship and LD was shown to significantly improve performance. Similar to others, the work further demonstrated that these accurate inference methods could be used to detect the IBD enrichment evident around disease-gene loci, highlighting the value of IBD detection in the mapping of disease susceptibility genes. Moltke et al. presented a Markov Chain Monte Carlo (MCMC) approach for the detection of IBD regions where segments of chromosomes are it iteratively partitioned into sets of identical descent [22]. In the above methods, there exists a tradeoff between accuracy and running time. Nonetheless, in most of the above methods, the complexity of the analysis in all these methods is quadratic in the number of individuals. Simply, every pair of individuals must be examined for relatedness. GERMLINE, by Gusev et al. aimed to reduce the time complexity of IBD inference at the cost of lower accuracy [14]. The GERMLINE method performs the IBD analysis on phased data. By populating hash tables with segments taken from the phased data, the method efficiently determines potential seeds of segments that are shared between individuals. These segments are then extended to determine if sufficient evidence exists to support IBD between specific pairs of individuals. As GERMLINE requires phased data in order to operate, the individuals are first phased using BEAGLE [6]. In a later work by Browning and Browning, fastIBD [5] was developed to efficiently determine IBD segments between pairs of individuals in large cohorts of thousands of samples in a feasible timeframe. Similar to GERMLINE, fastIBD employed a sliding window approach to allow efficient computation. Pairs of individuals sharing the same state in fastIBD's factorial HMM are considered in the evaluation of subsequent windows; shared segments are extended for pairs of individuals with a high probability of IBD. While GERMLINE provides a more time-efficient solution, previous work has shown the method to have a reduced ability to detect more ancient IBD segments in comparison to more accurate methods such as fastIBD. As phasing can be prohibitive when analyzing extremely large datasets, Henn et al. developed a method aimed at detecting larger IBD segments based on reverse-homozygous positions that does not require phasing [15]. While providing an efficient approach for IBD detection, the method is tuned to detect larger IBD segments, in order to achieve required specificity.

While advances in IBD detection have been made in recent years, accurately detecting IBD in large cohorts remains a challenge. As the cost of genotyping decreases, the number of genotyped individuals is increasing rapidly, and the genotyping density is growing to include millions of markers per sample. Since many of the accurate methods investigate all pairs of individuals for relatedness, the analysis complexity grows quadratically with the number of individuals in a studied sample. Such challenges require that IBD detection methods have high computational efficiency. More importantly, since the vast majority of examined pairs of individuals are unlikely to be related, an IBD detection method must exhibit extremely high specificity in order to avoid reporting an overwhelming number of false positives.

In this paper we present PARENTE, a novel method for the detection of IBD that exhibits high accuracy, and can be efficiently used for the analysis of large genotyped cohorts. PARENTE employs a variant of a likelihood-ratio test along with local thresholding to achieve significantly higher accuracy than the current state of the art. Our method can be applied directly on genotype data, without needing to first phase the genotypes, a step that can be computationally-intensive. The primary goal of our method is to efficiently detect which pairs of individuals in large corhorts are *related* to one another, in feasible time. This is done by finding pairs of individuals that share at least one IBD segment greater than x cM in size. Once these related pairs are identified, one can determine specific IBD segment boundaries as a post-processing step using a more complex IBD detection method of higher computational cost. We further show that PARENTE can also be directly used for the localization of the IBD segments within the related pairs, providing highly accurate results. PARENTE was able to successfully detect pairs of related individuals sharing a 6 cM IBD segment (the expected average IBD segment size for 7th cousins) with 90% sensitivity at a 5×10^{-5} false positive rate. In the more challenging case of a 4 cM shared segment, it detects related pairs with 86% sensitivity at a 8×10^{-3} false positive rate, which represents a 28% relative increase in sensitivity compared to fastIBD, a state-of-the-art method. Finally, we observed that PARENTE is an order of magnitude faster than fastIBD, as well. These results highlight the relevance of our method for the accurate and efficient analysis of large cohorts.

2   Methods

The PARENTE model employs a window-based approach, whereby multiple consecutive markers are grouped together and their joint probability is estimated given a hypothesized IBD state. Subsequently, the probabilities of multiple non-overlapping windows are merged via a naive Bayes model, producing the probability for the assumed IBD state in a given block of pre-defined length. The block lengths are derived from a target timespan covering common ancestors of interest, and the required accuracy as driven by the application.

Given N individuals sampled over M biallelic markers, let G be defined as the genotype matrix. We use $g_{i,j} \in \{0, 1, 2\}$ to denote the major allele count observed in the jth marker of the ith individual, and g_i as the vector corresponding to all M genotyped markers sampled for individual i. The measured genotypes G are assumed to have originated from a set of $2N$ underlying hidden haplotypes, denoted by the matrix H. The maternal and paternal alleles of the jth marker in the ith individual are marked as $h_{i,j}^m \in \{0, 1\}$ and $h_{i,j}^p \in \{0, 1\}$, respectively, corresponding to the major allele count in each. More broadly, however, we use h_j^* as a symbol to signify one of the alleles at the jth marker, corresponding to one of the population haplotypes comprising an individual's genotype. We use f_j to denote the major allele frequency of the jth marker in the sampled population. The M markers covering the genome are partitioned into a set of consecutive windows $W = \{w_1, ..., w_{\frac{M}{k}}\}$, each of size k. We use $m(w)$ to denote the indices of the k consecutive markers within the wth window, and $g_{i,m(w)}$ as the partial genotyping vector for individual i corresponding to these k markers. Finally, we define a block $B = \{w_t, ..., w_{t+k-1}\}$ as a set of consecutive windows.

For a target IBD block length l (in cM), the PARENTE method is defined as follows. All $\binom{N}{2}$ pairs of individuals are enumerated. For each pair of individuals, the genome is scanned by sliding a block B across each chromosome, where each block B starts from one of the $\frac{M}{k}$ possible window positions. The examined block B includes all successive windows that contain markers that are at most l cM away from the first marker of the first window in that block. For each such block B and pair of individuals i, i', an aggregated block score $\Lambda_B(g_i, g_{i'})$ is defined as follows:

$$\Lambda_B(g_i, g_{i'}) = \sum_{w \in B} \log s_w(g_{i,m(w)}, g_{i',m(w)}) \tag{1}$$

where $s_w(g_{i,m(w)}, g_{i',m(w)})$ is a window-specific score, computed using the genotypes of the two examined individuals i, i' within an examined window w. We call a pair of individuals i and i' to be IBD in block B whenever $\Lambda_B(g_i, g_{i'}) > T_B$, where T_B is a pre-defined threshold associated with block B. We compute this score for each block in the genome and call a pair of individuals to be *related* if any block in the genome is called to be IBD. The threshold T_B is defined such that the false-positive rate is controlled to a desired level. The block score $\Lambda_B(g_i, g_{i'})$ can be efficiently computed along the genome of two individuals. As blocks are scanned, window-scores corresponding to windows that are no longer part of the newly examined block B' are subtracted from the current block score $\Lambda_{B'}(g_i, g_{i'})$, and the window-scores corresponding to newly joining windows are simply added.

In the remainder of this section we derive two instantiations for the score function $s_w(g_{i,m(w)}, g_{i',m(w)})$. We first derive a score function s_w using a likelihood-ratio approach. We continue by deriving an embedded likelihood-ratio score which corrects for the reduced performance stemming from windows exhibiting high variance in the likelihood-ratio score. Finally, we will describe how the block-specific score threshold T_B is defined. In the Results section, we show that higher variance is associated with windows that have reduced ability to distinguish between genotypes originating from related individuals from those originating from unrelated individuals.

2.1   Likelihood Ratio Test

To efficiently detect IBD, we first develop a likelihood ratio-test (LRT) variant of our method. Within a sliding block comparing two individuals' genotypes, we contrast the probability that the they are IBD in the block against the probability that they are are not IBD. The LRT score is computed by estimating the likelihood of the individuals' genotypes within each block under two models, namely a model M_{IBD} corresponding to the hypothesis the two examined individuals are related, and a model $M_{\overline{\mathrm{IBD}}}$ corresponding to the hypothesis the two individuals are unrelated.

As suggested by Equation 1, for both M_{IBD} and $M_{\overline{\mathrm{IBD}}}$, we model the genotypes within a block B using a naive Bayes approach whereby all windows are independent given the IBD status of the two examined individuals within B. The probabilities of the genotypes within each window $w \in B$ comprising an examined block B are considered separately, and the product of these probabilities defines the probability of the observed genotypes within the examined block (or as a sum, under our log formulation). Namely, given a block of interest B, and the genotype of two examined individuals g_i and $g_{i'}$, the window-specific score in Equation 1 is defined as:

$$s_w^{\mathrm{LR}}(g_{i,m(w)}, g_{i',m(w)}) = \frac{p_{M_{\mathrm{IBD}}}(g_{i,m(w)}, g_{i',m(w)})}{p_{M_{\overline{\mathrm{IBD}}}}(g_{i,m(w)}, g_{i',m(w)})} \qquad (2)$$

Under the assumption that the sampled markers are in linkage equilibrium, meaning that the alleles within a window are not associated, the genotype probabilities under the two models are given by:

$$p_{M_{\mathrm{IBD}}}(g_{i,m(w)}, g_{i',m(w)}) = \prod_{j \in m(w)} p_{M_{\mathrm{IBD}}}(g_{i,j}, g_{i',j}) \qquad (3)$$

$$p_{M_{\overline{\mathrm{IBD}}}}(g_{i,m(w)}, g_{i',m(w)}) = \prod_{j \in m(w)} p_{M_{\overline{\mathrm{IBD}}}}(g_{i,j}, g_{i',j}).$$

The probability of the genotype pair $g_{i,j}, g_{i',j}$ under our two models is then defined as:

$$p_{M_{\mathrm{IBD}}}(g_{i,j}, g_{i',j}) = \sum_{h_j^1, h_j^2, h_j^3} p(g_{i,j}|h_j^1, h_j^2) \cdot p(g_{i',j}|h_j^1, h_j^3) \cdot p(h_j^1) \cdot p(h_j^2) \cdot p(h_j^3) \qquad (4)$$

$$p_{M_{\overline{\mathrm{IBD}}}}(g_{i,j}, g_{i',j}) = \sum_{h_j^1, h_j^2, h_j^3, h_j^4} p(g_{i,j}|h_j^1, h_j^2) \cdot p(g_{i',j}|h_j^3, h_j^4) \cdot p(h_j^1) \cdot p(h_j^2) \cdot p(h_j^3) \cdot p(h_j^4)$$

where $p(h_j^*) = f_j^{h_j^*} \cdot (1-f_j)^{(1-h_j^*)}$ as determined by the allele frequency at marker f_j. The probability $p(g_{i,j}|h_j^1, h_j^2)$ that the genotype $g_{i,j}$ was sampled given the underlying haplotypes h_j^1 and h_j^2, must accommodate for genotyping errors. We define $p(g_{i,j}|h_j^1, h_j^2)$ as follows:

$$p(g_{i,j}|h_j^1, h_j^2) = \begin{cases} 1 - \epsilon & g_{i,j} = h_j^1 + h_j^2 \\ \frac{\epsilon}{2} & \text{otherwise} \end{cases} \qquad (5)$$

where the parameter ϵ is tuned to capture the amount of expected genotyping error. Finally, to accommodate for missing data, we set the likelihood ratio at a marker to 0.5 if either genotype is missing.

We note that in the above model, the individuals can share at most a single haplotype. We further note that under the assumption of linkage equilibrium, the equivalent of a block LRT score $\Lambda_B(g_i, g_{i'})$ can be directly computed without windows by using the sums of log of the genotype probabilities, as defined by Equation 4. We utilize the window-based s_w formulation described in Equation 2 to facilitate our description of an extension that accounts for local score variability, which we now derive.

2.2   Embedded Likelihood Ratio Test

The model described thus far provides an efficient approach to identifying pairs of individuals that share a common ancestor, and in particular to detecting specific regions that are IBD. While alleviating some of the performance-related challenges that are evident when examining large cohorts by providing a computationally feasible approach, the model is sensitive to windows exhibiting highly variable scores. Namely, for each block, the window-score of a small sub-set of windows plays a critical role in the determination of the final block score. It is the high variability of such windows that limits the performance of the likelihood-ratio based test.

One approach that corrects for the detrimental impact of high-variance windows is based on the direct examination of window-level performance. The distribution of window-score can be examined given the genotypes from unrelated individuals, and contrasted against the distribution of the window-score given genotypes from related individuals. By contrasting these distributions, it is possible to detect and control for the impact of highly-variable windows. Specifically, to apply such a correction, we treat the LR described by Equation 2 as a random variable $S_w^{LR} = s_w^{\text{LR}}(g_{i,m(w)}, g_{i',m(w)})$. We then define two Gaussian models for the distribution of S_w^{LR}, one corresponding to the distribution of the score under related individuals, and a second corresponding to the distribution of the score given unrelated individuals:

$$S_w^{LR}|IBD \sim N(\mu_{w,\text{IBD}}, \sigma_{w,\text{IBD}}), \quad S_w^{LR}|\overline{\text{IBD}} \sim N(\mu_{w,\overline{\text{IBD}}}, \sigma_{w,\overline{\text{IBD}}}). \qquad (6)$$

Our modified score, which we term *embedded likelihood-ratio* (ELR), is finally defined as:

$$s_w^{ELR}(g_{i,m(w)}, g_{i',m(w)}) = \frac{P(S_w^{LR} = s_w^{\mathrm{LR}}(g_{i,m(w)}, g_{i',m(w)}) | \mathrm{IBD})}{P(S_w^{LR} = s_w^{\mathrm{LR}}(g_{i,m(w)}, g_{i',m(w)}) | \overline{\mathrm{IBD}})}. \tag{7}$$

In total, 4 additional parameters define our new model. Namely, the mean μ and standard deviation σ of the normal distributions used to approximate the behavior of our initial score s_B^{LR} under observations originating from related and unrelated individuals. In order to estimate these parameters, phased data is used to simulate related and unrelated individuals, yielding the means to compute empirical estimates for the score distributions. The phased haplotypes can be either generated from datasets containing trios, or via computationally-phased individuals. It is important to note that current phasing methods offer a sufficiently low switch-error rate such that their performance should have a negligible effect when considering haplotypes within a window of moderate size.

2.3   Genotyping-Error Function

In Equation 5 we describe the probability of genotypes given the hidden underlying haplotype. The conditional probability $p(g_{i,j}|h_j^1, h_j^2)$ derived accounts for genotyping error. While providing a more realistic model, it can in fact reduce the statistical power when failing to reject unrelated individuals. The lower power stems from the fact the impact of reverse-homozygous genotypes is reduced; such observations can be attributed to sampling errors rather than indication of unrelatedness under the realistic model. One can increase the penalty under such scenarios by controlling the genotyping error parameter ϵ. Our method strives to reduce the amount of false-positive pairs detected. Thus, we extend our method by introducing a genotyping-error function that increases the contrast between IBD and non-IBD segments. Specifically, when estimating the model parameters, we use ϵ as the genotyping error rate, whereas during inference, we replace ϵ in Equation 5 with a function $\phi(\epsilon) = v \cdot \epsilon$, where v is a scaling factor. In the Result section, we used $v = \frac{1}{100}$.

2.4   Likelihood-Ratio Test Threshold

When applying likelihood-ratio tests, thresholds are selected so as to control the false-positive rate. Specifically, the distribution of the test is examined under examples originating from the null distribution, and a threshold is selected to guarantee an expected performance in terms of false-positives. It is common to select a single, global threshold to control for the global proportion of type I errors. However, as each block in our method contains windows of different score distribution, a local, block-specific threshold T_B can be applied to improve the performance. In our method, we explore the distribution of $\Lambda_B(g_i, g_{i'})$ given the genotypes of unrelated individuals for each block, thus accommodating to the local behavior of our score. Given a training set of unrelated pairs and their corresponding block scores $D_{b,\overline{IBD}}$, we define the block threshold as:

$$T_B = \max(D_{b,\overline{IBD}}) + c\sigma_{D_{b,\overline{IBD}}} \tag{8}$$

where $\sigma_{D_{b,\overline{IBD}}}$ is the standard deviation observed in the block-scores, and c scales the margin defined by the standard deviation. In our experiments, we use values between -1.5 and 2.5 for the scaling-factor c.

In the Results section, we demonstrate that the combination of ELR and a block-specific threshold T_B provides superior performance in comparison to current state-of-the-art methods.

3   Results

The performance of PARENTE was evaluated using simulated data. We show that PARENTE has a superior accuracy performance when compared against fastIBD, which is considered state-of-the-art method for the accurate and efficient detection of IBD. We further explore the relative contribution to performance stemming from the use of the likelihood-ratio approach (LRT), the embedded LRT (ELRT) approach, and finally the use of a local threshold versus a global threshold. As a note on notation, for the remainder of this paper, we present the *window score* as $\log s_w(g_{i,m(w)}, g_{i',m(w)})$ instead of $s_w(g_{i,m(w)}, g_{i',m(w)})$.

Constructing Training and Testing Datasets. To train and evaluate the performance of PARENTE, we used the phased data from three Asian populations of the the HapMap Phase III panel [3]: Han Chinese in Beijing, China (CHB); Japanese in Tokyo, Japan (JPT); and Chinese in Metropolitan Denver, Colorado (CHB). Our experiments used polymorphic SNPs from the long arm of human chromosome 1. We randomly partitioned the unrelated individuals from these populations into a set of 154 training haplotypes and a set of 366 testing haplotypes. To create a larger dataset of unrelated individuals, we used the original haplotypes to generate *composite haplotypes* by simulating mosaics of the original haplotypes using an approach similar to [8]. Briefly, to generate a composite haplotype, we considered every 0.2 cM segment across the chromosome; for each segment, we copied the corresponding segment from one of the original haplotypes chosen uniformly at random. Due to the random process, some longer segments of two composite haplotypes were copied from the same original haplotype. Therefore, we removed 36 composite haplotypes that had more than 0.8 cM of contiguous sequence that was generated from the same original haplotype as another composite haplotype. A total of 500 composite training haplotypes and 1, 000 composite testing haplotypes were generated. In all of our experiments we use these composite haplotypes for training and testing. Thus, henceforth, we will refer to these composite haplotypes as simply training and testing haplotypes.

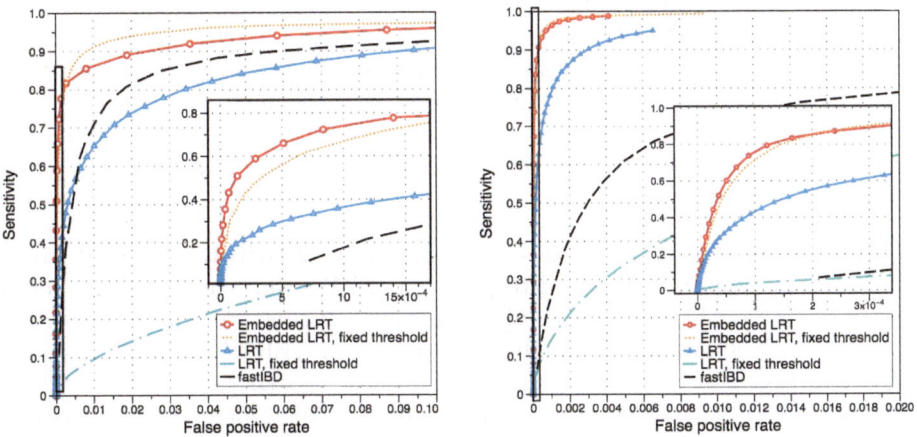

Fig. 1. (a) Performance of PARENTE for detecting related pairs of individuals sharing 4 cM IBD segments in comparison to fastIBD. PARENTE was applied using three different strategies: LRT, LRT with local thresholding, and ELRT. The magnified inset highlights PARENTE's superior performance when considering the high-specificity range. (b) Performance of PARENTE for detecting IBD segments compared to fastIBD. The same experiments from (a) were used, but the sensitivity and false positive rate were calculated based on the number of SNPs in IBD and non-IBD segments. Similarly, the magnified inset highlights PARENTE's superior performance in the high-specificity range.

Simulations to Evaluate Performance. To evaluate and characterize the performance of PARENTE, we created simulated pairs of related individuals that shared a single IBD segment of a specific size, ranging between 3 and 8 cM. We used a bootstrap approach to measure accuracy, using 100 trials per experiment, averaging the results of all trials within an experiment. For each trial, we simulated 80 pairs of related individuals by generating 80 pairs of composite individuals and inserting one shared IBD segment of a given size at a random position along the chromosome. After genotypes were copied and IBD was injected, a genotypic error rate of $\epsilon = 0.005$ was applied, changing the genotype call to one of the other two genotypes with equal probability. We designated the first simulated individual of each pair to be a *query* individual and the second individual as the *database* individual. Then we used PARENTE to predict whether IBD existed between each query individual and all database individuals by labeling a pair as IBD if at least one block had a score passing the block-specific threshold. We calculated sensitivity as the number of IBD pairs correctly predicted out of 8,000 true IBD pairs per experiment, and false positive rates as the number of non-IBD pairs incorrectly predicted as IBD out of the 632,000 non-IBD pairs per experiment.

When aiming to detect IBD segments of a particular length L (in cM), we defined the blocks to have the largest size possible l such that $L - 0.5 \leq l \leq L - 0.1$. We used block sizes slightly smaller than the target IBD segment size to account issues related to block-boundary, stemming from the varying density of the SNP array and the fact that blocks start at window boundaries (and not at arbitrary SNPs). This was done to increase the likelihood that at least one block fit completely within the any arbitrary IBD segment of length L.

In all our experiments, we used a window size of $k = 20$ SNPs per window, and simulated a single 4 cM segment for each related pair of individuals, except where stated otherwise.

PARENTE's Accuracy and Comparison to Fastibd. Our goal was to produce a fast, accurate method to predict IBD. We thus compared the performance of PARENTE to fastIBD [5], an efficient IBD detection method. fastIBD was previously shown to have higher accuracy than GERMLINE [14], a scalable IBD detection platform, and comparable accuracy to BEAGLE's slower, high-accuracy IBD inference method [8]. We evaluated the performance of fastIBD on our simulated dataset using the default parameters and IBD detection thresholds ranging from 1×10^{-6} to 1×10^{-30}. Following fastIBD's authors recommendations, we ran fastIBD ten times with ten different seeds and aggregated the results by taking the minimum score observed at each position in any of the runs. We applied a size filter to the fastIBD predictions, only considering called segments longer than 1 cM, a value selected for yielding the best performance for fastIBD. fastIBD further recommends providing additional genotypes to aid in training fastIBD's internal haplotype model. Our experiments indicate that the use of additional haplotypes did not increase the performance (results not shown). As fastIBD infers IBD segments from all pairs in a given cohort, all the query and database individuals was provided simultaneously, while only considering calls that were made between query and database individuals, following PARENTE's mode of operation.

To compare the accuracy of PARENTE and fastIBD, we performed the simulations described above, measuring accuracy on detecting which pairs of individuals shared a simulated 4 cM IBD segment. The results shown in Figure 1a demonstrate that PARENTE has a significantly higher accuracy in comparison to fastIBD when detecting pairs of related individuals. This difference in sensitivity further grows at high-specificity levels, which is a crucial parameter when analyzing large cohorts. Note that the use of a local threshold for the ELRT provides superior high-specificity performance over a global threshold strategy. In the case of the LRT, the local threshold provides a large increase in sensitivity at all specificity levels. We further compared the performance of PARENTE and fastIBD in the task of accurately determining the location and boundaries of IBD segments (see Figure 1b). Our experiments demonstrate that PARENTE achieves higher per-SNP, per-pair accuracy when compared to fastIBD. We note that when running fastIBD for this analysis we did not enforce the called segment size filter, as fastIBD performed better when the filter was not applied. The sensitivity for each related pair of individuals was measured as the fraction SNPs in the simulated IBD segment successfully detected to be IBD. For all pairs in the experiment, we measured the false positive rate as the fraction of

SNPs not in IBD segments that were incorrectly called as IBD. Since blocks can overlap in PARENTE, we labeled a SNP as IBD if it belonged to any block that had a score above the threshold.

We characterized PARENTE performance on a range of simulated IBD segment sizes from 3 cM to 8 cM, as depicted in Figure 2. These results show that PARENTE excels at high-specificity detection of IBD segments. For instance, PARENTE was able to successfully detect 8 cM IBD segments with 94% sensitivity and nearly zero false positive rate, and 6 cM IBD segments with 90% sensitivity and a 5×10^{-5} false positive rate.

As efficiency is key in the analysis of large cohorts, we measure execution time. In our experiments, the running time for PARENTE was approximately 10 times less than that of fastIBD. Specifically, PARENTE was able to process ~15 individual pairs per second on our trials of 6,400 pairs. Note that we measured running time in pairs per second as fastIBD analyzes all pairs within a cohort, whereas PARENTE was run on all pairings between query and database individuals.

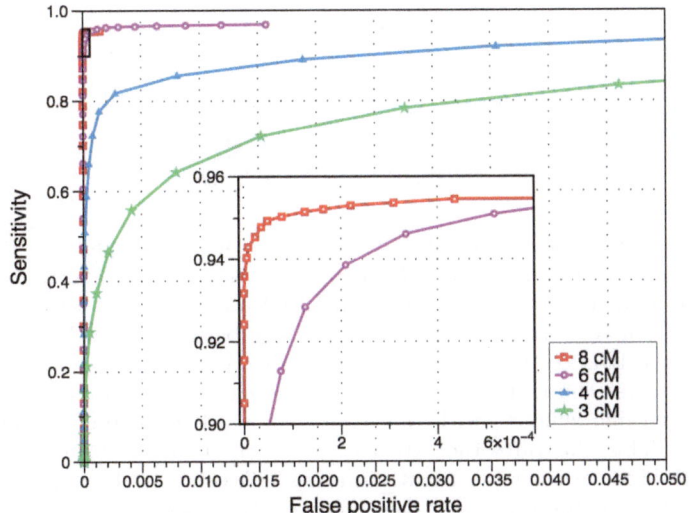

Fig. 2. Performance of PARENTE for detecting related pairs of individuals sharing IBD segments of various sizes. The magnified inset shows PARENTE's high sensitivity achieved at near-zero false positive rates for larger IBD segments.

Training PARENTE's Model and Thresholds. In order to compute our embedded LRT score, P_{IBD} and $P_{\overline{\text{IBD}}}$ first need to be evaluated for every window w. Simulated pairs of related and unrelated individuals was used for this process (see Equations 6,7). Simulated pairs of related individuals' genotypes were simulated so that each pair shared one entire haplotype along the chromosome. Specifically, each pair of related genotypes was generated by randomly selecting one haplotype from the training data to be shared by both genotypes as well as a unique haplotype for each genotype so that three distinct haplotypes were sampled. Pair of unrelated genotypes were simulated by randomly choosing four distinct training haplotypes, using two of the haplotypes for one genotype and the remaining two haplotypes for the second genotype. A total of $2,000$ pairs of related genotypes and $2,000$ pairs of unrelated genotypes were generated. For each window w and each pair of related and unrelated genotypes, we computed the LRT score assuming a genotyping error rate of $\epsilon = 0.005$; we then fit window-specific normal distributions to the scores of related and unrelated pairs resulting in $(\mu_{w,\mathrm{IBD}}, \sigma_{w,\mathrm{IBD}})$ and $(\mu_{w,\overline{\mathrm{IBD}}}, \sigma_{w,\overline{\mathrm{IBD}}})$, respectively.

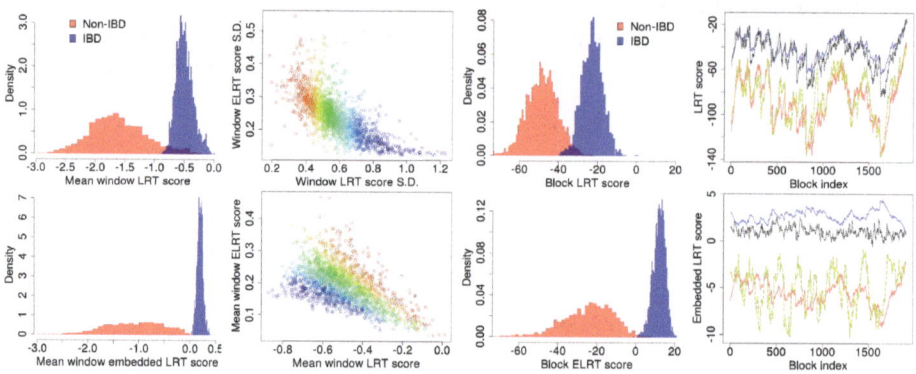

Fig. 3. (a) For each window, the mean window score of the IBD and non-IBD training data was computed; the histogram of these means is shown for the LRT and ELRT scores. When compared to the LRT score, the ELRT score has more separation between the IBD and non-IBD distributions, the boundary between them becomes centered at zero, and the IBD score variance is reduced. (b) Mean and standard deviation of window LRT scores and ELRT scores for IBD training data was computed. Each point represents a specific window, with the same color used to denote the same window in both plots. This illustrates the extent to which the ELRT reduces the variance of windows with high-variance, low-negative-mean LRT scores. (c) For a particular block, a histogram of the scores observed in the training data are shown. As with windows, the ELRT block scores feature better separation between IBD and non-IBD individuals, with a boundary close to zero. (d) Scores and thresholds across a chromosomal segment based on training data. The red line represents the mean score for non-IBD training data and the dark blue line represents the mean score for IBD training data. The yellow dashed line is the score for a single unrelated pair at each block. The dotted black line shows a local, block-specific threshold. This figure illustrates the consistent and improved separation between IBD and non-IBD score distributions at blocks across the chromosome for the ELRT over the LRT.

Embedded LRT and Local Thresholds. We computed the LRT and ELRT scores for windows and blocks for the unrelated and related training data and examined their properties in order to explore the differences between the ELRT and LRT strategies. Figure 3a shows the distribution of the average window-score for IBD and non-IBD segments. The figure demonstrates three notable properties of the ELRT, when compared to the LRT. First and foremost, there is greater separation between the scores of IBD and non-IBD segments; second, the boundary between the scores of IBD and non-IBD segments is very close to zero, suggesting well calibrated scores; third, the variance of the scores of IBD segments is controlled. To understand the role of the ELRT's reduction variance of the IBD window scores, we plotted the mean and standard deviation of the window scores for ELRT versus LRT (see Figure 3b). Note that the ideal score distribution for IBD segments would have a high mean and low variance in order to serve as a reliable predictor for the IBD state. Therefore, these plots clearly demonstrate that ELRT controls for windows that are unreliable predictors of IBD. Specifically, the windows with high variance and low negative mean LRT scores (the blue and violet points in the figure) are mapped to lower variance ELRT scores. We note that even though there is a negative trend between the average LRT scores and average ELRT scores, the ELRT scores stay above zero, the apparent boundary between IBD and non-IBD scores. The ELRT advantages at the window level translate to the block level, as seen in Figure 3c. This greater block score separation consequently allows PARENTE to achieve higher accuracy when using the embedded LRT score. In Figure 3d, the mean of these distributions can be seen for many blocks along chromosome, demonstrating the stability of the increased separation of the ELRT across the chromosome. This figure also shows the high variation in the block thresholds in for the LRT, which explains why the LRT's performance increases significantly when using block-specific thresholds compared to a global threshold.

Accuracy Performance Characteristics. Finally, we conducted additional experiments aimed at characterizing the performance of PARENTE. Specifically, we examined the effect of genotyping errors, the use of the genotyping-error function $\phi(\epsilon)$, and the effect of varying the window size k. First, we explored PARENTE's performance with and without $\phi(\epsilon)$, assessing differences in accuracy. When using $\phi(\epsilon)$ with the scaling factor $v = \frac{1}{100}$, PARENTE's sensitivity increased from 75% to 86% at the 1% FPR level. The improvement in sensitivity further increased at the 0.1% FPR level, from 45% when using ϵ to 73%, when $\phi(\epsilon)$ was applied. Next, we demonstrated that PARENTE is robust to changes in the window size parameter. IBD pairs were inferred on simulations with 4 cM injected IBD segments for a window size of 10, 20, and 30 SNPs per window. When using the LRT score, PARENTE's sensitivity changed less than 0.5% at the 0.1% FPR level. The differences were due to the fact that block boundaries were generated to begin and end at window boundaries, resulting in block definitions that were slightly different given the window size. As noted earlier, the varying windows size does not effect the LRT score, as the window-based model is equivalent to the direct computation of the score at the block level. Simply, the LRT score of a block can be equivalently computed by summing the individual SNP LRT scores or the window LRT scores. When using the embedded LRT score,

PARENTE's sensitivity varied by less than 2% at the 0.01% FPR level across the different window sizes. These differences can be attributed to differences in the window models as well as block boundary differences. Finally, we explored the extent to which genotyping errors affected PARENTE's performance. To this end, we repeated the simulations but introduced genotyping errors at different rates: 1%, 0.5%, and 0%. The model parameters ϵ and $\phi(\epsilon)$ were unchanged from previously described experiments, being set to $\epsilon = 0.005$ and $\frac{\epsilon}{100}$, respectively. We found that at the 0.1% FPR level, the sensitivity increased from 66% to 74% to 76% for the 1%, 0.5%, and 0% error rates, respectively. These results illustrate that PARENTE is robust to a realistic range of error rates of less than 0.5%.

4   Discussion

To improve computational efficiency when applying the described scoring functions, the log window score $\log s_w(g_{i,m(w)}, g_{i',m(w)})$ can be pre-computed for all possible pairs of genotypes for every window. For instance, with a window size of 5 SNPs, each window requires only $\frac{(3^5)(3^5+1)}{2} = 29,646$ values per window. The block score $\Lambda_b(g_i, g_{i'})$ can then be computed efficiently by retrieving and summing these values.

The model presented here assumes markers within each window are in linkage equilibrium. One approach to satisfy this assumption is via marker pruning using tools such as PLINK [25]. Alternatively, our model can be extended so as to incorporate the LD evident between neighboring markers. Previous work has shown that modeling LD can improve the performance of IBD methods [4].

In our work, the applied block-specific threshold strategy was based on the observed scores of unrelated pairs in the training data. The rationale behind this approach was to extremely control for false positions, since we aim to identify IBD in extremely large cohorts. Therefore, we calculated the threshold based on the maximum and variance of the observed training scores and a provided constant, c (see Equation 8). The default value $c = 0$ yielded a threshold with good performance (82% sensitivity at a 3×10^{-3} FPR for the embedded LRT); c can be adjusted to achieve the preferred tradeoff between specificity and sensitivity. We have observed that the margin between the related and unrelated distributions varies between blocks (see Figure 3). One may be able to increase sensitivity without loss of specificity by increasing the thresholds at blocks where the margin is large. In future work, we aim to explore additional stronger thresholding schemes in order to increase PARENTE's accuracy.

PARENTE makes the assumption that IBD segments along the genome are independent of one another, which holds true for distant relatives with relatively small IBD segments (eg 5 cM) that are expected to have at most one shared IBD segment. The assumption may not hold true for closely-related individuals, which are expected to share several IBD segments. However, due to the close relationships in these scenarios, these IBD segments also tend to be very large. Because PARENTE can accurately detect individual small IBD segments, it can

Table 1. As window size increases, a Gaussian distribution fits window LRT scores better. Given a window size (SNPs per window), the Kolmogorov−Smirnov test was performed on the scores of the training data for each window along the chromosomal segment. The mean p-value of all the windows is reported here.

	SNPs per window				
	3	**5**	**10**	**15**	**20**
Mean non-IBD KS p-value	7e-12	4e-11	8e-7	5e-6	2e-5
Mean IBD KS p-value	1e-9	4e-5	0.003	0.008	0.017

also detect each individual larger IBD segment, without needing to take into account that several large IBD segments may appear across the genome.

Our model uses a normal approximation of the LRT score distribution in order to compute the ELRT scores. With a window size of 20 SNPs per window, as used in our experiments, the LRT score distributions of most windows reasonably follow a Gaussian distribution. Naturally, however, for smaller window sizes (such as 3 SNPs window), most windows had score distributions that does not fit a Gaussian distribution. The poor approximation of the LRT score via a Gaussian distribution resulted in reduced performance (results not shown). We quantified window LRT score normality across various window sizes by using a Kolmogorov−Smirnov (KS) test on the related and unrelated training LRT scores for each window. The mean p-value of all the windows along the chromosome was computed. Table 1 shows these results, illustrating that the approximation using a Gaussian distribution provides a better fit as the window size increases. These observations indicate that it may be worthwhile to explore alternative parametric and empirical distributions for LRT, evaluating their impact on PARENTE's accuracy, especially when using small window sizes.

In this paper we presented PARENTE, a novel method for the accurate and efficient detection of IBD. Our results demonstrate that PARENTE has a superior accuracy in comparison to previous state-of-the-art methods, especially when set to control for extremely low false-positive rates. Furthermore, the methods efficiency enables the analysis of large-cohorts sampled over dense marker sets. As larger dataset are collected and sampled at an increasingly higher resolution via next-generation sequencing [21, 28], efficient methods such as PARENTE that can operate on non-phased genotype data become vital for their analysis. PARENTE is publicly and freely available at *http://parente.stanford.edu/*.

Acknowledgments. This material is based upon work supported by the National Science Foundation Graduate Research Fellowship under Grant No. DGE-1147470. Any opinions, findings, and conclusions or recommendations expressed in this material are those of the authors and do not necessarily reflect the views of the National Science Foundation. This work is also supported by a grant from the Stanford-KAUST alliance for academic excellence. We would like to thank Kelly Gilbert for helpful feedback in preparing the manuscript and two anonymous reviewers for many helpful comments.

References

[1] Abecasis, G.R., Cherny, S.S., Cookson, W.O., Cardon, L.R.: Merlin–rapid analysis of dense genetic maps using sparse gene flow trees. Nat. Genet. 30(1), 97–101 (2002)

[2] Alkuraya, F.S.: Homozygosity mapping: one more tool in the clinical geneticist's toolbox. Genet. Med. 12(4), 236–239 (2010)

[3] Altshuler, D.M., Gibbs, R.A., Peltonen, L., Dermitzakis, E., Schaffner, S.F., Yu, F., Bonnen, P.E., De Bakker, P.I.W., Deloukas, P., Gabriel, S.B., et al.: Integrating common and rare genetic variation in diverse human populations. Nature 467(7311), 52–58 (2010)

[4] Bercovici, S., Meek, C., Wexler, Y., Geiger, D.: Estimating genome-wide ibd sharing from snp data via an efficient hidden markov model of ld with application to gene mapping. Bioinformatics 26(12), i175–i182 (2010)

[5] Browning, B.L., Browning, S.R.: A fast, powerful method for detecting identity by descent. American Journal of Human Genetics 88(2), 173–182 (2011)

[6] Browning, S., Browning, B.: Rapid and accurate haplotype phasing and missing-data inference for whole-genome association studies by use of localized haplotype clustering. Am. J. Hum. Genet. 81(5), 1084–1097 (2007)

[7] Browning, S., Thompson, E.: Detecting Rare Variant Associations by Identity by Descent Mapping in Case-control Studies. Genetics 190, 1521–1531 (2012)

[8] Browning, S.R., Browning, B.L.: High-Resolution Detection of Identity by Descent in Unrelated Individuals. American Journal of Human Genetics 86(4), 526–539 (2010)

[9] Carey, V.J.: Mathematical and statistical methods for genetic analysis (2nd ed.). kenneth lange. Journal of the American Statistical Association 100, 712 (2005)

[10] Conrad, D.F., Keebler, J.E.M., DePristo, M.A., Lindsay, S.J., Zhang, Y., Casals, F., Idaghdour, Y., Hartl, C.L., Torroja, C., Garimella, K.V., Zilversmit, M., Cartwright, R., Rouleau, G.A., Daly, M., Stone, E.A., Hurles, M.E., Awadalla, P., for the 1000 Genomes Project: Variation in genome-wide mutation rates within and between human families. Nature Genetics (2011)

[11] Elston, R., Stewart, J.: A general model for the analysis of pedigree data. Hum. Hered. 21, 523–542 (1971)

[12] Ghahramani, Z., Jordan, M.I., Smyth, P.: Factorial hidden markov models. In: Machine Learning. MIT Press (1997)

[13] Gudbjartsson, D.F., Thorvaldsson, T., Kong, A., Gunnarsson, G., Ingolfsdottir, A.: Allegro version 2. Nature Genetics 37(10), 1015–1016 (2005)

[14] Gusev, A., Lowe, J.K., Stoffel, M., Daly, M.J., Altshuler, D., Breslow, J.L., Friedman, J.M., Pe'er, I.: Whole population, genome-wide mapping of hidden relatedness. Genome Research 19, 318–326 (2009), doi:10.1101/gr.081398.108

[15] Henn, B.M., Hon, L., Macpherson, J.M., Eriksson, N., Saxonov, S., Pe'er, I., Mountain, J.L.: Cryptic distant relatives are common in both isolated and cosmopolitan genetic samples. PLoS ONE 7(4), e34267 (2012)

[16] Ingólfsdóttir, A., Gudbjartsson, D.: Genetic Linkage Analysis Algorithms and Their Implementation. In: Priami, C., Merelli, E., Gonzalez, P., Omicini, A. (eds.) Transactions on Computational Systems Biology III. LNCS (LNBI), vol. 3737, pp. 123–144. Springer, Heidelberg (2005)

[17] Kyriazopoulou-Panagiotopoulou, S., Kashef Haghighi, D., Aerni, S.J., Sundquist, A., Bercovici, S., Batzoglou, S.: Reconstruction of genealogical relationships with applications to phase iii of hapmap. Bioinformatics 27(13), i333–i341 (2011)

[18] Lander, E.S., Green, P.: Construction of multilocus genetic maps in humans. Proceedings of the National Academy of Sciences 84, 2363–2367 (1987)

[19] Li, M.-H., Strandén, I., Tiirikka, T., Sevón-Aimonen, M.-L., Kantanen, J.: A comparison of approaches to estimate the inbreeding coefficient and pairwise relatedness using genomic and pedigree data in a sheep population. PLoS ONE 6(11), e26256 (2011)

[20] Markianos, K., Daly, M.J., Kruglyak, L.: Efficient multipoint linkage analysis through reduction of inheritance space. Am. J. Hum. Genet. 68(4), 963–977 (2001)

[21] 1000 Genomes Project. A map of human genome variation from population-scale sequencing. Nature 467(7319),1061–1073 (2010)

[22] Moltke, I., Albrechtsen, A., Thomas, Nielsen, F.C., Nielsen, R.: A method for detecting IBD regions simultaneously in multiple individuals with applications to disease genetics. Genome Research 21(7), 1168–1180 (2011)

[23] Nalls, M.A., Simon-Sanchez, J., Gibbs, J.R., Paisan-Ruiz, C., Bras, J.T., Tanaka, T., Matarin, M., Scholz, S., Weitz, C., Harris, T.B., Ferrucci, L., Hardy, J., Singleton, A.B.: Measures of autozygosity in decline: Globalization, urbanization, and its implications for medical genetics. PLoS Genet 5(3), e1000415 (2009)

[24] Ott, J.: Analysis of Human Genetic Linkage. The Johns Hopkins series in contemporary medicine and public health. Johns Hopkins University Press (1999)

[25] Purcell, S., Neale, B., Todd-Brown, K., Thomas, L., Ferreira, M.A., Bender, D., Maller, J., Sklar, P., de Bakker, P.I., Daly, M.J., Sham, P.C.: PLINK: a tool set for whole-genome association and population-based linkage analyses. American Journal of Human Genetics 81(3), 559–575 (2007)

[26] Rabiner, L.R.: A tutorial on hidden markov models and selected applications in speech recognition. Proceedings of the IEEE, 257–286 (1989)

[27] Ralph, P., Coop, G.: The geography of recent genetic ancestry across Europe (July 2012)

[28] WTCCC. Genome-wide association study of 14,000 cases of seven common diseases and 3,000 shared controls. Nature 447(7145), 661–678 (2007)

Learning Natural Selection from the Site Frequency Spectrum

Roy Ronen[1], Nitin Udpa[1], Eran Halperin[2,3,4], and Vineet Bafna[5,*]

[1] Bioinformatics Graduate Program, University of California, San Diego, CA, USA
[2] International Computer Science Institute, Berkeley, CA, USA
[3] The Blavatnik School of Computer Science, Tel-Aviv University, Tel-Aviv, Israel
[4] Department of Molecular Microbiology & Biotechnology, Tel-Aviv University, Tel-Aviv, Israel
[5] Department of Computer Science & Engineering, University of California, San Diego, CA, USA
vbafna@cs.ucsd.edu

Abstract. Genetic adaptation to external stimuli occurs through the combined action of mutation and selection. A central problem in genetics is to identify loci responsive to specific selective pressures. Over the last two decades, many tests have been proposed to identify genomic signatures of natural selection. However, the power of these tests changes unpredictably from one dataset to another, with no single dominant method. We build upon recent work that connects many of these tests in a common framework, by describing how positive selection strongly impacts the observed site frequency spectrum (SFS). Many of the proposed tests quantify the skew in SFS to predict selection. Here, we show that the skew depends on many parameters, including the selection coefficient, and time since selection. Moreover, for each of the different regimes of positive selection, informative features of the scaled SFS can be learned from simulated data and applied to population-scale variation data. Using support vector machines, we develop a test that is effective over all selection regimes. On simulated datasets, our test outperforms existing ones over the entire parameter space. We apply our test to variation data from *Drosophila melanogaster* populations adapted to hypoxia, and identify new loci that were missed by previous approaches, but strengthen the role of the Notch pathway in hypoxia tolerance.

Natural selection works by preferentially expanding the pool of beneficial (*fit*) alleles. At the genetic level, the increased fitness may stem either from a *de novo* mutation that is beneficial in the current environment, or from a new environmental stress leading to increased relative fitness of standing variation. Over time, haplotypes carrying these variants start to dominate the population, causing reduced genetic diversity. This process, known as a 'selective sweep', is mitigated by recombination and is therefore mostly observed in the vicinity of the beneficial allele. Improving our ability to detect the genomic signatures of

---

* Corresponding author.

M. Deng et al. (Eds.): RECOMB 2013, LNBI 7821, pp. 230–233, 2013.
© Springer-Verlag Berlin Heidelberg 2013

natural selection is crucial for shedding light on genes responsible for adaptation to selective constraints, including disease.

Following Fu [1], let ξ_i denote the number of polymorphic sites at frequency i/n in a sample of size n. The *site frequency spectrum* (SFS) vector ξ, and the scaled SFS vector ξ', are defined as:

$$\xi = [\xi_1, \xi_2, \ldots, \xi_{n-1}] \qquad \xi' = [1\xi_1, 2\xi_2, \ldots, (n-1)\xi_{n-1}] \qquad (1)$$

In a constant sized population evolving neutrally, it has been shown [1] that $E(\xi_i) = \theta/i$ for all $i = (1, \ldots, n-1)$. This implies that each ξ'_i $(= i\xi_i)$ is an unbiased estimator of θ, and that the scaled SFS ξ' is uniform in expectation. However, for populations evolving under directional selection this is not the case. Individuals carrying a favorable allele are preferentially chosen to procreate with probability $\propto 1 + s$, where s is the selection coefficient. As a result, the frequency of the favored allele and of those linked to it rises exponentially with parameter s, eventually reaching fixation at a rate dependent on s. Not surprisingly, directional selection has a dramatic effect on the scaled SFS. Near the point of fixation, the scaled SFS is characterized by an abundance of very high frequency alleles, and a near-absence of intermediate frequency alleles. Notably, the scaled SFS of regions evolving under directional selection differs from that of regions evolving neutrally even in the pre-fixation and post-fixation regimes. Many tests of neutrality have been proposed based on the site frequency spectrum [2–4]. To a first approximation, these tests operate by quantifying the 'skew' in the SFS of a given population sample, relative to the expected under neutral conditions. A subset of these tests do this by comparing different estimators of the population scaled mutation rate $\theta = 4N_e\mu$, where μ is the mutation rate and N_e the effective population size.

Under neutrality, any weighted linear combination of ξ' yields an unbiased estimator of θ. Thus, known estimators such as Tajima's θ_π and Fay & Wu's θ_H [2, 3] can be re-derived simply by choosing appropriate weights [5]. Since different estimators of θ are affected to varying extents by directional selection, many tests of neutrality, such as Tajima's D and Fay & Wu's H, are based on taking the difference between two estimators. These, also, can be defined as weighted linear combinations of ξ'. In both cases, the expected value of (D, H) is 0 under neutral evolution, but < 0 for populations evolving under directional selection. A potential caveat of these tests is that although the scaled SFS changes considerably with time under selection (τ), and with the selection coefficient (s), the test statistic applies a fixed weight function. It is therefore not surprising that the performance of these tests varies widely depending on the values of s and τ.

Here, rather than inferring selection using fixed summary statistics (such as θ-based tests) on the scaled SFS, we propose inferring it directly using supervised learning. Specifically, we use Support Vector Machines (SVMs) trained on data from extensive forward simulations of the Wright-Fisher model under various parameters. Being uniform in expectation under neutrality, the scaled SFS provided a natural choice of features to learn from. We considered the relative importance of features for classifying neutrality from different regimes of positive selection, and characterized commonalities in these features across the parameter space. Rather than a fixed weight function (model) that only performs well under certain regimes of selection, we were able to learn multiple weight functions of the scaled SFS, each corresponding to a different regime of natural selection, and each providing optimal performance in its respective regime. Combined, these resulted in a test that improves over existing methods when applied to simulated data. Using this as foundation, we develop an algorithmic framework, *SFselect*, with which we can apply these principles to real population polymorphism data. Additionally, we develop a similar approach, *XP-SFselect*, for cross population testing based on the two-dimensional SFS [6–8].

We applied XP-SFselect to data obtained from in-laboratory selection experiments on Drosophila melanogaster in hypoxic (4% O_2) conditions [9]. In that study, we used existing methods to identify elements of the Notch pathway as evolving under positive selection. Here, in addition to the previously identified regions, we show the Notch gene itself to be affected by a selective sweep, further implicating the Notch pathway as playing a crucial role in hypoxia tolerance.

Although there have been recent applications of machine learning to SFS-based and LD-based summary statistics for inferring selection [10–12], to the best of our knowledge, our study represents the first attempt to apply supervised learning directly to the scaled SFS to this end.

References

[1] Fu, Y.X.: Statistical properties of segregating sites. Theor. Popul. Biol. 48, 172–197 (1995)

[2] Tajima, F.: Statistical method for testing the neutral mutation hypothesis by DNA polymorphism. Genetics 123, 585–595 (1989)

[3] Fay, J.C., Wu, C.I.: Hitchhiking under positive Darwinian selection. Genetics 155, 1405–1413 (2000)

[4] Chen, H., Patterson, N., Reich, D.: Population differentiation as a test for selective sweeps. Genome Res. 20, 393–402 (2010)

[5] Achaz, G.: Frequency spectrum neutrality tests: one for all and all for one. Genetics 183, 249–258 (2009)

[6] Chen, H., Green, R.E., Paabo, S., Slatkin, M.: The joint allele-frequency spectrum in closely related species. Genetics 177, 387–398 (2007)

[7] Gutenkunst, R.N., Hernandez, R.D., Williamson, S.H., Bustamante, C.D.: Inferring the joint demographic history of multiple populations from multidimensional SNP frequency data. PLoS Genet. 5, e1000695 (2009)

[8] Nielsen, R., Hubisz, M.J., Hellmann, I., Torgerson, D., Andres, A.M., Albrechtsen, A., Gutenkunst, R., Adams, M.D., Cargill, M., Boyko, A., Indap, A., Bustamante, C.D., Clark, A.G.: Darwinian and demographic forces affecting human protein coding genes. Genome Res. 19, 838–849 (2009)

[9] Zhou, D., Udpa, N., Gersten, M., Visk, D.W., Bashir, A., Xue, J., Frazer, K.A., Posakony, J.W., Subramaniam, S., Bafna, V., Haddad, G.G.: Experimental selection of hypoxia-tolerant Drosophila melanogaster. Proc. Natl. Acad. Sci. U.S.A. 108, 2349–2354 (2011)

[10] Lin, K., Li, H., Schlötterer, C., Futschik, A.: Distinguishing positive selection from neutral evolution: Boosting the performance of summary statistics. Genetics 187(1), 229–244 (2011)
[11] Pavlidis, P., Jensen, J.D., Stephan, W.: Searching for footprints of positive selection in whole-genome snp data from nonequilibrium populations. Genetics 185(3), 907–922 (2010)
[12] Kern, A.D., Haussler, D.: A population genetic hidden markov model for detecting genomic regions under selection. Molecular Biology and Evolution (2010)

Considering Unknown Unknowns - Reconstruction of Non-confoundable Causal Relations in Biological Networks

Mohammad Javad Sadeh, Giusi Moffa, and Rainer Spang

Institute of Functional Genomics,
Computational Diagnostics Group, University of Regensburg,
Josef Engertstr. 9, D-93053 Regensburg. Germany
{mohammed.sadeh,giusi.moffa,rainer.spang}@klinik.uni-r.de
https://genomics.uni-regensburg.de/site/spang-group

Abstract. Our current understanding of cellular networks is rather incomplete. We miss important but sofar unknown genes and mechanisms in the pathways. Moreover, we often only have a partial account of the molecular interactions and modifications of the known players. When analyzing the cell, we look through narrow windows leaving potentially important events in blind spots. Network reconstruction is naturally confined to what we have observed. Little is known on how the incompleteness of our observations confounds our interpretation of the available data.

Here we ask the question, which features of a network can be confounded by incomplete observations and which cannot. In the context of nested effects models, we show that in the presence of missing observations or hidden factors a reliable reconstruction of the full network is not feasible. Nevertheless, we can show that certain characteristics of signaling networks like the existence of cross talk between certain branches of the network can be inferred in a non-confoundable way. We derive a test for inferring such non-confoundable characteristics of signaling networks. Next, we introduce a new data structure to represent partially reconstructed signaling networks. Finally, we evaluate our method both on simulated data and in the context of a study on early stem cell differentiation in mice.

Keywords: Biological networks, Network reconstruction, Nested Effects Models, Hidden variables.

1 Introduction

In February 2002, Donald Rumsfeld, the then US Secretary of Defense, stated at a Defense Department briefing: "There are known knowns. There are things we know that we know. There are known unknowns. That is to say, there are things that we now know we dont know. But there are also unknown unknowns. There are things we do not know we dont know" [1]. The concept of unknown unknowns

M. Deng et al. (Eds.): RECOMB 2013, LNBI 7821, pp. 234–248, 2013.
© Springer-Verlag Berlin Heidelberg 2013

is eminent to many fields of research. In the context of biological networks, known knowns make up our literature knowledge on physical and functional interactions of signaling molecules. Known unknowns might be what our current research projects are about, but unknown unknowns are those cellular mechanisms that we do not even anticipate today. They can be best appreciated in a historic perspective: Today, the role of many micro RNAs and epigenetic modifications of chromatin structure are known known mechanisms in many pathways. For other instances they are still in the realm of known unknowns. But going back 15 years in history they were unknown unknowns. Models of signaling pathways did not include them and the vast majority of molecular biologists did not anticipate the important role they play.

Once unknown unknowns become known, two different scenarios can occur: (i) The new observations can add to our understanding of a network or (ii) they can fundamentally change our perspective of the networks. In scenario (i) the network becomes more nodes and edges but the already existing parts of the network do not change. In scenario (ii), we learn that our old working hypothesis of the network was confounded by the mechanisms we were not aware of. The hidden effects of unknown unknown players made the interplay of the known players appear different than they really are.

This raises the question what of our current understanding of biological networks can be confounded by hidden mechanisms and what can not. We believe the question can only be addressed meaningfully in the context of a formal statistical network reconstruction framework, like Bayesian networks [2], gaussian graphical networks [3], boolean networks [4], or nested effects models [5].

In these frameworks, unknown unknowns are a set of hidden nodes. Together with the observed nodes they form a directed large biological network. The edges of the network encode causal relations. This means that if there is a directed edge from A to B, then perturbing A leads to changes in B. We call this large network the ground truth network (GTN). In practice it is almost always unknown. Observed and modeled is only a subset of the GTN nodes resulting in a "current state of the art network" (CSAN). This network only connects observed nodes. Importantly, in the GTN the hidden nodes can affect the observed nodes. A CSAN is reconstructed correctly, if it is identical to the subnetwork that the observed nodes form in the GTN. However, such a subnetwork does not need to exist [6]. Clearly, in such a case every network built exclusively from observed nodes is incorrect.

The problem of hidden nodes in network analysis has been recognized for a long time, e.g. in causal inference theory [7]. In the context of Bayesian network reconstruction, the structural EM algorithm can be used [8] to account for some missing observations. Moreover, the concept of structural signatures facilitates the detection and approximate location of a hidden variable in a network [6]. However, latent variable approaches are only practical if the number of hidden nodes is small, an assumption that is questionable, and it is often violated in the domain of molecular biology.

The reconstruction of a correct subnetwork from very incomplete observations might be too ambitious. Alternatively, one can strive for deriving features of a network that are correct, no matter what is going on outside of the observation window. Colombo et al. [9] introduced the concept of partial ancestral graphs (PAG) extending work in [10]. A PAG describes the common causal features of all directed acyclic graphs (DAGs) that can not be reliably distinguished if one accounts for possible effects of hidden nodes. The PAG is not a fully resolved network. Its information content lies in the features of the networks it excludes, since this exclusion is guaranteed not to be an artifact caused by hidden nodes. The inference is not confoundable. Colombo et al. [9] describe a computationally efficient algorithm that allows for the asymptotically consistent estimation of sparse high dimensional PAGs. A charming feature of the method is that it works exclusively using observational data. Practical drawbacks are the limited biological interpretability of general Bayesian networks learned on gene expression data and missing to exploiting functional information revealed in cell perturbation experiments. In fact, applications to molecular biology have not been reported to date.

Here we consider the concept of a partial but not confoundable network reconstruction in the context of nested effects models (NEMs) [5]. NEMs differ from the more general networks designed by Colombo *et al.* in two ways: (i) They are learned from interventional data, (ii) All edges except for those involving leaf nodes encode deterministic information flow, e.g. local transition probabilities are zero or one. NEMs assume that the cellular information flow is deterministic, stochasticity only comes in via noisy observations [11]. These features make not confoundable network inference simpler and allow for straightforward applications in systems biology.

The paper is organized as follows. In the context of nested effects models, we show that in the presence of unknown unknowns, network reconstruction can be flawed. We introduce a simple edge by edge partial network reconstruction algorithm called *Non Confoundable Network Analysis* (No-CONAN) to derive non confoundable network properties. In analogy to PAGs, we define a data structure that encodes the partially resolved networks (pNEM for partial nested effects model). On simulated data, we demonstrate the performance of our algorithm and in a first application to embryonic stem cell differentiation in mice, we demonstrate that taking unknown unknowns into account changes our account of real biological networks.

2   Method

Nested Effects Models. For self-containedness, we briefly review nested effects models as introduced in [5]. NEMs learn upstream/downstream relations in non-transcriptional signaling pathways from the nesting of transcriptiopnal downstream effects when perturbing the signaling genes. In a nutshell: NEMs infer that a gene A operates upstream of a gene B in a pathway, if the downstream effects resulting from silencing gene B are a noisy subset of those resulting from silencing gene A. Following [5] we call the perturbed genes in the signaling pathway S-genes and the genes that show expression changes in response to perturbation of E-genes.

A NEM is a directed and possibly cyclic network that connects the S-genes representing the flow of information in the underlying signaling pathway. E-genes can be linked to single S-genes forming leaf nodes of the network. The directed edges linking E-genes to S-genes can be estimated together with the S-gene network [12] or they can be taken by bayesian maximum a posterioiri approach [13] or they can be treated as nuisance parameters that are integrated out [5,14,15].

The underlying data consists of gene expression profiles of gene silencing assays and corresponding controls. Typically, a pathway is stimulated both in cells where it is intact (controls) and in cells where it is partially disrupted by silencing one of its S-genes. If the silencing of an S-gene blocks the flow of information from the pathway initiating receptor to the E-gene, the E-gene no longer changes expression in response to stimulation. In the language of nested effect models the E-gene shows a silencing effect with respect to the S-gene and the crucial assumption is that E-genes must attach to at most one S-gene. In each experiment, one S-gene is silenced by RNAi and silencing effects on thousands of E-genes are measured. The expression data D_{ij} is assumed to be discretized to 0 and 1, with a 1 indicating that a silencing effect of S_j was observed on E_i. Signal propagation within the pathways is assumed to be deterministic, hence the silencing of S_j is expected to produce silencing effects in all E-genes downstream of S_j. Consequently, every network topology is associated with an expected data pattern across all silencing assays: the silencing scheme [5]. If the network is acyclic, the silencing scheme defines a partial order relation on the S-genes reflecting the expected nesting of downstream effects. Noise comes into play at the level of observations. NEMs allow for both false positive and false negative observations accounting for them by fixed rates α and β in the likelihood. Hence, NEMs aim to detect a noisy subset relation in the observations D_{ij} and represent it as a directed network, where the directed edges can be interpreted as upstream/downstream relations of genes in the pathway. Clearly if a gene A is upstream of a gene B and B is upstream of a gene C, then A must be upstream of C as well. This is reflected in the likelihood equivalence of all networks that have the same transitive closure. In other words NEMs are a degenerate type of Bayesian networks where all non-leaf nodes are not observable and all edges between S-genes are associated with 0/1 local probabilities [11]. While the first property makes network reconstruction hard, the second is a simplification that renders network reconstruction practical again.

Several extensions of NEMs exist including networks that distinguish activation and repression [16], likelihoods allowing continuous silencing data [12,17], dynamic models [15,18], and models including direct observations of S-gene activation [19]. The methodology is implemented in a bioconductor package *nem* and cran package *nessy* [20]. A comprehensive review and evaluation can be found in [21].

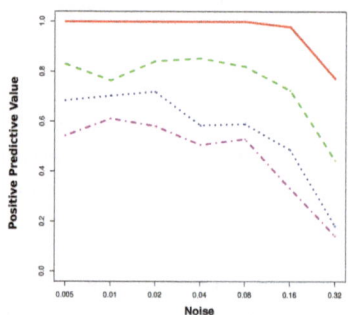

Fig. 1. In simulations hidden nodes compromise network reconstruction: Shown is the accuracy of standard NEM based network reconstructions if hidden nodes are present. The x-axis shows the degree of noise used in the stimulations. The y-axis shows the positive predictive value of reconstructed edges of the subnetwork of observed nodes. The different lines correspond to different numbers of hidden confounders (red 0, green 4, blue 8, purple 12).

Hidden Nodes Compromise NEM Based Network Reconstruction. For general Bayesian networks it is well known that hidden nodes can confound the reconstruction of networks [9,22]. Here we show that this problem still exists for the more specialized NEM. We generated data for networks that include both observed and unobserved nodes and reconstruct the subnetwork of observed nodes: We generated 100 random networks of 4 nodes and extended them by $n = 0, 4, 8, 16$ additional hidden nodes. Artificial silencing data was generated for the extended networks as described previously [5]. Only the data for the 4 observable nodes was used to reconstruct 4 node networks. These were compared to the corresponding subnetworks of the larger networks. The extended networks represent the ground truth signaling pathway while the 4 node subnetworks represent the small window through which we observe it. Figure 1 shows positive predicted values of network reconstruction (y-axis) for different noise levels (x-axis). The red line corresponds to network reconstruction without hidden nodes, while the green, blue and purple lines refer to 4, 8 and 12 additional hidden nodes respectively. We observe a marked decrease in network reconstruction performance when hidden nodes can confound the flow of information of the observed nodes.

Alien Silencing Patterns Are the Clue to a Non-confoundable Network Analysis. We analyze all pairs of S-genes S_1 and S_2 separately using only the data from silencing S_1 and S_2. Since our analysis will be non-confoundble by genes outside of the observation window, it will also not be affected by the the remaining S-genes that we voluntarily did not take into account.

For a pair of genes S_1 and S_2 we distinguish five possible upstream/downstream relations summarized in Figure 2A. (R1) S_1 is upstream of S_2, (R2) S_1 is downstream of S_2, (R3) S_1 and S_2 lie in a feedback loop in which case they are both up and downstream of each other indicated by the double arrow, (R4) S_1 and S_2 lie in independent modules of the network and do not interact with each other at all, and (R5) S_1 and S_2 are in different branches of a signaling network but jointly regulate at least one possibly hidden S-gene H. The five relations are encoded by the different edge types:

$$\mathcal{R} := \{R_1, \cdots, R_5\} = \{S_1 \to S_2, S_1 \leftarrow S_2, S_1 \leftrightarrow S_2, S_1 \cdots S_2, S_1 \to H \leftarrow S_2\} \quad (1)$$

With only two S-genes, an E-gene can show 4 different silencing patterns: It responds to both perturbations (1,1), only to one of them (1,0) and (0,1) or to none (0,0). Each upstream/downstream relation induces an expected subset of these 4 patterns. For example, in relation (R1) an E-gene can be unconnected to both S_1 and S_2 in which case it does not show a silencing effect neither when silencing S_1 nor when silencing S_2, yielding the expected pattern (0,0). It can be attached to S_1 in which case it is expected to show an effect when silencing S_1 but not when silencing the downstream gene S_2, yielding the expected pattern (1,0). And last, it can be linked to S_2 and show silencing effects both when silencing S_1 and S_2 yielding the pattern (1,1). Figure 2A gives the set of expected silencing patterns for all five upstream/downstream relations. Note that only relation R5 can produce all 4 possible silencing patterns. For the remaining relations at least one pattern is not expected. We call these unexpected patterns *alien patterns*.

We next investigated the possible influence of hidden nodes on the sets of expected and alien patterns (Figure2B). There are nine possible positions of a hidden confounder. The silencing patterns associated with these positions are shown in Figure2B. The most important observation is that any position of hidden confounders in the network does not change the sets of expected and alien silencing patterns (Figure2B). Note that in R4 the hidden node marked in red produces the alien pattern of R4. However, we have accounted for this problem by distinguishing the two relations R4 and R5 from the beginning. The conclusion that no alien patterns can occur through confounding facilitates our non-confoundabe analysis: If the observation of an alien pattern can not be through confounding effects it must be due to noise in the observation. Note that the assumption of deterministic signal propagation is crucial here. In relation R1 we assume that a perturbation of S_1 is deterministically propagated to S_2, which rules out the silencing pattern (0,1).

The Accumulation of Alien Patterns Is Evidence against Respective Upstream/Downstream Relations. For a pair of S-genes we can systematically consider all five upstream/downstream relations and see whether they conform with the observed data. Each of the relations R1-R4 has at least one alien pattern. Every observation of an E-gene that displays this alien pattern is evidence against the respective relation. Few alien patterns can occur due to observation noise but a large number of alien patterns is unlikely. We will set up a test to detect significantly high occurrences of alien patterns.

Binary NEMs [5] model observation noise by a false positive rate α, the probability that an observed effect is a noise artifact and a false negative rate β,

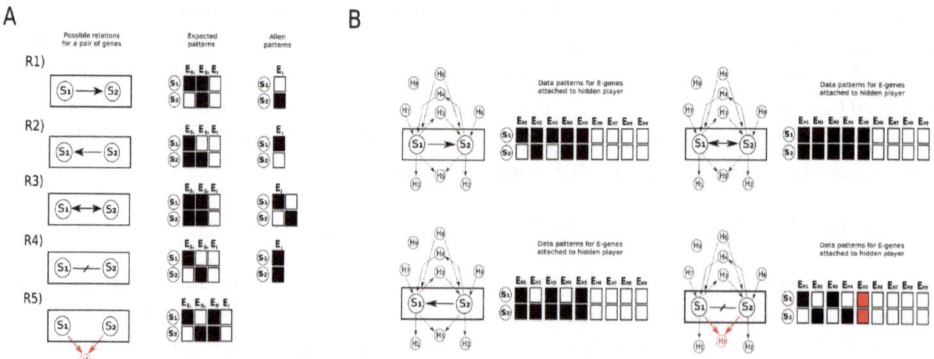

Fig. 2. Pairwise upstream/downstream relations and their alien patterns: A
Shown are the five possible possible relations R1,..., R5 together with their expected
silencing patterns and their alien patterns. **B** Hidden nodes are introduced in all possible configurations and the expected patterns of E-genes attached to the hidden nodes
are shown. In R4 the Hidden note marked in red produces the alien pattern of R4.
Node that this constellation leads to the constellation in R5.

the probability that we miss a true silencing effect. Further, the occurrence of
observation errors is assumed to be independent across E-genes. We can derive
limits for the probability that a certain number k of alien patterns occur given
a relation $R \in \mathcal{R}$ For example if $R = R1$ we have

$$P(K \geq k | S_1 \longrightarrow S_2) \leq \sum_{i=k}^{n} \binom{n}{i} \gamma_{R1}^i (1 - \gamma_{R1})^{n-i}, \tag{2}$$

where k is the observed number of alien patterns, n the total number of E-genes
and γ_{R1} an upper bound for the probability of observing the alien pattern.
If R1 holds true, the alien pattern (0,1) needs to be produced by noise from
one of the three expected patterns (1,0), (1,1) and (0,0). Starting from (1,0)
requires both a false positive and a false negative observation which happens
with probability $\gamma_1 = \alpha \cdot \beta$, starting from (1,1) we need one true positive and
one false negative observation which occurs with probability $\gamma_2 = \beta \cdot (1 - \beta)$.
Finally, generating the alien pattern (0,1) from (0,0) requires one true negative
and one false positive observation and occurs with probability $\gamma_3 = (1 - \alpha) \cdot \alpha$.
Setting $\gamma_{R1} = \max(\gamma_1, \gamma_2, \gamma_3)$ yields the bound (2). Similarly we obtain:

$$P(K \geq k | R) \leq \sum_{i=k}^{n} \binom{n}{i} \gamma_R^i (1 - \gamma_R)^{n-i} \tag{3}$$

for all $R \in \mathcal{R} - \{R_5\}$. Here γ_R is a bound for the probability of observing the
alien pattern of R. If some of the above probabilities become sufficiently small,
we gather evidence against the respective relations. We exclude a relation R if
and only if

$$P(K \geq k | R) < \kappa, \tag{4}$$

where κ is a calibration parameter that is set to 0.05 in all applications below. Note that $R5$ can not be rejected since it does not have an alien pattern.

Partial Network Reconstruction. If $R5$ is correct and we can reject relations $R1$-$R4$ leaving only relation $R5$ as compatible with the data, we have fully resolved the relation of S_1 and S_2. In cases where $R5$ is incorrect, the best we can achieve is that all but one relation from $R1$-$R4$ is rejected leaving us with one edge type and the always existing possibility that $R5$ is true. However, also this does not need to be the case. It is possible that we can not reject several relations leaving us with higher uncertainties on the true structure of the signaling network. We do not further resolve the network but confine ourselves to describing what we know and what we don't know. To do this we introduce the new data structure of a partial Nested Effects Model (pNEM). A pNEM is a graph connecting all S-genes however using a variety of different edge types. Each edge type is describing a set of relations that could not be rejected. This language of edge types is summarized in Figure 3. For example, if we exclude all relations except R5, there is no edge between S_1 and S_2. If we reject all relations except $R3$, $R4$ and $R5$ we draw a red double sided edge, and so on. Sixteen different edge types are needed to encode our partial network knowledge. In the next section we show an example of a pNEM.

Equation (4) has the form of a statistical test. When choosing κ sufficiently small we bound the probability of excluding a correct relation. The null hypothesis is that the tested relation is correct and that all observed alien patterns are due to noise alone. However, a small κ also leads to poorly resolved networks with only few excluded relations. This raises the issue of the power of the test. An edge between two S-genes is well resolved if the true relation generates many E-genes with silencing patterns that are alien to many alternative relations. For example if the relation $S_1 \rightarrow S_2$ holds true, every E-gene that is attached to S_1 and produces the expected pattern (1,0) produces evidence against the competing relations $S_1 \leftarrow S_2$ and $S_1 \leftrightarrow S_2$ since (1,0) is alien to both these relations, but not against the relation $S_1 \cdot \cdot S_2$, since (1,0) is not alien to it. However, E-genes attached to S_2 with the expected pattern (1,1) produce evidence against $S_1 \cdot \cdot S_2$. If we have enough E-genes of both types we will be able to reject all relations except the correct one and the not-rejectable relation $R5$. Inspecting Figure 2 points to a problem with edges that are of the type $S_1 \leftrightarrow S_2$, since in this constellation only the patterns (1,1) and (0,0) are produced but none of the alien patterns of the two directed relations $R1$ and $R2$. Since NEMs operate on transitively closed networks the relation $S_1 \leftrightarrow S_2$ is indicative of genes involved in a feedback loop. In other words our method is not capable of reliably detecting feedback loops; a non-circular constellation can often not be ruled out. Nevertheless, our method is valid also for biological networks with feedback loops. It does not produce spurious results in this case, but reports that it can not resolve the loop reliably. If in contrast the true network is not cyclic, our method has the potential to exclude a loop reliably.

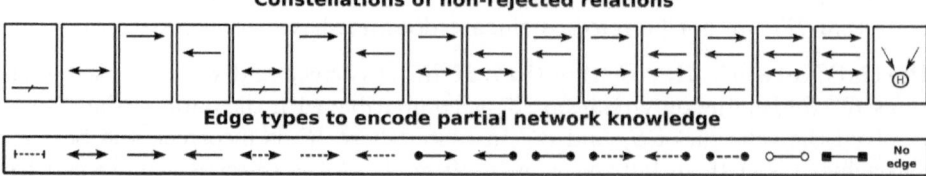

Fig. 3. The pNEM code: The top row of boxes shows combinations of relations and the bottom boxes show the corresponding edge types we use to encode that none of the edges in the set could be excluded by No-CONAN (The color are used to distinguish the same edge types in different constellations).

3   Simulation Experiments

We test the performance of No-CONAN in the context of simulation experiments using artificial data. In such simulations the true state of the network is known unlike in biological scenarios. Moreover, the artificial data fully conforms to all assumptions of NEMs, which is certainly not the case for real biological data.

Example. We start with an example to illustrate how No-CONAN works. Consider the GTN in Figure 4. Note that it has only one hidden node, but this node is in a central position of the network. We attach a total of 350 E-genes uniformly to the S-genes and generate artificial data using a moderate noise levels of 0.15 for both false negative and false positive observations as described previously [5]. Then, using only data for the observable nodes, we reconstruct the network using triplet search in a standard NEM approach [14] and using No-CONAN. Figure 5 compares the NEM to the pNEM. The NEM incorrectly predicts a feedback loop like structure. The pNEM in contrast did not do this. It actually resolved the relation between S_5 and S_6 as $R5$ thus predicting the existence of the hidden node at that position. Also all other predicted relations are correct, with the exception of S_6 and S_3, where the pNEM is undecided whether a directed relation exists (incorrect) or not (correct).

Evaluations. We generate 100 random networks of size 8 and generate data for these networks using noise levels varying between 0.005 (very low) and 0.32 (very high). We attach a total of 100 E-genes uniformly to the S-genes of the network and add another 900 E-genes which are unrelated to the networks. These have an expected silencing pattern of (0,0) but display occasional silencing effects due to noise. The number of unrelated E-genes might effect the power of the testing. We then run No-CONAN on every pair of nodes in each of 100 networks and reject all relations possible using $\kappa = 0.05$. The results are organized according to the true underlying relations in Figure 5A. Each of the five plots corresponds to one true relation. The x-axis shows the different noise levels while the y-axis shows the relative frequency of rejecting the different relations, which are marked by different colors. For example the left most plot corresponds to all situations where the true relation between nodes is $S_1 \longrightarrow S_2$. Rejection rates for this

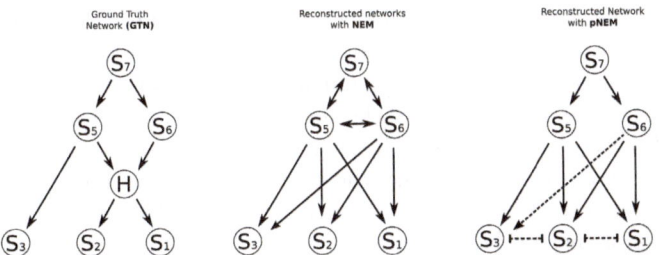

Fig. 4. Performance comparison: The left part of the figure shows a GTN that has one hidden node. We attach a total of 350 E-genes uniformly to the S-genes and generate artificial data using a moderate noise levels of 0.15 for both false negative and false positive observations. The plots in the middle and on the right side of the figure show the reconstructed network using triplet search in a standard NEM approach and pNEM, respectively only for the observable nodes. The NEM incorrectly predicts a feedback loop like structure. In contrast the pNEM resolved the relation between S_5 and S_6 as $R5$ thus predicting the existence of the hidden node at that position. Also all other predicted relations are correct, with the exception of S_6 and S_3, where the pNEM is undecided whether a directed relation exists (incorrect) or not (correct).

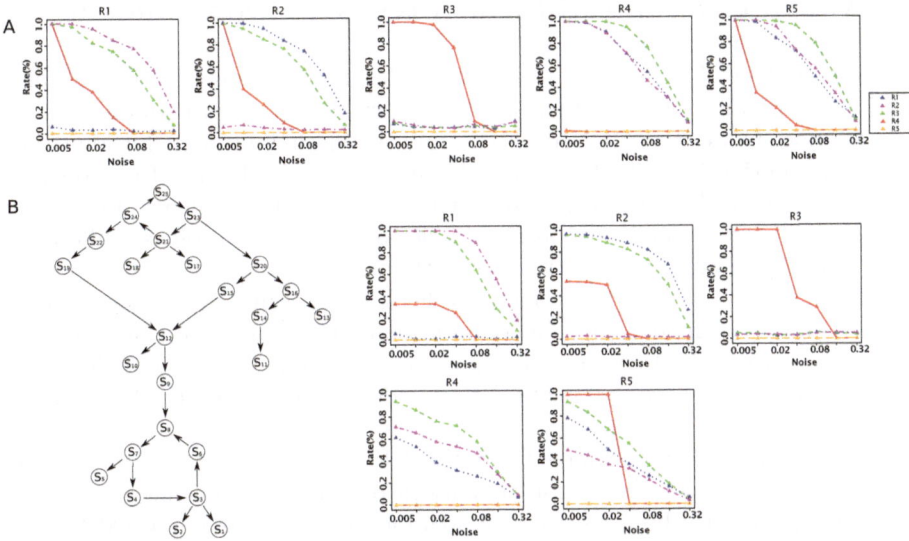

Fig. 5. Simulation results: A Small network simulations. Each of the five plots corresponds to one true relation. The x-axis shows the different noise levels. The y-axis shows the relative frequency of rejecting the relations (R1: blue, R2: purple R3: green, R4: red and R5: orange) **B** Simulations on a large network. Left: The ground truth network. Right: Performance plots that organized like in **A**.

relation are marked in blue and we can see that the relation does not get falsely rejected even for very high noise levels. In contrast the 3 competing relations marked in red, purple and green are virtually always rejected except for very high noise levels and even for maximal noise we reject them in about half of the cases. We do similarly well for the next two relations. If the true relation is the feedback loop $S_1 \leftrightarrow S_2$, we still get hardly any false positive rejections but loose almost all power in rejecting the two directed relations. As described in the previous section, this is expected since a feedback loop does not produce the alien patterns of these relations.

Finally we examine the performance of No-CONAN in the context of the larger 25 node network shown in Figure 5B. Note that the network contains two feedback loops, one towards the root and another one close to a leaf. The crucial difference from the smaller networks is the ratio of S-genes inside the observation window (2 in both cases) and those outside of it (6 vs. 23). In fact, this unfavorable ratio of observed versus unobserved nodes compromises the resolution of the pNEMs generated by No-CONAN. Importantly, we still hardly ever falsely reject a correct relation. However, except for very low noise levels rejection rates of incorrect relations go down. Nevertheless, except for very high noise levels we reject substantial fractions of relations thus partially learning the structure of the network.

4   An Application to Murine Stem Cell Development

We test No-CONAN in a study on molecular mechanisms of self-renewal in murine embryonic stem cells (ESCs). Ivanova et al. [23] down-regulated six factors (Nanog, Oct4, Sox2, Esrrb, Tbx3, and Tcl1) that need to be jointly expressed in murine ESCs to keep the cells in a self-renewal state. In response to the interventions the cells go into differentiation and the resulting shifts in the transcriptome were monitored in time series of expression profiles. Differentiation includes the successive destruction of the self-renewal network. This process of differentiation has been previously modeled twice using nested effect models [15] and [18]. Both models have in common that they are dynamic nested effects models exploiting the temporal information of the time series but differ in the likelihood functions used. None of them considered the possibility of unobserved factors.

In the NEM framework Nanog, Oct4, Sox2, Esrrb, Tbx3, and Tcl1 are S-genes, whereas genes showing expression changes in response to silencing are E-genes. A silencing effect is observed if the expression of an E-gene is pushed from its level in self-renewing cells to its level in differentiated cells. We preprocess and discretize the data as described in [15]. Then we run No-CONAN on the data of the last time point of all time series. Figure 6 A shows the pNEM produced from No-CONAN while C and D are the transitive closures of the networks derived in [15] and [18].

Notably, many edges of the pNEM are optimally resolved and often agree with those in the two previous models. E.g. the linear backbone of the network

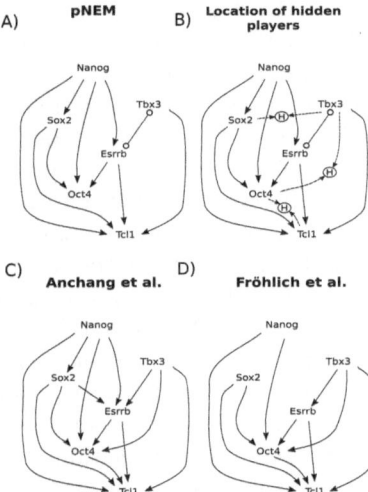

Fig. 6. Murine stem cell self-renewal network: **A** shows the pNEM produced from No-CONAN. **B** shows prediction results from the observation that all relations except for R5 could be excluded for the respective pairs of genes. The pNEM predicts the existence of certain hidden nodes in positions marked in **B**. **C** and **D** are the transitive closures of the networks derived in [15] and [18]. Many edges of the pNEM are optimally resolved and often agree with those in the two previous models in **C** and **D**.

Nanog → Sox2 → Oct4 observed in the Anchang model [15] could be resolved unambiguously even when taking hidden confounders into account.

In contrast, the role of the remaining genes Tcl1, Tbx3 and Esrrb could not be determined unambiguously with the available observations. For example, our pNEM proclaims that there is an interaction between Esrrb and Tbx3 but can not determine its nature. It could be a feedback loop as well as any directed edge depending on how a potential unknown gene is influencing the process. Moreover, the pNEM differs from the two NEMs in that it predicts the existence of certain hidden nodes in positions marked in Figure 6 B. These predictions result from the observation that all relations except for R5 could be excluded for the respective pairs of genes. In summary, non-confoundable analysis sustains a previous hypothesis on the role of Nanog, Sox2, and Oct4 interactions in stem cell differentiation but also points to possible ambiguities with respect to the role of Tcl1, Tbx3 and Esrrb.

5   Discussion

We introduced No-CONAN, a novel method that partially reconstructs the upstream/downstream relations of non-transcriptional signaling networks from interventional data. The method is set in the framework of nested effects models but has the additional feature that its inference can not be confounded by hidden nodes. The key idea is the definition of alien silencing patterns that can not be confounded by unobserved nodes. The output of No-CONAN is not a fully resolved network but a pNEM: A network of upstream/downstream relations where for some pairs of nodes several relations remain conformable with the data. The information in a pNEM lies in the upstream/downstream relations that it excludes. A pNEM encodes what we know but also what we can not know unless we can be sure that we have observed all nodes of a network. The uncertainties left with certain edges are the price we have to pay to ensure that our results are non-confoundable by mechanisms outside the window of observations.

No-CONAN is reliable in that it does not produce false information by rejecting correct relations. By construction, No-CONAN has two limitations affecting its power in resolving the network. It can never reject the relation $S_1 \to H \leftarrow S_2$, since this relation has no alien silencing patterns. Moreover, No-CONAN has very little power in resolving a true feedback loop since feedback does not produce the alien patterns of the two directed relations. Nevertheless, No-CONAN is generating new non confoundable insights into network structures by rejecting many though not all incorrect relations.

Partial network reconstruction is a relatively new concept in network analysis. It can be seen as a safeguard against possibly severe confounding effects caused by unobserved mechanisms. Clearly, such a non-confoundable analysis is only valid within the formal context of a network model. In this paper we used the framework of nested effect models. The assumptions of nested effect models might be incorrect in certain applications as is true for every modeling framework. We believe that no formal analysis can safeguard against this. However, the concept of unknown mechanisms can be represented in many formal frameworks, and simulations can mimic our partial observation of a true underlying network. Here we represented all the unknown unknowns of biology as unobserved nodes of a nested effects models and strived for extracting as much information on the full nested effects models from the incomplete data we obtained from looking through a narrow window.

Donald Rumsfeld continued his speech by saying "If I know the answer I'll tell you the answer, and if I don't, I'll just respond, cleverly". We do not know whether a pNEM is a "clever" response but it aims to be a realistic and an honest one. The partial networks aims to encode what we know that we know, but it also encodes what we can not know for certain, unless we are absolutely sure that we have a complete account of all biological mechanisms affecting cell signaling.

Acknowledgement.   This   work   was   supported   by   BMBF   grants (EraSys:0315714B) and the Bavarian Genomenetwork BayGene. We thank Achim Tresch for fruitful discussions and Florian Markowetz, Christian Hundsrucker, Julia C. Engelmann and Claudio Lottaz for carefully proof reading the manuscript.

References

1. Rumsfeld, D.: Dod news briefing-secretary rumsfeld and gen. myers. us department of defence (2002)
2. Friedman, N., Linial, M., Nachman, I., Pe'er, D.: Using bayesian networks to analyze expression data. J. Comput. Biol. 7(3-4), 601–620 (2000)
3. Schäfer, J., Strimmer, K.: An empirical bayes approach to inferring large-scale gene association networks. Bioinformatics 21, 754–764 (2005)
4. Saez-Rodriguez, J., Alexopoulos, L.G., Epperlein, J., Samaga, R., Lauffenburger, D.A., Klamt, S., Sorger, P.K.: Discrete logic modelling as a means to link protein signalling networks with functional analysis of mammalian signal transduction. Mol. Syst. Biol. 5, 331 (2009)
5. Markowetz, F., Bloch, J., Spang, R.: Non-transcriptional pathway features reconstructed from secondary effects of rna interference. Bioinformatics 21, 4026–4032 (2005)
6. Elidan, G., Ninio, M., Friedman, N., Shuurmans, D.: Data perturbation for escaping local maxima in learning. In: Proceedings of the National Conference on Artificial Intelligence, pp. 132–139. AAAI Press, MIT Press, Menlo Park, Cambridge (1999, 2002)
7. Pearl, J.: Causality: models, reasoning, and inference, vol. 47. Cambridge Univ. Press (2000)
8. McLachlan, G., Krishnan, T.: The EM algorithm and extensions, vol. 274. Wiley, New York (1997)
9. Colombo, D., Maathuis, M., Kalisch, M., Richardson, T.: Learning high-dimensional directed acyclic graphs with latent and selection variables. Arxiv preprint arXiv:1104.5617 (2011)
10. Richardson, T., Spirtes, P.: Ancestral graph markov models. The Annals of Statistics 30(4), 962–1030 (2002)
11. Zeller, C., Fröhlich, H., Tresch, A.: A bayesian network view on nested effects models. EURASIP J. Bioinform. Syst. Biol., 195272 (2009)
12. Tresch, A., Markowetz, F.: Structure learning in nested effects models. Stat. Appl. Genet. Mol. Biol. 7(1), Article9 (2008)
13. Niederberger, T., Etzold, S., Lidschreiber, M., Maier, K.C., Martin, D.E., Fröhlich, H., Cramer, P., Tresch, A.: Mc eminem maps the interaction landscape of the mediator. PLoS Comput. Biol. 8, e1002568 (2012)
14. Markowetz, F., Kostka, D., Troyanskaya, O.G., Spang, R.: Nested effects models for high-dimensional phenotyping screens. Bioinformatics 23, i305–i312 (2007)
15. Anchang, B., Sadeh, M.J., Jacob, J., Tresch, A., Vlad, M.O., Oefner, P.J., Spang, R.: Modeling the temporal interplay of molecular signaling and gene expression by using dynamic nested effects models. Proc. Natl. Acad. Sci. U S A 106, 6447–6452 (2009)
16. Vaske, C.J., House, C., Luu, T., Frank, B., Yeang, C.-H., Lee, N.H., Stuart, J.M.: A factor graph nested effects model to identify networks from genetic perturbations. PLoS Comput. Biol. 5, e1000274 (2009)
17. Fröhlich, H., Fellmann, M., Sültmann, H., Poustka, A., Beißbarth, T.: Estimating large-scale signaling networks through nested effect models with intervention effects from microarray data. Bioinformatics 24, 2650–2656 (2008)
18. Fröhlich, H., Praveen, P., Tresch, A.: Fast and efficient dynamic nested effects models. Bioinformatics 27, 238–244 (2011)

19. Bender, C., Henjes, F., Fröhlich, H., Wiemann, S., Korf, U., Beißbarth, T.: Dynamic deterministic effects propagation networks: learning signalling pathways from longitudinal protein array data. Bioinformatics 26(18), i596–i602 (2010)
20. Fröhlich, H., Beißbarth, T., Tresch, A., Kostka, D., Jacob, J., Spang, R., Markowetz, F.: Analyzing gene perturbation screens with nested effects models in r and bioconductor. Bioinformatics 24, 2549–2550 (2008)
21. Fröhlich, H., Tresch, A., Beißbarth, T.: Nested effects models for learning signaling networks from perturbation data. Biom. J. 51, 304–323 (2009)
22. Koller, D., Friedman, N.: Probabilistic graphical models: principles and techniques. The MIT Press (2009)
23. Ivanova, N., Dobrin, R., Lu, R., Kotenko, I., Levorse, J., DeCoste, C., Schafer, X., Lun, Y., Lemischka, I.R.: Dissecting self-renewal in stem cells with rna interference. Nature 442, 533–538 (2006)

Inference of Tumor Phylogenies with Improved Somatic Mutation Discovery

Raheleh Salari[1], Syed Shayon Saleh[1], Dorna Kashef-Haghighi[1], David Khavari[1], Daniel E. Newburger[2], Robert B. West[3], Arend Sidow[3,4], and Serafim Batzoglou[1,*]

[1] Department of Computer Science, Stanford University,
353 Serra Mall, Stanford, CA 94305
[2] Biomedical Informatics Training Program, Stanford University,
353 Serra Mall, Stanford, CA 94305
[3] Department of Pathology, Stanford University School of Medicine,
300 Pasteur Drive, Stanford, CA 94305
[4] Department of Genetics, Stanford University School of Medicine,
300 Pasteur Drive, Stanford, CA 94305
serafim@cs.stanford.edu

Abstract. Next-generation sequencing technologies provide a powerful tool for studying genome evolution during progression of advanced diseases such as cancer. Although many recent studies have employed new sequencing technologies to detect mutations across multiple, genetically related tumors, current methods do not exploit available phylogenetic information to improve the accuracy of their variant calls. Here, we present a novel algorithm that uses somatic single nucleotide variations (SNVs) in multiple, related tissue samples as lineage markers for phylogenetic tree reconstruction. Our method then leverages the inferred phylogeny to improve the accuracy of SNV discovery. Experimental analyses demonstrate that our method achieves up to 32% improvement for somatic SNV calling of multiple related samples over the accuracy of GATK's Unified Genotyper, the state of the art multisample SNV caller.

Keywords: tumor phylogeny, cancer evolution, genetic variations.

1   Introduction

Next-generation genome sequencing technologies have provided a means to identify and characterize the large number of mutations present in a human tumor. It is now widely known that cancer genomes are highly mutated by several mechanisms, which can lead to short mutations such as single nucleotide variants (SNVs), structural changes such as copy-number variations, or complex patterns of mutation such as chromothripsis. Pairwise comparison between the genetic landscape of late-stage tumors and matched normal tissue has provided a first understanding of the

---

* Corresponding author.

M. Deng et al. (Eds.): RECOMB 2013, LNBI 7821, pp. 249–263, 2013.

mutational state of cancer genomes [3,5,14,20-21,27-28]. However, the development of effective treatment hinges upon deeper investigation of tumor progression pathways, which can be efficiently conducted by analyzing multiple tumors originating from the same neoplastic progenitor. This kind of study, especially of related tumors at different stages of development, reveals mutations that drive cancer progression and helps to identify early-stage tumors that can turn into cancerous tissues. Recently, several groups have pursued this goal by building cancer-specific phylogenetic trees that illustrate the order, timing, and rates of genomic mutation based on sequencing multiple tumor samples within a patient [4,21]. Using exome sequencing data of nephrectomy specimens and their metastases, Gerlinger et al [8] constructed tumor phylogenetic trees for two patients. Newburger et al [19] performed whole genome deep sequencing of multiple breast cancer tumors from six patients and built trees that relate the tissue samples within each patient. The construction of phylogenetic trees in such studies has particular clinical relevance; in addition to pinpointing drug targets that arise in the aggressive late-stage tumors, it allows researchers to choose drug targets from among the earliest mutagenic events, common to all cancerous lesions. Targeting these events treats early neosplasms as well as late-stage tumors, thereby removing the reservoir for recurrence of the cancer.

SNV calling in tumor samples is essential for cancer characterization (diagnosis, identifying driver mutations, etc.), but current SNV callers for cancer remain highly inaccurate. Specialized tumor-normal SNV calling methods very effectively leverage the fact that the tumor and normal samples are genetically very similar [7,13,18,23]. However, they don't currently take complex relationships into account. Multisample callers are able to use more general, population-level information to improve variant calling across many samples [1,6,16], but this approach wastes a tremendous amount of information when samples are known to be related. For example, GATK's Unified Genotyper [6,16], the state of the art multisample SNV caller, sums the likelihoods of SNVs across the samples to take into account the recurrence of true SNVs, but it ignores the general pattern of true SNVs between samples that reveals their phylogenetic relation. In a parallel track, several papers used techniques to infer phylogeny [19,31] across multiple samples, but they never used this information to improve their variant calls. An ideal method should both infer phylogeny and apply that information. Here we devised a method that extends the advantages of tumor-normal callers to multiple samples with complex but unknown relationships to accurately reveal both the phylogenetic relationship and identify genetic variants.

Our method uses somatic point mutations as markers to construct tumor phylogeny trees. We then use these trees to perform error correction. The basic assumption of our method is that the single nucleotide variations (SNVs) follow the perfect phylogeny model, which assumes that mutations cannot recur in separate samples independently by chance and that recombination events do not happen between generations. These assumptions are reasonable in the context of cancer genomics. Following the perfect phylogeny assumption, true variant calls are tree compatible. Thus, phylogeny trees in which tumor samples are leaves can be constructed by using a character-based phylogenetic inference method. However, noisy sequence data, sequence alignment errors, and biases in mutation caller methods introduce both false positive and false negative SNV calls. As a result, identified SNVs usually contain several conflicts. In data from our breast cancer genomic evolution study [19] we observed that up to 20% of phylogenetically informative SNV calls are incompatible. In order to proceed with tree construction the incompatibility of data must be resolved or at least minimized. To rescue these mutations we propose an elegant algorithmic approach that benefits from the samples' phylogenetic relation; a valuable piece of information not used by any existing SNV calling method.

Our conflict resolution strategy is based on two approaches: editing mutations and identifying subclones. In several cases the alternative allele frequency of conflicting SNVs in different samples provides information for editing the mutations to reconcile it with the expected phylogenetic relation. In addition to variant calling errors, heterogeneity of samples can also result in conflicting SNVs, since a heterogeneous tumor can contain several subpopulations, each possessing its own genetic variations and progression stages. In this situation, the conflicting mutation profiles represent a true biological state, and we wish to identify subclones in order to update phylogenetic tree. Although there are some unmixing approaches for separating cell populations in tumor data [25,31], none can be simply adapted to next generation genome sequencing data of solid tumors. As a part of conflict resolution process we identify conflicts caused by subclones. Coupling it with our mutation editing process enabled us to identify several subclonal mutations, which was previously possible only through ultra deep sequencing [4,9].

In summary, our algorithm infers the tumor phylogeny tree from multisample genotype information retrieved by GATK (or another SNV caller). Using the phylogenetic information inferred from the majority of SNVs, the algorithm resolves conflicts among the remaining SNVs. The main contribution of this paper, in addition to tumor phylogeny tree construction, is to improve the accuracy of SNV calls by resolving these conflicts. Conflict resolution is a highly sensitive process, especially with larger numbers of samples. We measured the performance of our method at each step in a comprehensive simulation study. The simulation analysis demonstrates that our algorithm constructs highly accurate phylogenenetic trees while also achieving an average accuracy of 89% in reassigning conflicting mutations. The fast and efficient conflict resolution step improves the accuracy of GATK by up to 32% when assessing whether a multisample SNV call produced the correct mutation status for every sample. These results strongly suggest that the method can benefit several cancer sequencing applications that involve multiple, related tumors.

2   Methods

Given genotype information inferred from sequencing data of multiple samples, we aim to construct a perfect phylogeny tree that supports the maximum number of genotypes. Here we utilize somatic SNVs as genotypes. Let $S = \{s_1, s_2, ..., s_n\}$ be the set of n samples, and $G = \{g_1, g_2, ..., g_m\}$ be the universal set mutations in one or more samples. Each mutation g_j is characterized by a binary profile $\{g_{1j}g_{2j}...g_{i,j}... g_{nj}\}$ where $g_{i,j}$ is 1 if and only if g_j is called in the ith sample. We group all mutations with the same profile into a mutation group. Therefore, over m' distinct mutation groups, mutation matrix $M_{n \times m'}$ is defined such that each row represents a sample and each column represents a mutation group. Note that there are no duplicate columns, and each column has at least one entry that is 1. Two columns (or mutation groups) in M are said to be in conflict if and only if the two columns contain three rows with the pairs 1,1; 0,1; and 1,0. We first describe the outline of the algorithm in subsection 2.1 and discuss the details of each step in the following subsections.

2.1   Overview of Our Algorithm

Our algorithm first constructs a consensus perfect phylogeny tree based on the maximum number of compatible mutations. Then, in the conflict resolution process, it reconciles conflicting mutations with the tree by: i) editing the profile of these mutations, or ii) extending the tree by identifying significnt subclones. Once resolved, previously conflicting mutations are added to the tree.. The general steps of the algorithm are shown in Algorithm 1.

Algorithm 1. Tumor Phylogeny Inference

input : Genotype information of somatic mutations for multiple samples

1.   M ← Mutation Matrix
     // find the largest number of compatible mutations (section 2.2 and 2.3)
2.   G ← Weighted Conflict Graph
3.   Find the Maximal Independent Set *MIS* in G
     // use Gusfield's Algorithm [12]
4.   Construct the perfect phylogeny tree for *MIS*
     // conflict resolution
5.   **while** ∃ resolvable conflict **do**
         // edit mutations (section 2.4)
6.   **for** all conflicting mutations g **do**
7.   **if** ∃ evidence for g to be called in a non-conflicting mutation group **then**
8.   Move g to the most prominent group
9.   **end**
10.   **end**
11.   // identify subclones (section 2.5)
12.   **if** ∃ significant conflicting mutation group **then**
13.   Identify the subclones and add them to the tree
14.   **end**
15. **end**

In order to find the largest number of compatible mutations we build the weighted conflict graph G for mutation groups. The concept of a conflict graph was first introduced by Gusfield et al. [11] to explore the incompatibility of data. A maximal independent set of G is the maximum number of compatible mutations. However, a mutation group should be significantly large to be considered in the phylogeny tree. In the following subsection, we explain how to decide whether a mutation group is significant. Using Gusfield's algorithm [12], an efficient character-based method for perfect phylogeny tree construction, the consensus phylogeny tree for significant mutation groups in maximal independent is constructed.

The next step is iterative conflict resolution to enhance the tree and improve the accuracy of mutation profiles. If there is significant evidence for a conflicting mutation to be called in a non-conflicting mutation group already in the tree, we move the mutation to that group. To do so, we need to edit the binary profile of the mutation by changing the mutation call status for specific samples (see subsection 2.4). Significant conflicting mutation groups can be the result of mixture samples; where a mixture sample consists of multiple genetically distinct populations. Multiple subclones with distinct progression stages and phylogeny paths impose a DAG instead of tree for phylogenetic relationship between samples. Therefore, it is essential to identify the mixture sample and break it into several leaves representing its subclones. This procedure is explained in subsection 2.5.

2.2   Significant Mutation Group

In a phylogeny tree, where samples are leaves, each node corresponds to a mutation group, which contains mutations called in all samples belonging to the subtree rooted at that node. A tree with k non-root nodes supports k different somatic mutation groups. Given the total number of somatic mutations m, there are $(m + k - 1)!/m!\,(k - 1)!$ ways to distribute mutations into k groups. The probability of all groups having at least x mutations or more is:

$$\mathbb{P}(\text{size of all mutation groups} \geq x) = \frac{(m - xk + k - 1)!\,m!}{(m - xk)!\,(m + k - 1)!}$$

$$= \prod_{i=1}^{k-1} \frac{m - xk + i}{m + i} \approx \left(\frac{m - xk}{m}\right)^{k-1} \tag{1}$$

The approximation of the probability is obtained by the fact that number of mutation groups is much smaller than the total number of mutations in practice. Equation 1 is in fact the p-value of the event where the number of mutations in each group is larger than x. Using equation 1, one can choose x based on a significance threshold.

2.3   Maximum Number of Compatible Mutations

In order to find the maximum number of compatible mutations we build a weighted conflict graph. We first run Gusfield's algorithm on mutation matrix M to find all pairwise conflicts between mutation groups. We then build the weighted conflict graph $G = (V, E, W)$ as follows.

- V is a set of nodes, where each node represents a mutation group.
- E is a set of edges, where (v, u) exists if mutation groups v and u are in conflict.
- W is a set of node weights, where for each node it is equal to size of the mutation group.

The problem of finding maximum number of conflict-free mutations can be modeled as a maximal independent set problem on conflict graph G. The problem is known to be NP-complete and searching the exact solution is time consuming. The greedy algorithm works as follows: at each step, choose the uncovered node with the highest weight and remove all of its neighbors. As we discussed earlier, mutation groups should be large enough to be considered valid. Therefore, at each step we set k to the number of selected nodes and test if the new mutation group is large enough. The algorithm adds the mutation group to the solution set if it is sufficiently large. Otherwise it stops. In the rest of the paper, mutation groups that are in the solution set are denoted as valid groups and others as conflict groups.

2.4   Editing Mutation Profiles

Evidence of presence. The probability of seeing a base $\in \{A, G, T, C\}$ at each locus is $(1 - e)$ for the true underlying allele and $e/3$ otherwise, where e is the error rate at the base. For sample i and mutation j, let d_{ij} be the total coverage, and k_{ij} be the alternative allele coverage. The probability of not having genotype g_{ij}, i.e. all observed alternative allele bases result from sequencing error, is:

$$\mathbb{P}_{ij} = \mathbb{P}_{ij}\left(k_{ij}, d_{ij} \mid g_j \text{ is not a mutation in sample } s_i\right) = \binom{d_{ij}}{k_{ij}} (\frac{e_{ij}}{3})^{k_{ij}} (1 - e_{ij})^{d_{ij}-k_{ij}} \quad (2)$$

We compute the p-value of observing k_{ij} alternate allele bases at total coverage d_{ij} , assuming mutation g_j is not in sample s_i. Let the null hypothesis be that there is no mutation at the locus and all read bases are results of sequencing error.

$$p - value(k_{ij}, d_{ij}) = \sum_{k=k_{ij}}^{d_{ij}} \binom{d_{ij}}{k} (\frac{e_{ij}}{3})^k (1 - e_{ij})^{d_{ij}-k} \quad (3)$$

If the p-value is less than a chosen significance level we reject the null hypothesis and therefore there is evidence of presence for mutation g_j in sample s_i.

Target group. For each conflict group, all valid groups within a specific edit distance are potential target groups. By computing the p-value of evidence of presence for a conflicting mutation, we decide if the mutation is editable to a potential target group. A mutation can be editable to more than one potential target group. Each editing suggests a possible error pattern - XOR of source and target profiles determines which specific samples contain error. Our objective here is to edit the profile of as many mutations as possible while the number of distinct error patterns is minimized. Our problem of choosing target groups for conflicting mutations can be formulated as a classical set cover problem. Let X ={g_z} be the set of all mutations that can be moved to at least one target group. Denote Y as the set of subsets of X, where each subset represents the mutations that can be edited to the same potential target group. We look for minimum number of target groups, minimum elements of Y, which cover all mutations in X. The problem is known to be NP-complete and searching for the exact solution is not feasible. We applied the standard greedy algorithm: at each stage, choose the target group that contains the largest number of uncovered mutations. Figure 1 represents a simple example with four samples and set of conflicting mutations that can be moved to valid groups.

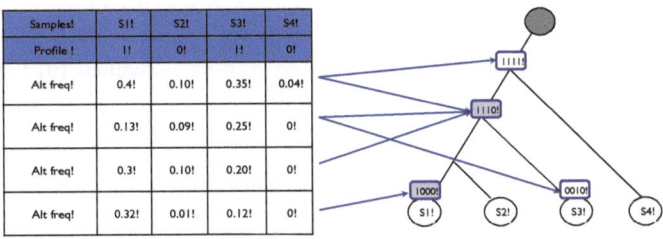

Samples!	S1!	S2!	S3!	S4!
Profile !	1!	0!	1!	0!
Alt freq!	0.4!	0.10!	0.35!	0.04!
Alt freq!	0.13!	0.09!	0.25!	0!
Alt freq!	0.3!	0.10!	0.20!	0!
Alt freq!	0.32!	0.01!	0.12!	0!

Fig. 1. Illustration of the profile editing strategy. Phylogeny tree of tumor samples s_1-s_4 is shown on the right. Group 1010 is a conflict group with four editable mutations. The arrow between each mutation and node of the tree represents evidence of presence for the mutation in the corresponding group. Groups {1110, 1000} are selected as target groups by the set cover algorithm.

2.5   Identification of Subpopulations

We used large conflict groups identified in section 2.3 to generate subclonal populations. There are several issues that our algorithm resolves: i) identification of mixture samples, ii) building a phylogeny of subclones, and iii) estimating proportion of each subclone. Algorithm 2 presents the pseudocode of our method that identifies subclones and adds them to the tree. It is not guaranteed to find the optimal tree. To identify the mixture samples, we first look for a sample involved in the maximum number of significant conflict groups. This sample is marked as α, and its corresponding conflict groups are marked as active conflicts. We extend α to a subtree if all leaves in the subtree are involved in all active conflicts. The next step searches for a set of samples with the same phylogeny history as a subclone of α. We find the ancestry of α such that all involving samples in active conflicts are its descendants. The highest subtree of the ancestry where all of its leaves are involved in active conflics is chosen as the subtree of interest, β. Then, the algorithm adds subclones to the tree where the phylogeny path of subclones is supported by both the previous mutation groups in the tree and the current active conflicts (Figure 2). We run our editing mutation profiles strategy to improve quality of mutations in subclone (Line 23). Since a lower number of reads covering subclones resulted in high false negative rates in mixture samples, our editing procedure is able to rescue several mutations missed in subclones.

Algorithm 2. Identification of Subclone

1.   **for** sample s $\in$ S **do**
2. $\text{cost}_1(s) \leftarrow \Sigma$ size of significant conflict groups that s shares
3.   **end**
4. $\alpha \leftarrow$ s with maximum cost_1
5.   set of active conflicts $\{\theta\} \leftarrow$ all conflict groups contributed to $\text{cost}_1(\alpha)$
6.   **for** r $\in$ ancestries of α **do**
7.   **if** all descendant samples of r are involved in all active conflicts $\{\theta\}$ **then**
8. $\alpha \leftarrow r$
9.   **else**
10. $\text{cost}_2(r) \leftarrow \Sigma$ size of active conflict θ,
              w . r is root of the deepest subtree that includes all involving samples of θ
11.   **end**
12. **end**
13. $\rho \leftarrow$ r with maximum cost_2
14. Assign each conflict group θ contributed to $\text{cost}_2(\rho)$ to root of a subtree such that union of
      samples in α and subtree is equal to the set of all involving samples of θ
15. **If** no θ assigned **then**
16.   Choose next best α, and **goto 5**
17. **end**
18. $\beta \leftarrow$ the highest subtree of ρ with assigned active conflict
19. **if** α is child of ρ **then**
20.   Add subclone of β to α // Figure 2a
21. **elseif** β is child of ρ **then**
22.   **while** β has a child v with an assigned θ **do**
23.   Run editing mutations approach for mutation group of v to target group θ
24.   **if** θ > mutation group of v **then**
25. $\beta \leftarrow v$
26.   **else goto 28**
27.   **end**
28.   Add subclone of α to β // Figure 2b
29. **else**
30.   Build a subtree for subclones of α and β using active conflict groups
31.   Add the subtree to ρ // Figure 2c
32. **end**
33. Estimate the proportion of subclones in each sample
34. Scale the depth coverage for each subclone regarding to its proportion
35. Update M

When subpopulations are identified, we can estimate of their size. Consider two subclones of a sample and the two nodes in tree where the phylogeny paths of subclones are first separated. These nodes refer to two mutation groups representing private subclonal mutations. The ratio between the average alternative allele frequencies of these mutation groups in that sample can give us an estimate of the proportion of subclones. Subsequently, depth coverage for private mutations should be scaled according to this proportion. Finally mutation matrix M (as well as SNV profiles) is updated by replacing the row for sample s with rows for all new subclones.

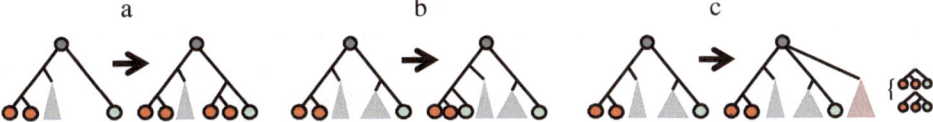

Fig. 2. Illustration of adding subpopulations to tree. Algorithm 2 identified leaf colored by green as α, node colored by gray as ρ, and leaves colored by orange as in subtree β. In case a, α is child of ρ, therefore β is inserted to α. In case b, α shares conflicts with all nodes from ρ to β, therefore α is inserted to β. In case c, a subtree (colored pink) including leaves in β and α is built and inserted to ρ. On the right side possible subtrees of subclones of β and α are shown.

3 Data

3.1 Simulation

To assess the performance of our method we developed a simulator to generate short read data for complex phylogeny trees. Using dwgsim [30], our program introduces somatic point mutations for all nodes of a given phylogeny tree and simulates paired end read sequencing data for each sample that include mutations of sample's phylogeny path. To accurately simulate the tumor development, tapered alternative allele frequency is modeled in the trees. We also simulate trees with mixture samples by combining reads produced for subclones.

We simulated 120 random trees for 3 to 10 samples; 40 of them with one or two mixed samples. For each node of tree we set a random number of somatic mutations within the range of 200 to 2000. The average alternative allele frequency is (f_1-node depth*f_2), where f_1 is the initial frequency rate (set to be 50%) and f_2 is the decreasing rate (set to be 5%). We ran all of our simulation cases on chromosome 22 of a diploid version of the NA12878 genome, built from the 1000 Genomes Project [24,29]. Short reads are produced in Illumina standard format with length 100, base error rate 2% and coverage 15x.

3.2 GATK Pipeline

We aligned the simulated short reads for each sample to the human genome (hg19) using BWA (0.5.9) [15]. The BAM files were then processed via base quality recalibration, duplicate marking, and local realignment following GATK's best practice workflow for variant detection (v3). Read pairs with identical coordinates and orientations were marked as duplicates using Picard MarkDuplicates tool and were ignored in the subsequent analysis. GATK's local realignment step was done to realign sample-level reads around known indels in order to minimize the number of mismatches. We used recommended indel sets from Mills et al. [17] and the 1000 Genomes Project from GATK resource bundle. We ran tools CountCovariates and TableRecalibration from GATK for the base quality recalibration using the standard set of covariates. GATK's Unified Genotyper multisample SNV caller was used on the realigned and recalibrated bam files to detect the non-reference sites among the samples and assign genotypes to each sample. In this step, the minimum phred-scale confidence thresholds at which variants should be emitted and called were both set to 30. Minimum base quality was set to 20. To reduce false positive rate, SNVs with the average coverage per sample less than half of the total coverage were discarded from the final set of identified variants.

4   Results

In our experiments we used p-value < 0.01 for deciding on the minimum size for a valid mutation group, and p-value < 0.1 for evidence of presence. We limited ourselves to edit mutation profiles with edit distance less than 3, and to one round of conflict resolution. Since mixture samples have a huge effect on the complexity of the problem, for better performance analyses we report the accuracy of the method for samples with and without subpopulations separately. Table 1 shows the summary of our simulation study on random trees without mixture samples. Somatic SNVs are achieved after filtering germline SNV calls from GATK output. Phylogenetically informative SNVs are all somatic SNVs except those called only in a single sample. Conflicting SNVs are in the minimum set chosen by the algorithm explained in subsection 2.3. The accuracy of GATK is defined as the ratio of somatic SNVs called with a correct profile to all somatic SNVs called by GATK. A given SNV profile is correct if only if the SNV is called correctly in all samples. The accuracy of the tree is measured as the accuracy of nodes in tree; a node is correct if it represents a true somatic mutation group.

Table 1. Summary of simulation study in samples without subpopulation. All values are the average over test cases with same number of samples.

Tumor Samples	Somatic SNVs	Phylogenetically informative SNVs	Ratio of Conflicting SNVs	GATK Accuracy	Tree Accuracy	Runtime (sec)
3	1534	349	0.10	0.72	1.00	0.50
4	3216	1203	0.11	0.79	0.98	0.82
5	2087	1504	0.22	0.70	0.97	1.38
6	2342	1505	0.19	0.72	0.99	3.16
7	3982	3258	0.30	0.55	0.98	7.90
8	3479	2716	0.21	0.66	0.98	10.02
9	3415	2642	0.19	0.64	0.95	10.50
10	3673	2972	0.26	0.59	0.96	16.48

As suggested by these results our method is highly accurate and efficient in tree reconstruction. More precisely, in our simulated trees when a sufficient number of SNVs was called in a true mutations group the algorithm did not miss that. However, there were examples of false positive mutation groups that followed a general scenario. Consider three samples $\{s_1, s_2, s_3\}$ where the only true SNVs are private mutations and germline mutations. Sequencing errors cause GATK miscalls some of the germline SNVs in a group shared by two samples. In most cases the false discovery rate was low, and the algorithm marked these kinds of groups as conflict and moved false SNVs to the correct group in the conflict resolution step. However, there were a few cases where GATK false discovery rate was quite high. Consequently, since there was no incompatibility in data our algorithm could not recognize the false SNVs as a conflict group. As a result, the node representing the false mutation group was added to tree. Obviously, reducing sequencing error rate or increasing sequencing coverage easily prevents this scenario. We assessed the accuracy of our editing mutations approach for those SNVs mapped to either germline or simulated mutations. Note that our method edits false somatic SNVs to germline mutation groups as well. The accuracy and precision for editing mutations are measured as follows:

$$Accuracy = \frac{\#\ SNVs\ edited\ to\ the\ correct\ group}{\#\ conflict\ SNVs} \tag{6}$$

$$Precision = \frac{\#\ SNVs\ edited\ to\ the\ correct\ group}{\#\ edited\ conflict\ SNVs} \tag{7}$$

Figure 3 presents the accuracy as well as the ratio of improvement in GATK accuracy by moving SNVs to the correct groups. Improvement over GATK accuracy is measured as the fraction of correct SNVs produced by our algorithm and the fraction of correct SNVs produced by the original GATK output. The results confirm that our editing mutations strategy is indeed effective with average accuracy 86% and average precision 92% on all simulated trees. Similar to GATK's multisample SNV

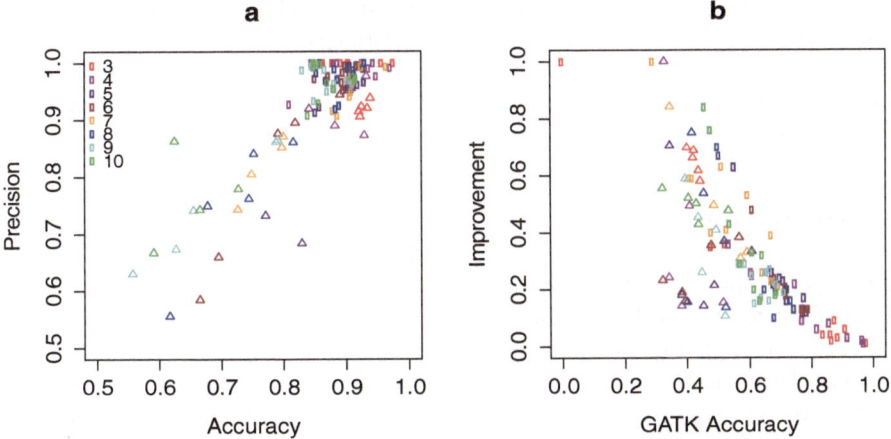

Fig. 3. Accuracy and precision of editing mutations. Each marker represents a simulated tree; circles for trees without mixture samples and triangles for trees with mixture samples. Markers are colored according to the number of samples in tree.

caller, our performance is dependent on the number of samples. The improvement over GATK accuracy was up to 32% in average, while the average GATK accuracy for somatic SNVs was only 67% on test cases without subpopulation, and 44% on cases with subpopulations.

Table 2. Simulation results for subpopulation detection. All values are the average over test cases with same number of samples.

Tumor Samples	GATK Accuracy	GATK Accuracy Subclonal SNVs	Improvement over GATK	Error Rate of Est. Size	Tree Accuracy
3	0.42	0.38	0.65	0.12	1.00
4	0.39	0.36	0.41	0.14	0.85
5	0.45	0.34	0.31	0.14	0.93
6	0.43	0.28	0.26	0.14	0.88
7	0.48	0.33	0.30	0.13	0.86
8	0.46	0.27	0.51	0.11	0.80
9	0.46	0.21	0.36	0.12	0.83
10	0.42	0.24	0.50	0.18	0.89

To evaluate the performance of our algorithm for subclone detection, we investigated the accuracy of predicted subclones in 40 simulated trees. Table 2 presents several measures including improvement over GATK, error rate of estimation of size of subclones, and accuracy of tree reconstruction. Despite the very low accuracy of GATK specifically for SNVs in mixture samples, only 30% on average, our method is able to identify subclones in 86% of cases. Since our conditions for adding subclones are quite restrictive, the false positive rate in our 120 simulated trees was zero. Our editing mutations strategy on average achieved a 42% improvement over GATK accuracy by recovering mutations in subclones. These results demonstrate the effectiveness of using phylogenetic relation between samples for subclone identification.

5   Discussion

In this paper we presented the first approach for tumor phylogeny tree reconstruction with conflict resolution for somatic mutations. Our algorithm first constructs a consensus perfect phylogeny tree based on the maximum number of non-conflicting mutations. Then, in iterative conflict resolution steps, it integrates more mutations into the tree by either editing the mutation profile or identifying significant subclones. Our conflict resolution approach results in a significant improvement in the accuracy of called somatic mutations. More specifically, our simulation analyses confirm that our conflict resolution step improves the accuracy of GATK's state of the art multisample SNV caller by up to 32%.

SNVs are not the only genetic changes whose study helps elucidate cancer evolution. There are additional mechanisms such as copy-number variations and complex structural variations involved in tumor development. CNVs are of particular interest because they represent some of the earliest mutagenic events, and thus are important drug targets. Although experimental and computational approaches for somatic structural variation detection are not yet mature, recent progress in copy number variation discovery motivates us to include CNVs in our phylogeny tree method in future work. Once phylogeny trees have been built based upon somatic mutations, we propose to map aneuploidy events onto them. To determine the order of aneuploidy events in phylogeny trees constructed based on somatic mutations, a statistical test can be employed[1]. This approach works if CNVs are conflict free. In the future, we would like to extend our conflict resolution strategy for aneuploidy events.

Reliable detection of mutations in subclones suffers from low sequence coverage. Although with our method we find evidence for subclones and we can find their path in phylogeny at a certain level of accuracy, the power of our method for detecting rare subclonal variants is still limited by genotype information. A deep resequencing analysis can be used to validate the phylogeny path of heterogeneous samples discovered by our method.

Acknowledgments. RS was supported by NSERC postdoctoral fellowship (PDF). DKH was supported by a STMicroelectronics Stanford Graduate Fellowship. SS and DK were supported by Stanford CURIS program. DEN was supported by training grant from NIH/NLM and a Bio-X Stanford Interdisciplinary Graduate Fellowship. This work was funded by a grant from KAUST to SB, and the Sequencing Initiative of the Stanford Department of Pathology to RW and AS.

References

1. Bansal, V., et al.: Accurate detection and genotyping of SNPs utilizing population sequencing data. Genome Res. 20, 537–545 (2010)
2. Beroukhim, R., et al.: The land-scape of somatic copy-number alteration across human cancers. Nature 463, 899–905 (2010)
3. Bignell, G.R., et al.: Signatures of mutation and selection in the cancer genome. Nature 463, 893–898 (2010)
4. Campbell, P.J., et al.: Subclonal phylogenetic structures in cancer revealed by ultra-deep sequencing. Proc. Natl. Acad. Sci. U S A 105(35), 13081–13086 (2008)
5. Chapman, M.A., et al.: Initial genome sequencing and analysis of multiple myeloma. Nature 471, 467–472 (2011)
6. DePristo, M., et al.: A framework for variation discovery and genotyping using next-generation DNA sequencing data. Nature Genet. 43, 491–498 (2011)
7. Ding, J., et al.: Feature based classifiers for somatic mutation detection in tumour-normal paired sequencing data. Bioinformatics 28(2), 167–175 (2012)
8. Gerlinger, M., et al.: Intratumor heterogeneity and branched evolution revealed by multiregion sequencing. N. Engl. J. Med. 366, 883–892 (2012)
9. Gerstung, M., et al.: Reliable detection of subclonal single-nucleotide variants in tumour cell populations. Nature Communications 3 (2011)

---

[1] The algorithm is presented in our parallel study [19].

10. Greenman, C., et al.: Patterns of somatic mutation in human cancer genomes. Nature 446, 153–158 (2007)
11. Gusfield, D., Eddhu, S., Langley, C.: Efficient Reconstruction of Phylogenetic. Networks with Constrained Recombination. In: Proc. IEEE CSB (2003)
12. Gusfield, D.: Efficient algorithms for inferring evolutionary trees. Networks 21, 19–28 (1991)
13. Larson, D.E., et al.: SomaticSniper: Identification of Somatic Point Mutations in Whole Genome Sequencing Data. Bioinformatics 28(3), 311–317 (2012)
14. Ley, T.J., et al.: DNA sequencing of a cytogenetically normal acute myeloid leukaemia genome. Nature 456, 66–72 (2008)
15. Li, H., Durbin, R.: Fast and accurate short read alignment with Burrows-Wheeler Transform. Bioinformatics 25, 1754–1760 (2009)
16. McKenna, A., et al.: The Genome Analysis Toolkit: a MapReduce framework for analyzing next-generation DNA sequencing data. Genome Res. 20, 1297–1303 (2010)
17. Mills, R.E., Luttig, C.T., Larkins, C.E., Beauchamp, A., Tsui, C., Pittard, W.S., Devine, S.E.: An initial map of insertion and deletion (INDEL) variation in the human genome. Genome Res. 16, 1182–1190 (2006)
18. muTect: A Reliable and Accurate Method for Detecting Somatic Mutations in Next Generation Cancer Genome Sequencing, https:// confluence.broadinstitute.org/display/CGATools/MuTect
19. Newburger, D.E., et al.: Genome Evolution during Progression to Breast Cancer (submitted)
20. Nik-Zainal, S., et al.: Mutational Processes Molding the Genomes of 21 Breast Cancers. Cell 149, 979–993 (2012)
21. Nik-Zainal, S., et al.: The life history of 21 breast cancers. Cell 149, 994–1007 (2012)
22. Pleasance, E.D., et al.: A comprehensive catalogue of somatic mutations from a human cancer genome. Nature 463, 191–196 (2010)
23. Roth, A., et al.: JointSNVMix: A Probabilistic Model For Accurate Detection of Somatic Mutations in Normal/Tumour Paired Next Generation Sequencing Data. Bioinformatics 28(7), 907–913 (2012)
24. Rozowsky, J., et al.: Allseq: analysis of allele Specific Expression and Binding in a Network Framework. Mol. Sys. Bio. (2011)
25. Schwartz, R., Schackney, S.E.: Applying unmixing to gene expression data for tumor phylogeny inference. BMC Bioinformatics 11, 42 (2010)
26. Shah, S., et al.: Mutational evolution in a lobular breast tumour profiled at single nucleotide resolution. Nature 461(7265), 809–813 (2009)
27. Stratton, M.R.: Exploring the genomes of cancer cells: progress and promise. Science 331, 1553–1558 (2011)
28. Stratton, M.R., Campbell, P.J., Futreal, P.A.: The cancer genome. Nature 458, 719–724 (2009)
29. The 1000 Genomes Project Consortium, et al.: A map of human genome variation from population-scale sequencing. Nature 467, 1061–1073 (2010)
30. Whole Genome Simulation, http://sourceforge.net/apps/mediawiki/dnaa/index.php
31. Zhang, G., et al.: Development of a phylogenetic tree model to investigate the role of genetic mutations in endometrial tumors. Oncol. Rep. 25(5), 1447–1454 (2011)
32. Zhang, Y., et al.: Molecular Evolutionary Analysis of Cancer Cell Lines. Mol. Cancer Ther. 9(2), 279–291 (2010)

Appendix: Significant Subset of Mutations

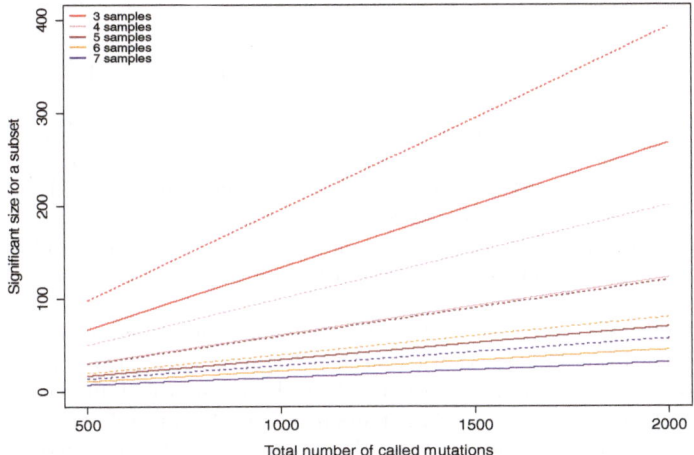

Fig. S1. Minimum size for a large enough mutation group. Solid lines presents values computed with p-value < 0.1 and dotted lines presents values computed with p-value < 0.01.

Abstract: Using the Fast Fourier Transform to Accelerate the Computational Search for RNA Conformational Switches

Evan Senter[1], Saad Sheikh[2], Ivan Dotu[1], Yann Ponty[3], and Peter Clote[1]

[1] Biology Department, Boston College, Chestnut Hill MA 02467 USA
{clote,ivan.dotu}@bc.edu, evansenter@gmail.com
[2] Computer Science Department, University of Florida, Gainesville FL 32601 USA
sheikh@cise.ufl.edu
[3] Laboratoire d'Informatique, Ecole Polytechnique, F-91128 Palaiseau Cedex, France
yann.ponty@lix.polytechnique.fr

Abstract. We describe the broad outline of a new thermodynamics-based algorithm, FFTbor, that uses the fast Fourier transform to perform polynomial interpolation to compute the Boltzmann probability that secondary structures differ by k base pairs from an arbitrary reference structure of a given RNA sequence. The algorithm, which runs in quartic time $O(n^4)$ and quadratic space $O(n^2)$, is used to determine the correlation between kinetic folding speed and the *ruggedness* of the energy landscape, and to predict the location of riboswitch expression platform candidates. The full paper appears in *PLoS ONE* (2012) 19 Dec 2012. A web server is available at http://bioinformatics.bc.edu/clotelab/FFTbor/.

Keywords: RNA secondary structure; partition function; fast Fourier transform; Lagrange interpolation.

In [2], we developed a dynamic programming algorithm, RNAbor, which simultaneously computes for each integer k, the Boltzmann probability $p_k = \frac{Z_k}{Z}$ of the subensemble of structures whose base pair distance to a given *initial*, or *reference*, structure S^* is k.[1] RNAbor stores the value of the (partial) partition functions $Z_k(i,j)$ for all $1 \leq i \leq j \leq n$ and $0 \leq k \leq n$, each of which requires quadratic time to compute. Thus it follows that RNAbor runs in time $O(n^5)$ and space $O(n^3)$, which severely limits its applicability to genomic annotation. This restriction is somewhat mitigated by the fact that in [1], we showed how to use sampling to efficiently approximate RNAbor in cubic time $O(n^3)$ and quadratic space $O(n^2)$, *provided* that the starting structure S^* is the minimum free energy (MFE) structure. We expect that a more efficient version of RNAbor could be used in applications in genomics and synthetic biology, to detect potential conformational switches – RNA sequences containing two or more (distinct) metastable structures.

---

[1] Here Z denotes the partition function, defined as the sum of all Boltzmann factors $\exp(-E(S)/RT)$, over all secondary structures S of a given RNA sequence, R denotes the universal gas constant and T absolute temperature. Similarly Z_k denotes the sum of all Boltzmann factors of all structures S, whose base pair distance to the initial structure S^* is exactly k.

M. Deng et al. (Eds.): RECOMB 2013, LNBI 7821, pp. 264–265, 2013.
© Springer-Verlag Berlin Heidelberg 2013

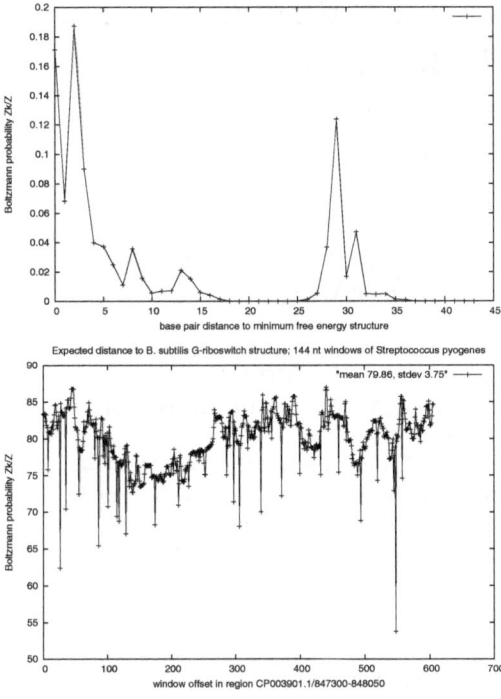

Fig. 1. *(Top)* Output of `FFTbor` on the 46 nt spliced leader conformational switch of *Leptomonas collosoma*, where reference structure S^* is taken to be the minimum free energy structure. *(Bottom)* Expected base pair distance $\sum_k k \cdot Z_k/Z$ from the reference structure of the guanine riboswitch of *Bacillus subtilis*, depicted in Figure 1A of [3]. `FFTbor` was run on all 144 nt windows of CP003901.1/847300-848050, comprising the 5′ untranslated region of the XPT gene (guanosine monophosphate reductase, with coding region at CP003901.1/848026-848607) of the unrelated organism *Streptococcus pyogenes A20*. `FFTbor` detects the guanine riboswitch at position 847848, where expected base pair distance to S^* is minimized (53.79) corresponding to a Z-score of -6.95. This prediction corresponds well with the Rfam prediction at nearby position 847844.

In this abstract, we announce a radically different algorithm, `FFTbor`, that uses polynomial interpolation to compute the coefficients $p_0, \ldots, p_{n-1}$ of the polynomial $p(x) = p_0 + p_1 x + \cdots + p_{n-1} x^{n-1}$, where p_k is defined by $p_k = \frac{Z_k}{Z}$. Due to severe numerical instability issues in both the Lagrange interpolation formula and in Gaussian elimination, we employ the Fast Fourier Transform (FFT) to compute the inverse Discrete Fourier Transform (DFT) on values $y_0, \ldots, y_{n-1}$, where $y_k = p(\omega^k)$ and $\omega = e^{2\pi i/n}$ is the principal nth complex root of unity. This gives rise to an improved version of `RNAbor`, denoted `FFTbor`, which runs in time $O(n^4)$ and space $O(n^2)$ on a single processor, and in time $O(n^3)$ on a theoretical n-cores processor or cluster (e.g. using `OpenMP`). Figure 1 (top) depicts the *rugged* energy landscape typical of a conformational switch, while Figure 1 (bottom) depicts expected base pair distance, of each size 144 window in the 5′-UTR of *S. pyogenes*, to the XPT riboswitch structure of *B. subtilis*.

References

1. Clote, P., Lou, F., Lorenz, W.A.: Maximum expected accuracy structural neighbors of an RNA secondary structure. BMC Bioinformatics 13(suppl. 5), S6 (2012)
2. Freyhult, E., Moulton, V., Clote, P.: Boltzmann probability of RNA structural neighbors and riboswitch detection. Bioinformatics 23(16), 2054–2062 (2007)
3. Serganov, A., Yuan, Y.R., Pikovskaya, O., Polonskaia, A., Malinina, L., Phan, A.T., Hobartner, C., Micura, R., Breaker, R.R., Patel, D.J.: Structural basis for discriminative regulation of gene expression by adenine- and guanine-sensing mRNAs. Chem. Biol. 11(12), 1729–1741 (2004)

MethylCRF, an Algorithm for Estimating Absolute Methylation Levels at Single CpG Resolution from Methylation Enrichment and Restriction Enzyme Sequencing Methods

Michael Stevens[1,2], Jeffrey B. Cheng[3], Mingchao Xie[1],
Joseph F. Costello[4,*], and Ting Wang[1,2,*]

[1] Department of Genetics, Center for Genome Sciences and Systems Biology,
Washington University School of Medicine, MO 63108
[2] Department of Computer Science and Engineering, Washington University in St. Louis,
MO 63130
[3] Department of Dermatology, University of California San Francisco, CA 94143
[4] Brain Tumor Research Center, Department of Neurosurgery, Helen Diller Family
Comprehensive Cancer Center, University of California San Francisco, CA 94143
jcostello@cc.ucsf.edu, twang@genetics.wustl.edu

Keywords: DNA methylation, Conditional Random Fields, Epigenomics.

1 Introduction

We introduce MethylCRF, a novel Conditional Random Fields-based algorithm to integrate methylated DNA immunoprecipitation (MeDIP-seq) and methylation-sensitive restriction enzyme (MRE-seq) sequencing data to predict DNA methylation levels at single CpG resolution. MethylCRF was benchmarked for accuracy against Infinium arrays, RRBS, whole-genome shotgun-bisulfite (WGBS) sequencing and locus specific-bisufite sequencing on the same DNA. MethylCRF transformation of MeDIP/MRE was equivalent to a biological replicate of WGBS in quantification, coverage and resolution, providing a lower cost and widely accessible strategy to create full methylomes.

2 Methods

Sequencing-based DNA methylation profiling methods provide an unprecedented opportunity to map complete DNA methylomes. These include whole genome bisulfite sequencing (WGBS, MethylC-seq[1] or BS-seq[2]), Reduced-Representation Bisulfite-Sequencing (RRBS) [3], and enrichment-based methods such as MeDIP-seq[4,5], MBD-seq[6] and MRE-seq[5]. There are few complete single nucleotide DNA methylome

---

[*] Corresponding authors.

M. Deng et al. (Eds.): RECOMB 2013, LNBI 7821, pp. 266–268, 2013.
© Springer-Verlag Berlin Heidelberg 2013

maps of humans due to the high cost of producing such methylomes using whole genome bisulfite sequencing based methods. In contrast, many more lower-cost DNA methylomes of either lower resolution or lower coverage have been generated across diverse biological and disease states.

We developed a combined computational and experimental strategy to produce single CpG resolution DNA methylomes of all 28 million CpGs for human at a fraction of the cost of whole genome bisulfite sequencing method. Our computational model "methylCRF" is based on Conditional Random Fields and models the conditional probability of the variables of interest, (in our case CpG methylation levels) given the predictor values (e.g., MRE-seq, MeDIP-seq, and genomic context). The complete model contains features including MeDIP-seq and MRE-seq measurements covering individual CpGs, distance between neighboring CpGs, and genomic annotations including CpG islands, genes, repeats, and evolutionary conservation of DNA sequences. Using this model, we integrate MeDIP-seq and MRE-seq data to predict DNA methylation at single CpG level, similar to what whole genome bisulfite sequencing can do. However, the cost of our two assays combined is less than 10% of that of whole genome bisulfite.

3   Results

MethylCRF was benchmarked for accuracy against Infinium arrays, RRBS, WGBS sequencing and locus specific-bisufite sequencing on the same DNA. MethylCRF transformation of MeDIP/MRE was equivalent to a biological replicate of WGBS in quantification, coverage and resolution. We used conventional bisulfite-cloning-sequencing strategy to validate several loci where our predictions do not agree with whole genome bisulfite data, and in majority of the cases (11 out of 12) methylCRF predictions of methylation level based on MeDIP-seq and MRE-seq agree better with validated results than does the whole genome bisulfite sequencing. Therefore, methylCRF provides a lower cost and widely accessible strategy to create full DNA methylomes.

4   Conclusions

Our results suggest that methylCRF is an effective statistical framework capable of integrating two fundamentally different sequencing-based DNA methylation assays, MeDIP-seq and MRE-seq, to predict genome-wide, single CpG resolution methylome maps. The concordance of our methylCRF predictions with WGBS falls within the range of concordance between two WGBS experiments on similar cells. MethylCRF will thus significantly increase the value of high-coverage DNA methylomes produced using much less expensive methods, and provide a general statistical framework for integrating contributions from various types of DNA methylation data regardless of their coverage, resolution, and nature of their readout.

References

1. Lister, R., et al.: Human DNA methylomes at base resolution show widespread epigenomic differences. Nature 462 (2009)
2. Laurent, L., et al.: Dynamic changes in the human methylome during differentiation. Genome Res. 20, 320–331 (2010)
3. Meissner, A., et al.: Genome-scale DNA methylation maps of pluripotent and differentiated cells. Nature 454, 766–770 (2008)
4. Weber, M., et al.: Chromosome-wide and promoter-specific analyses identify sites of differential DNA methylation in normal and transformed human cells. Nat. Genet. 37, 853–862 (2005)
5. Maunakea, A.K., et al.: Conserved role of intragenic DNA methylation in regulating alternative promoters. Nature 466, 253–257 (2010)
6. Serre, D., Lee, B.H., Ting, A.H.: MBD-isolated Genome Sequencing provides a high-throughput and comprehensive survey of DNA methylation in the human genome. Nucleic Acids Res. 38, 391–399 (2009)

Counting Motifs in the Entire Biological Network from Noisy and Incomplete Data
(Extended Abstract)

Ngoc Hieu Tran[1], Kwok Pui Choi[1,2,*], and Louxin Zhang[2,3,*]

[1] Department of Statistics and Applied Probability
[2] Department of Mathematics
[3] Graduate School for Integrative Sciences and Engineering
National University of Singapore, Singapore
{stackp,matzlx}@nus.edu.sg

Small over-represented motifs in biological networks are believed to represent essential functional units of biological processes. A natural question is to gauge whether a motif occurs abundantly or rarely in a biological network. Given that high-throughput biotechnology is only able to interrogate a portion of the entire biological network with non-negligible errors, we develop a powerful method to correct link errors in estimating undirected or directed motif counts in the entire network from noisy subnetwork data.

Consider a network $\mathcal{G}(V, E)$ with n nodes, where V is the set of nodes and E the set of links. Let $\mathcal{G}^{\mathrm{obs}}\left(V^{\mathrm{obs}}, E^{\mathrm{obs}}\right)$ be an observed subnetwork of $\mathcal{G}$, and let n^{obs} be the number of nodes in $\mathcal{G}^{\mathrm{obs}}$. For an arbitrary motif $\mathcal{M}$ with m nodes, let $N_{\mathcal{M}}$ and $N_{\mathcal{M}}^{\mathrm{obs}}$ respectively denote the number of occurrences of $\mathcal{M}$ in $\mathcal{G}$ and $\mathcal{G}^{\mathrm{obs}}$. Assuming that $\mathcal{G}^{\mathrm{obs}}$ and n are known, our interest is to estimate $N_{\mathcal{M}}$ from the observed subnetwork $\mathcal{G}^{\mathrm{obs}}$. Following [3,5], we modeled an observed subnetwork as the outcome of a uniform node sampling process in the following sense: each node from V is independently sampled with probability p, $0 < p < 1$, and the subgraph induced from E by the sampled nodes is the observed subnetwork $\mathcal{G}^{\mathrm{obs}}$. When $\mathcal{G}^{\mathrm{obs}}$ is free from link errors, $\widehat{N}_{\mathcal{M}}$ given in Eqn. (1) is an asymptotically unbiased and consistent estimator for $N_{\mathcal{M}}$. For realistic estimation, this estimator, however, is inaccurate due to link errors in real datasets. Modelling spurious and missing links as a random process with false positive rate r_+ and false negative rate r_-, we corrected the bias and adjusted $\widehat{N}_{\mathcal{M}}$ to $\widetilde{N}_{\mathcal{M}}$ in Eqn. (2):

$$\widehat{N}_{\mathcal{M}} = \frac{\binom{n}{m}}{\binom{n^{\mathrm{obs}}}{m}} N_{\mathcal{M}}^{\mathrm{obs}}, \tag{1}$$

$$\widetilde{N}_{\mathcal{M}} = \frac{1}{(1 - r_+ - r_-)^s} \left(\widehat{N}_{\mathcal{M}} - \widetilde{W}_{\mathcal{M}} \right), \tag{2}$$

where s is the number of links in $\mathcal{M}$ and $\widetilde{W}_{\mathcal{M}}$ a function of n, r_-, r_+, and $\widetilde{N}_{\mathcal{M}'}$ for all proper sub-motifs $\mathcal{M}'$ of $\mathcal{M}$. We derived the bias-corrected estimator $\widetilde{N}_{\mathcal{M}}$ for triad and quadriad motifs in directed and undirected networks.

---

* Corresponding authors.

M. Deng et al. (Eds.): RECOMB 2013, LNBI 7821, pp. 269–270, 2013.

Next, we re-estimated the interactome size (i.e., the number of interactions) in the entire protein-protein interaction (PPI) networks of *S. cerevisiae*, *C. elegans*, *H. sapiens*, and *A. thaliana*. We used PPI datasets produced from yeast two-hybrid (Y2H) experiments [1,4,6,7]. We also obtained the quality parameters of those datasets from the original papers. Our estimators using the false positive and false negative rates yielded similar results to those previously obtained in [1,4,6,7] by considering the precision and sensitivity. We further estimated the number of triangles in each of the interactomes. We found that although the *A. thaliana* interactome is about 1.38 times that of the human interactome, it contains far fewer triangles than the human interactome does. The triangle density of the human and *C. elegans* interactomes are similar and 2.5 times that of the *A. thaliana* and *S. cerevisiae*. The human interactome is only 12.4 times that of the *S. cerevisiae* interactome, and yet the number of triangles of the former is about 125 times that of the latter, that is 2.5 times of what is expected.

Recently, the transcription factor (TF) regulatory network of forty-one human cell and tissue types were obtained from genome-wide in vivo DNaseI footprints map [2]. Surprisingly, we found that there is a very strong positive linear correlation between the counts in the TF networks of different cell types even for the triad and quadriad motifs that are topologically very different.

In conclusion, by taking into account of spurious and missing link rates, we have developed for the first time a powerful method for estimating the number of motif occurrences in the entire network from noisy subnetwork data. Such a method is important because exact motif enumeration is possible only if the network is completely known, which is often not the case for biological networks. The dynamics of human PPI and TF networks uncovered in the present paper are consistent with biological intuition about the complexity of life.

References

1. *Arabidopsis* Interactome Mapping Consortium: Evidence for network evolution in an *Arabidopsis* interactome map. Science 333, 601–607 (2011)
2. Neph, S., et al.: Circuitry and dynamics of human transcription factor regulatory networks. Cell 150, 1274–1286 (2012)
3. Rottger, R., Ruckert, U., Taubert, J., Baumbach, J.: How little do we actually know? - On the size of gene regulatory networks. IEEE/ACM Trans. Comput. Biol. Bioinform. 9, 1293–1300 (2012)
4. Simonis, N., et al.: Empirically controlled mapping of the *C. elegans* protein-protein interactome network. Nature Methods 6, 47–54 (2009)
5. Stumpf, M.P.H., et al.: Estimating the size of the human interactome. Proc. Natl. Acad. Sci. U.S.A. 105, 6959–6964 (2008)
6. Venkatesan, K., et al.: An empirical framework for binary interactome mapping. Nature Methods 6, 83–90 (2009)
7. Yu, H., et al.: High-quality binary protein interaction map of the yeast interactome network. Science 322, 104–110 (2008)

Extracting Structural Information from Residual Chemical Shift Anisotropy: Analytic Solutions for Peptide Plane Orientations and Applications to Determine Protein Structure[*]

Chittaranjan Tripathy[1], Anthony K. Yan[1,2],
Pei Zhou[2], and Bruce Randall Donald[1,2,**]

[1] Department of Computer Science, Duke University, Durham, NC 27708
[2] Department of Biochemistry, Duke University Medical Center, Durham, NC 27710
brd+recomb13@cs.duke.edu

Abstract. Residual dipolar coupling (RDC) and residual chemical shift anisotropy (RCSA) provide orientational restraints on internuclear vectors and the principal axes of chemical shift anisotropy (CSA) tensors, respectively. Mathematically, while an RDC represents a single sphero-conic, an RCSA can be interpreted as a *linear combination* of *two* sphero-conics. Since RDCs and RCSAs are described by a molecular alignment tensor, they contain inherent structural ambiguity due to the symmetry of the alignment tensor and the symmetry of the molecular fragment, which often leads to more than one orientation and conformation for the fragment consistent with the measured RDCs and RCSAs. While the orientational multiplicities have been long studied for RDCs, structural ambiguities arising from RCSAs have not been investigated. In this paper, we give exact and tight bounds on the number of peptide plane orientations consistent with multiple RDCs and/or RCSAs measured in one alignment medium. We prove that at most 16 orientations are possible for a peptide plane, which can be computed in closed form by solving a merely quadratic equation, and applying symmetry operations. Furthermore, we show that RCSAs can be used in the initial stages of structure determination to obtain highly accurate protein backbone global folds. We exploit the mathematical interplay between sphero-conics derived from RCSA and RDC, and protein kinematics, to derive quartic equations, which can be solved in closed-form to compute the protein backbone dihedral angles (ϕ, ψ). Building upon this, we designed a novel, sparse-data, polynomial-time divide-and-conquer algorithm to compute protein backbone conformations. Results on experimental NMR data for the protein human ubiquitin demonstrate that our algorithm computes backbone conformations with high accuracy from $^{13}C'$-RCSA or ^{15}N-RCSA, and N-H^N RDC data. We show that the structural information present in $^{13}C'$-RCSA and ^{15}N-RCSA can be extracted analytically, and used in a rigorous algorithmic framework to compute a high-quality protein backbone global fold, from a limited amount of NMR

[*] Grant sponsor: National Institutes of Health; Grant numbers: R01 GM-65982 and R01 GM-78031 to BRD, and R01 GM-079376 to PZ.
[**] Corresponding author

M. Deng et al. (Eds.): RECOMB 2013, LNBI 7821, pp. 271–284, 2013.
© Springer-Verlag Berlin Heidelberg 2013

data. This will benefit automated NOE assignment and high-resolution protein backbone structure determination from sparse NMR data.

1   Introduction

Nuclear magnetic resonance (NMR) spectroscopy is one of the most powerful experimental techniques for the study of macromolecular structure and dynamics, particularly for proteins in solution. NMR complements X-ray crystallography in that it can obtain structural information for proteins that are hard to crystallize, intrinsically disordered proteins [18,38], and denatured proteins [36]. NMR has also emerged as a major tool to probe protein-ligand interactions [20] under near physiological conditions, as well as to investigate invisible excited states in proteins and extract information on these minor conformers [3,4].

The NMR technique is based on the sensitivity of magnetic properties of the nuclei to its local chemical and electronic environment in the presence of a strong and static external magnetic field of the spectrometer. The observable for a nucleus, called its *chemical shift*, arises from the nuclear shielding effect caused by the local magnetic field, induced by the circulation of electrons surrounding the nucleus. This induced field can be described by a second-rank chemical shift (or shielding) anisotropy (CSA) tensor, which can be rewritten to correspond to the isotropic, anisotropic antisymmetric, and anisotropic symmetric parts.

In solution, due to isotropic molecular tumbling, the anisotropic parts of the CSA tensor average out to zero due to rotational diffusion, and only the remaining isotropic chemical shift, δ_{iso}, is observed. While isotropic chemical shifts play an increasingly important role in NMR structure elucidation and refinement [9, 45,59], and dynamics [1], our understanding of the relationship between structure and chemical shifts is still far from complete, especially in the context of proteins [68] and other macromolecules. The antisymmetric part of the CSA tensor often has a negligible effect on the relaxation rates, and hence can be ignored. The symmetric part of the CSA tensor is a traceless, second-rank tensor, usually represented by its three eigenvalues in the *principal order frame* and the orientations of its three principal axes (eigenvectors) with respect to the molecular frame.

Accurate knowledge of CSA tensors is essential to the quantitative determination and interpretation of dynamics, relaxation interference [33,46], residual chemical shift anisotropy (RCSA) [12,39,63], and NMR structure determination and refinement [15, 26, 27, 61]. In solid-state NMR, CSA tensors can be determined from powder patterns [32], or magic-angle spinning (MAS) spectra [60]. In solution NMR, the CSA tensors can be determined from relaxation and CSA-dipolar cross-correlation experiments [30, 62, 63], or from offsets in resonance peaks upon partial alignment [7]. The presence of an alignment medium introduces partial alignment in the molecules. Residual dipolar coupling (RDC) [34] can easily be extracted, often with high-precision, as the difference between the line-splittings in weakly aligned and isotropic buffer solutions. The small difference in chemical shifts observed under partially aligned conditions and isotropic conditions gives rise to the RCSA effect [39]. Techniques to measure RCSA include temperature-dependent phase transition of certain liquid crystals [12], varying concentration of the aligning medium [6], utilizing MAS to eliminate the effects of protein alignment relative to the magnetic field [21], and the recently introduced two-stage NMR tube method by Prestegard and coworkers [28].

Similar to RDC, RCSA contains rich and orientationally sensitive structural information [46] that complements other types of structural restraints such as the nuclear Overhauser effect (NOE) distance restraints and scalar couplings. Since amide nitrogen RCSA (^{15}N-RCSA) and carbonyl RCSA (^{13}C'-RCSA) can be measured to high precision [12, 63], they have been used as structural restraints for protein structure validation [14] and refinement [10, 26, 27, 53] during the *final* stages of traditional protein structure determination [5, 42]. However, to our knowledge, RCSAs have never been used in the *initial* stages of structure computation to compute the backbone global fold of a protein. Methods that primarily use RDCs in the initial stages of structure computation [16, 19, 47, 65] have been shown to have many advantages over traditional NOE-based structure determination protocols. Recently, in [35,45], Baker and Bax and coworkers have developed protocols within the ROSETTA [25] protein structure modeling framework, that use only backbone chemical shifts, RDCs, and amide proton NOE distances to compute high-quality protein backbone conformations. However, these approaches do not use structural restraints from RCSA data. Further, most of these approaches use stochastic search, and therefore, lack any algorithmic guarantee on the quality of the solution or running time.

In recent work from our laboratory [17, 51, 52, 56, 58, 64], polynomial-time algorithms have been proposed for high-resolution backbone global fold determination from a minimal amount of RDC data. This framework is called RDC-ANALYTIC. The core of the RDC-ANALYTIC suite is based on representing RDC and protein kinematics in algebraic form, and solving them analytically to obtain closed-form solutions for the backbone dihedrals and peptide plane orientations, in a divide-and-conquer framework to compute the global fold. These algorithms have been used in [57, 65, 66] to develop new algorithms for NOE assignment, which led to the development of a new framework [65] for high-resolution protein structure determination, which was used prospectively to solve the solution structure of the FF Domain 2 of human transcription elongation factor CA150 (FF2) (PDB id: 2KIQ). Recently, we have developed a novel algorithm, POOL [51,52], within the RDC-ANALYTIC framework, to determine protein loop conformations from a minimal amount of RDC data. However, RDC-ANALYTIC did not exploit orientational restraints from RCSA data.

In this work, we show that orientational restraints from ^{13}C'-RCSAs or ^{15}N-RCSAs can be used in combination with N-H^N RDCs in an analytic, systematic search-based, divide-and-conquer framework to determine individual peptide plane orientations and protein backbone conformations. Our new algorithm is a part of the RDC-ANALYTIC framework, and is called RDC-CSA-ANALYTIC. Two demonstrations of applying RDC-CSA-ANALYTIC, (1) using ^{13}C'-RCSA and N-H^N RDC, and (2) using ^{15}N-RCSA and N-H^N RDC, to compute the global fold of ubiquitin, and promising results from the application of our algorithm on real biological NMR data, are presented below.

Furthermore, we pursued the fundamental question of determining the peptide plane orientations when 3 measurements are used, each of which is either an RCSA on a nucleus or an RDC on an internuclear vector on the peptide plane. This is important, because for perdeuterated proteins, RDCs are usually measured on N-H^N, C$^\alpha$-C$'$, and C$'$-N coplanar vectors. Further, ^{15}N-RCSA and ^{13}C$'$-RCSA can be interpreted with respect to the CSA tensor components on the peptide plane. Previously, Brüschweiler and coworkers [23] showed that it is possible to derive analytic expressions, containing transcendental functions, for the 16 possible peptide plane orientations using only RDCs. However, they only showed a lower bound on the number of solutions. In addition, their work did not consider orientational restraints from RCSAs. In this work, we derive closed-form analytic expressions for the peptide plane orientations from RCSAs and RDCs on coplanar vectors measured in one alignment medium. We prove that *at most* 16 orientations are possible for the peptide plane, which can be computed in closed form by solving a *quadratic equation*, and then applying symmetry operations. This is remarkable because for decades, all previous approaches required, at worst, solving equations involving transcendental functions, or at best, solving polynomial equations of degree 4 or higher. We give a $\Theta(1)$-time deterministic algorithm, 3PLANAR, to compute all possible peptide plane orientations.

2   Theory and Methods

2.1   Residual Dipolar Coupling

The residual dipolar coupling r between two spin-$\frac{1}{2}$ nuclei a and b, described by a unit internuclear vector $\mathbf{v}$, due to anisotropic distribution of orientations in the presence of an alignment medium, relative to a strong static magnetic field direction $\mathbf{B}$ is given by [16, 48, 49]

$$r = D_{\max}\mathbf{v}^T\mathbf{S}\mathbf{v}. \tag{1}$$

Here the dipolar interaction constant $D_{\max}$ depends on the gyromagnetic ratios of the nuclei a and b, and the vibrational ensemble-averaged inverse cube of the distance between them. $\mathbf{S}$ is the *Saupe order matrix* [40], or *alignment tensor* that specifies the ensemble-averaged anisotropic orientation of the protein in the laboratory frame. $\mathbf{S}$ is a 3×3 symmetric, traceless, rank 2 tensor with five independent elements [34, 48, 49]. Letting $D_{\max} = 1$ (i.e., scaling the RDCs appropriately), and considering a global coordinate frame that diagonalizes $\mathbf{S}$, often called the *principal order frame* (POF), Eq. (1) can be written as

$$r = S_{xx}x^2 + S_{yy}y^2 + S_{zz}z^2, \tag{2}$$

where S_{xx}, S_{yy} and S_{zz} are the three diagonal elements of a diagonalized alignment tensor $\mathbf{S}$, and x, y and z are, respectively, the x, y and z components of the unit vector $\mathbf{v}$ in a POF that diagonalizes $\mathbf{S}$. Note that $S_{xx} + S_{yy} + S_{zz} = 0$ because $\mathbf{S}$ is traceless. Since $\mathbf{v}$ is a unit vector, an RDC constrains the corresponding internuclear vector $\mathbf{v}$ to lie on the intersection of a concentric unit

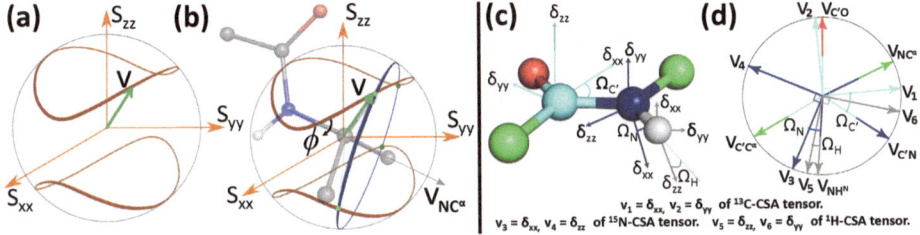

Fig. 1. *Left Panel.* (a) The internuclear vector **v** (green arrow) is constrained to lie on one of the two pringle-shaped RDC sphero-conic curves (brown) lying on a unit sphere. (b) The kinematic circle (blue), of the internuclear vector **v** (here $\mathbf{v}_{C^\alpha H^\alpha}$), around the axis $\mathbf{v}_{NC^\alpha}$, intersects the sphero-conic curves in at most four points (green dots) leading to a maximum of four possible orientations for **v**. The case is similar when ϕ is solved and a ψ-defining RDC is measured for an internuclear vector **v**, e.g., $\mathbf{v}_{NH^N}$. *Right Panel.* (c) Orientations of the principal components of ^{13}C-, ^{15}N- and ^{1}H-CSA tensors with respect to the peptide plane are shown in cyan, blue and gray, respectively. δ_{zz} is the most- and δ_{xx} is the least-shielded component. For each tensor, one of the components is approximately perpendicular to the peptide plane; therefore, the other two components lie on the peptide plane, and are completely defined by the angle Ω. The values of the angles $\Omega_{C'}$, Ω_N and Ω_H can be set to fixed values [10,12], e.g., 38°, 19° and 8° as reported in [12]. (d) The wagon wheel shows the CSA tensor components on the peptide plane along with the bond vectors drawn using C′ atom as the origin.

sphere and a quadric (Eq. (2)). This gives a pair of closed curves inscribed on the unit sphere that are diametrically opposite to each other (see Figure 1 (a, b)). These curves are known as *sphero-conics* or *sphero-quartics* [8,37]. Since $|\mathbf{v}| = 1$, Eq. (2) can be rewritten in the following form:

$$ax^2 + by^2 = c, \tag{3}$$

where $a = S_{xx} - S_{zz}$, $b = S_{yy} - S_{zz}$, and $c = r - S_{zz}$. Henceforth, we refer to Eq. (3) as the *reduced RDC equation*. For further background on RDCs and RDC-based structure determination, the reader is referred to [16,17,34,48,49].

2.2 Residual Chemical Shift Anisotropy

For a given nucleus, the difference in chemical shifts between the liquid crystalline phase (δ_{aniso}) and the isotropic phase (δ_{iso}) is the RCSA, and is given by [12, 14,39,63]

$$\Delta\delta = \delta_{\text{aniso}} - \delta_{\text{iso}} = \frac{2}{3} \sum_{i \in \{x,y,z\}} \langle P_2(\cos\theta_{ii}) \rangle \delta_{ii}, \tag{4}$$

where $P_2(\alpha) = (3\alpha^2 - 1)/2$ is the second Legendre polynomial, δ_{xx}, δ_{yy} and δ_{zz} are the principal components of the CSA tensor, and θ_{xx}, θ_{yy} and θ_{zz} are the respective angles between the principal axes of the traceless, second-rank CSA tensor and the magnetic field direction **B**. The angle brackets, $\langle \cdots \rangle$, denote ensemble averaging. After suitable algebraic manipulations, we can write Eq. (4) as

$$\Delta\delta = \lambda_1 \boldsymbol{\delta}_{xx}^T \mathbf{S} \boldsymbol{\delta}_{xx} + \lambda_2 \boldsymbol{\delta}_{yy}^T \mathbf{S} \boldsymbol{\delta}_{yy}, \tag{5}$$

where

$$\lambda_1 = \frac{1}{3}(2\delta_{xx} + \delta_{yy}) \tag{6}$$

and

$$\lambda_2 = \frac{1}{3}(2\delta_{yy} + \delta_{xx}) \tag{7}$$

are two constants since δ_{xx} and δ_{yy} are known experimentally. Eq. (5) therefore expresses the $\Delta\delta$ as a *linear combination* of two virtual RDC sphero-conics on two unit vectors $\boldsymbol{\delta}_{xx}$ and $\boldsymbol{\delta}_{yy}$ that can be realized on the peptide plane. This derivation applies, *mutatis mutandis*, to any two choices of unit vectors from $\{\boldsymbol{\delta}_{xx}, \boldsymbol{\delta}_{yy}, \boldsymbol{\delta}_{zz}\}$. Working in the POF of the molecular alignment tensor, we can write the above equation as

$$\Delta\delta = S_{xx}(\lambda_1 x_1^2 + \lambda_2 x_2^2) + S_{yy}(\lambda_1 y_1^2 + \lambda_2 y_2^2) + S_{zz}(\lambda_1 z_1^2 + \lambda_2 z_2^2), \tag{8}$$

where the unit vectors $\boldsymbol{\delta}_{xx} = (x_1, y_1, z_1)^T$ and $\boldsymbol{\delta}_{yy} = (x_2, y_2, z_2)^T$ in the POF of the molecular alignment tensor. Eq. (8) can be simplified to the following form

$$a(\lambda_1 x_1^2 + \lambda_2 x_2^2) + b(\lambda_1 y_1^2 + \lambda_2 y_2^2) = c, \tag{9}$$

where $a = S_{xx} - S_{zz}, b = S_{yy} - S_{zz}$, and $c = \Delta\delta - (\lambda_1 + \lambda_2)S_{zz}$. Henceforth, we refer to Eq. (9) as the *reduced RCSA equation*.

Figure 1 (c, d) shows the local structure of a peptide plane on which the principal components of ^{13}C-, ^{15}N- and ^{1}H-CSA tensors are realized. δ_{zz} and δ_{xx} are respectively the most- and least- shielded CSA tensor components. We denote ^{13}C'-RCSA, ^{15}N-RCSA and ^{1}H-RCSA by $\Delta\delta_{C'}$, $\Delta\delta_N$ and $\Delta\delta_H$, respectively.

2.3   The RDC-CSA-ANALYTIC Algorithm

RDC-CSA-ANALYTIC computes the backbone global fold of proteins using RDC and RCSA data in one alignment medium. Table 1 describes the RDC and RCSA types that RDC-CSA-ANALYTIC uses to compute the backbone dihedrals exactly and in closed form. A ϕ-*defining* RDC is used to compute the backbone dihedral ϕ, and a ψ-*defining* RDC or RCSA is used to compute the backbone dihedral ψ, in the increasing order of residue number. The input data to RDC-CSA-ANALYTIC include: (1) the primary sequence of the protein; (2) any combination of at least two RDCs or RCSAs per residue measured in one alignment medium; (3) a sparse set of NOEs; (4) secondary structure element (SSE) boundaries based on TALOS [13, 44] dihedral restraints; and (5) the rotamer library [31].

Previously, we have shown that when a ϕ-defining and a ψ-defining RDC are available for a residue, the corresponding values for ϕ and ψ can be computed by solving quartic equations [51, 52, 64]. RDC-CSA-ANALYTIC extends this to the cases when a ψ-defining RCSA is available in addition to a ϕ-defining RDC (see Proposition 1 below), e.g., when C$^\alpha$-C' or C$^\alpha$-H$^\alpha$ RDC, and ^{13}C'-RCSA or

Table 1. RDC-CSA-ANALYTIC uses a ϕ-defining RDC to compute the backbone dihedral ϕ, and a ψ-defining RDC or RCSA to compute the backbone dihedral ψ exactly and in closed form

ϕ-defining RDC	C^α-H^α, C^α-C', C^α-C^β
ψ-defining RDC/RCSA	N-H^N, C'-N, C'-H^N, $^{13}C'$-RCSA, ^{15}N-RCSA, ^{1}H-RCSA

^{15}N-RCSA data is available. However, in solution NMR, $^{13}C'$-RCSA and/or ^{15}N-RCSA can be measured, often with high precision, along with N-H^N RDC, for large and perdeuterated protein systems, for which C^α-H^α RDCs at the chiral C^α center cannot be measured, and C^α-C' RDCs measurements are often less precise. Therefore, it is important to be able to determine the global fold from these types of measurements. RDC-CSA-ANALYTIC algorithm specifically provides a solution to this problem. Here we solve the most general case when two ψ-defining RCSAs and/or RDCs are available for residues. Further, this includes the case when (only) two RCSAs per residue are available. It can be shown that (see the supporting information (SI) **Appendix A** available online [50]) one must solve a 32 degree univariate polynomial equation to solve for all possible (at most 32) (ϕ, ψ) pairs, which is a difficult computational problem.

However, for a given value of ϕ_i, the values of ψ_i can be computed by solving a quartic equation (see Proposition 1 below). Here we present a hybrid approach that employs a systematic search over ϕ combined with solutions to two quartic equations for ψ derived from two ψ-defining RDC/RCSA values r_1 and r_2, to compute the backbone dihedrals (ϕ, ψ) pairs. For each ϕ, sampled systematically from the Ramachandran map, let A and B (each of size ≤ 4) be the sets of all ψ values computed using r_1 and r_2, respectively. If $A \cap B \neq \emptyset$, then for a $\psi \in A \cap B$, the corresponding (ϕ, ψ) pair is a solution. However, in practice, there are two issues that need to be addressed. First, due to finite-resolution sampling of ϕ, and experimental errors in the RDC and RCSA data, the intersection of sets A and B can be an empty set, even though there exist $\psi_A \in A$ and $\psi_B \in B$ such that $|\psi_A - \psi_B| < \delta$, for some small delta $\delta > 0$ which depends on the resolution of sampling of ϕ. This issue can be addressed by choosing a suitable resolution α for systematic sampling of ϕ, and choosing a corresponding small value for δ. Both α and δ are input parameters to our algorithm. We use $\alpha = 0.2°$ and $\delta = 0.5°$. We choose a $\psi \in [\psi_A, \psi_B]$ when $|\psi_A - \psi_B| < \delta$. Further, our choice of ψ does not increase the RDC and RCSA RMSDs (i.e., the RMS deviation between the back-computed and experimental values) so much that they exceed user-defined thresholds; otherwise, the solution is discarded. Second, due to fine sampling of ϕ, often multiple pairs of (ϕ, ψ) cluster in a small region of the Ramachandran map. We cluster these solutions, and choose a set of representative candidates so that the complexity of the conformation tree search is not adversely affected.

A description of the core modules of RDC-CSA-ANALYTIC, and the inner working details are provided in the SI **Appendix B** available online [50].

The Analytic Step: Peptide Plane Orientations from N-H^N RDC, and ^{13}C-RCSA or ^{15}N-RCSA Measured in One Alignment Medium. To compute ψ_i for residue i, any of the ^{13}C′-RCSA, ^{15}N-RCSA or ^{1}H-RCSA can be used (see Table 1). Here we will use ^{13}C′-RCSA and derive the necessary mathematical tools for computing the dihedral ψ_i. Our derivation holds for ^{15}N-RCSA and ^{1}H-RCSA with minor modifications.

Proposition 1. *Given the diagonalized alignment tensor components S_{xx} and S_{yy}, the peptide plane P_i, the dihedral ϕ_i, and the ^{13}C′-RCSA $\Delta\delta_{C'}$ for residue i, there exist at most 4 possible values of the dihedral angle ψ_i that satisfy $\Delta\delta_{C'}$, and they can be computed exactly and in closed form by solving a quartic equation.*

Proof. The derivation below assumes standard protein geometry, which is exploited in the kinematics [52, 56]. Let the unit vector $\mathbf{v}_0 = (0, 0, 1)^T$ be the N-H^N bond vector of residue i in the local coordinate frame defined on the peptide plane P_i. Let $\mathbf{v}_1 = (x_1, y_1, z_1)^T$ and $\mathbf{v}_2 = (x_2, y_2, z_2)^T$ be the unit vectors defined with respect to the POF on the peptide plane P_{i+1}. We can write the forward kinematics relations between $\mathbf{v}_0$ and $\mathbf{v}_1$, and between $\mathbf{v}_0$ and $\mathbf{v}_2$ as follows:

$$\mathbf{v}_1 = \mathbf{R}_{i,\text{POF}} \, \mathbf{R}_l \, \mathbf{R}_z(\phi_i) \, \mathbf{R}_m \, \mathbf{R}_z(\psi_i) \, \mathbf{R}_r \, \mathbf{v}_0, \tag{10}$$

$$\mathbf{v}_2 = \mathbf{R}_{i,\text{POF}} \, \mathbf{R}_l \, \mathbf{R}_z(\phi_i) \, \mathbf{R}_m \, \mathbf{R}_z(\psi_i) \, \mathbf{R}'_r \, \mathbf{v}_0. \tag{11}$$

Here $\mathbf{R}_l$, $\mathbf{R}_m$, $\mathbf{R}_r$ and $\mathbf{R}'_r$ are constant rotation matrices. $\mathbf{R}_{i,\text{POF}}$ is rotation matrix of P_i with respect to the POF. $\mathbf{R}_z(\phi_i)$ is the rotation about the z-axis by ϕ_i, and is a constant rotation matrix since ϕ_i is known. $\mathbf{R}_z(\psi_i)$ is the rotation about the z-axis by ψ_i. Let $c = \cos\psi_i$ and $s = \sin\psi_i$. Using this in Eq. (10) and Eq. (11) and simplifying we obtain

$$x_1 = A_{10} + A_{11}c + A_{12}s, \qquad x_2 = A_{20} + A_{21}c + A_{22}s, \tag{12}$$

$$y_1 = B_{10} + B_{11}c + B_{12}s, \qquad y_2 = B_{20} + B_{21}c + B_{22}s, \tag{13}$$

$$z_1 = C_{10} + C_{11}c + C_{12}s, \qquad z_2 = C_{20} + C_{21}c + C_{22}s, \tag{14}$$

where A_{ij}, B_{ij}, C_{ij} for $1 \leq i \leq 2$ and $0 \leq j \leq 2$ are constants. Using Eq. (12) to Eq. (14) in the reduced RCSA equation (Eq. (9)), and simplifying we obtain

$$K_0 + K_1 c + K_2 s + K_3 cs + K_4 c^2 + K_5 s^2 = 0, \tag{15}$$

where K_i, $0 \leq i \leq 5$ are constants. Using half-angle substitutions

$$u = \tan(\frac{\psi_i}{2}), \quad c = \frac{1 - u^2}{1 + u^2}, \text{ and } s = \frac{2u}{1 + u^2} \tag{16}$$

in Eq. (15) we obtain

$$L_0 + L_1 u + L_2 u^2 + L_3 u^3 + L_4 u^4 = 0, \tag{17}$$

where L_i, $0 \leq i \leq 4$ are constants. Eq. (17) is a quartic equation that can be solved exactly and in closed form. For each real solution (at most four are possible), the corresponding ψ_i value can be computed using Eq. (16). $\qquad\square$

Corollary 1. *Given the diagonalized alignment tensor components S_{xx} and S_{yy}, the peptide plane P_i, the dihedral ϕ_i, and a ψ-defining RDC r for P_{i+1}, there exist at most 4 possible values of the dihedral ψ_i that satisfy r. The possible values of ψ_i can be computed exactly and in closed form by solving a quartic equation.*

Proof. The proof follows from Proposition 1 by setting $\lambda_1 = 1$, $\lambda_2 = 0$ in Eq. (9), and treating $\mathbf{v}_1$ as the vector for which the ψ-defining RDC r is measured.   □

2.4   The 3PLANAR Algorithm

We show that given any combination of three RCSAs and/or RDCs for internuclear vectors on a peptide plane (and in general, for any planar structural motif), there exist at most 16 possible orientations of the peptide plane that satisfy the three given orientational restraints. We further show that the 16 possible orientations can be computed in closed form by solving a *quadratic equation*. It is the only case where we have discovered a quadratic equation-based solution to constraints involving second-rank tensors, e.g., RDCs and RCSAs; all previous exact solutions to RCSA and/or RDC equations required solving quartic or higher degree equations. This we obtained by exploiting the symmetry of the equations in a novel way. Our main result is stated as the following proposition.

Proposition 2. *Given a rhombic alignment tensor, and 3 measurements, each of which is either an RCSA on a nucleus on the peptide plane P or an RDC on an internuclear vector on P, there exist at most 16 possible orientations for P that satisfy the 3 measurements, and these orientations can be written and solved in closed form by solving a quadratic equation.*

Proof. The proof is presented in the SI **Appendix C** available online [50].   □

Proposition 2, is incorporated into the 3PLANAR algorithm, which requires the following as input: (1) the diagonalized alignment tensor components (S_{yy}, S_{zz}); and (2) three orientational restraints, such as the CSA tensor parameters along with the RCSA values $(\delta_{xx}, \delta_{yy}, \Omega, \Delta\delta)$ and/or RDCs. It outputs all the possible oriented peptide planes consistent with the RDC and/or RCSA data.

3   Results and Discussion

3.1   Backbone Global Fold of Ubiquitin from Experimental RDC and RCSA

We applied our algorithm to compute the global fold of human ubiquitin. The protein ubiquitin has been a model system in many solution-state [6,7,12,14,24, 30,56] and solid-state [41,43] NMR studies. The solution structure of ubiquitin (PDB id: 1D3Z), and a 1.8 Å X-ray crystallographic structure of ubiquitin [55] (PDB id: 1UBQ), available in the PDB [2], were used as references. The experimental N-H^N RDC, ^{13}C′-RCSA and ^{15}N-RCSA data were obtained from the previously published work by Cornilescu and Bax [12]. We used the uniform

Table 2. Results on the alignment tensor computation, and RDC and RCSA data fit. (a) Experimental NMR data is from [12]. RMSD is the root-mean-square deviation between the back-computed and experimental values. (b) N-H^N RDC and ^{13}C'-RCSA, and (c) N-H^N RDC and ^{15}N-RCSA, were used to compute the global fold. (d) The alignment tensors for the global folds computed by RDC-CSA-ANALYTIC agree well with that of the reference NMR structure.

Model	RDC and RCSA^a used & RMSDs	Diagonalized Alignment Tensor S_{yy}, S_{zz}	Rhombicity^d (ρ)
1D3Z	N-H^N: 1.11 Hz	-2.31, 51.17	0.61
1D3Z	N-H^N: 1.17 Hz, ^{13}C'-RCSA: 6.85 ppb	-1.40, 50.57	0.63
1D3Z	N-H^N: 1.40 Hz, ^{15}N-RCSA: 10.08 ppb	-3.56, 49.40	0.57
RDC-CSA-ANALYTIC ^{b}	N-H^N: 1.21 Hz, ^{13}C'-RCSA: 7.38 ppb	-0.71, 51.11	0.65
RDC-CSA-ANALYTIC ^{c}	N-H^N: 1.13 Hz, ^{15}N-RCSA: 9.22 ppb	-3.98, 46.10	0.55

Table 3. Backbone RMSDs (Å) of SSE fragments computed by RDC-CSA-ANALYTIC. (a) NMR data is from [12]. (b) 12 H-bond information, and (c) 5 C$^\alpha$-C$^\alpha$ approximate distance restraints derived from NOEs [56] were used.

Data Used; Reference^a	α-helix I23–K33	β_1 Q2–T7	β_2 T12–V17	β_3 Q41–F45	β_4 K48–L50	β_5 S65–V70	β-sheet^b $\beta_{1,\ldots,5}$	Global Fold^c
N-H^N,^{13}C'-RCSA; NMR	0.27	0.24	0.35	0.16	0.19	0.20	0.71	1.04
N-H^N,^{13}C'-RCSA; X-ray	0.23	0.32	0.37	0.28	0.20	0.25	0.79	1.09
N-H^N,^{15}N-RCSA; NMR	0.26	0.51	0.42	0.28	0.31	0.30	0.93	1.31
N-H^N,^{15}N-RCSA; X-ray	0.25	0.61	0.43	0.36	0.32	0.35	0.99	1.38

average values of the principal components of the CSA tensors reported in [12]. Such an assumption has been used widely in the literature for protein structure refinement against RCSA data [10, 26, 27, 53]. Whenever residue-specific CSA tensors can be determined, as in [6, 7, 30, 62, 63], RDC-CSA-ANALYTIC can use those tensors. The NOE restraints and hydrogen bond information for ubiquitin were extracted from the NMR restraint file for the PDB id 1D3Z [14].

Since RCSA measurement is usually accompanied by that of N-H^N RDC, recorded for the same sample under the same alignment conditions [10, 12, 28], computing an accurate backbone global fold from this limited amount of data, as a first step in protein structure computation, is of considerable interest. Here we present results of backbone global fold computation by RDC-CSA-ANALYTIC using (1) N-H^N RDC and ^{13}C'-RCSA, and (2) N-H^N RDC and ^{15}N-RCSA.

The alignment tensor was computed from N-H^N RDC and ^{13}C'-RCSA/^{15}N-RCSA data by bootstrapping the computation with an ideal helix for the helical region I23–K33 of ubiquitin (see the SI **Appendix B** [50]), and was used subsequently in the global fold computation. As summarized in Table 2, the alignment tensors agree well with those computed for the reference NMR structure. It is worth noting that if the alignment tensor is estimated by other methods [11], RDC-CSA-ANALYTIC can use that as input to compute the backbone global fold.

RDC-CSA-ANALYTIC computed accurate backbone conformations from N-H^N RDC plus either ^{13}C'-RCSA or ^{15}N-RCSA data. As shown in Table 3, the backbone RMSDs between the computed SSEs and the reference structures are within 0.61 Å, and for about half of the cases they are less than 0.3 Å. The SSE back-

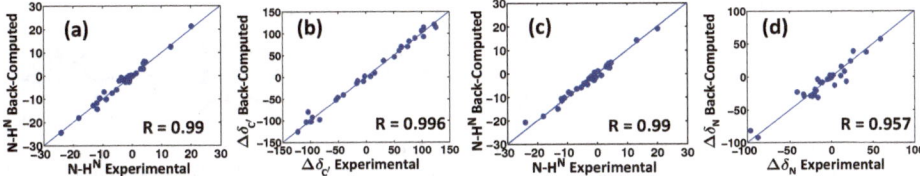

Fig. 2. Correlations between back-computed and experimental N-H^N RDCs and ^{13}C'-RCSA (a, b), and those for N-H^N RDC and ^{15}N-RCSA (c, d) shown for the global folds computed by RDC-CSA-ANALYTIC

Orientational Restraints:	N-H^N RDC and C'-RCSA		N-H^N RDC and N-RCSA	
Ref. Structure:	NMR	X-ray	NMR	X-ray
β-sheet: Q2-T7, T12-V17, Q41-F45, K48-L50, S65-V70	RMSD: 0.71 Å	RMSD: 0.79 Å	RMSD: 0.93 Å	RMSD: 0.99 Å
α-helix: I23-K33	RMSD: 0.27 Å	RMSD: 0.23 Å	RMSD: 0.26 Å	RMSD: 0.25 Å
Global-fold:	RMSD: 1.04 Å	RMSD: 1.09 Å	RMSD: 1.31 Å	RMSD: 1.38 Å

RDC-CSA-ANALYTIC   NMR: 1D3Z   X-ray: 1UBQ

Fig. 3. Overlay of the ubiquitin global fold computed by RDC-CSA-ANALYTIC using N-H^N RDC and ^{13}C'-RCSA or ^{15}N-RCSA versus the NMR and X-ray reference structures

bones computed using N-H^N RDC and ^{13}C'-RCSA data agree better with the reference structures than those computed using N-H^N RDC and ^{15}N-RCSA data and compared with the reference structures. Table 2 and Figure 2 show that the back-computed RDCs and RCSAs for the RDC-CSA-ANALYTIC-computed structures are in good agreement with their experimental counterparts. For the structure computed using N-H^N RDC and ^{15}N-RCSA, the ^{15}N-RCSA Pearson's correlation coefficient is 0.957, and for other three cases the correlation coefficients are 0.99 or more (see Figure 2). This is explained by the slightly better quality structure obtained using N-H^N RDC and ^{13}C'-RCSA data than that obtained using N-H^N RDC and ^{15}N-RCSA data. The β-sheet is computed using 12 hydrogen bond restraints in addition to the RDC and RCSA data. The α-helix (I23–K33) and the β-sheet for ubiquitin were packed using 5 approximate C$^\alpha$-C$^\alpha$ distances derived from NOEs using the method described in [56]. The top 1000 packed structures obtained from the packing of ubiquitin α-helix and β-sheet, computed using N-H^N RDC and ^{13}C'-RCSA data, have backbone RMSDs within the range 1.04–1.39 Å versus the reference NMR structure, and 1.09–1.42 Å versus the X-ray reference structure. The top 1000 packed structures obtained from the packing of ubiquitin α-helix and β-sheet, computed using

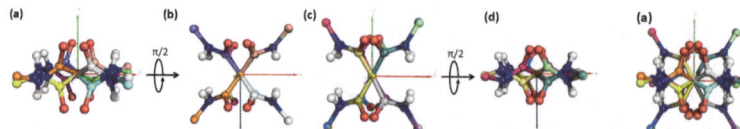

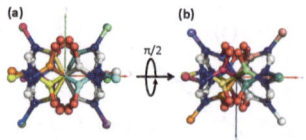

Fig. 4. The peptide plane orientations correspond to the two roots of the quadratic equations (Proposition 2) derived from C^α-C', C'-N and N-H^N RDCs measured in single alignment medium

Fig. 5. Visualization of all sixteen peptide plane orientations together

N-H^N RDC and ^{15}N-RCSA data have backbone RMSDs within the range 1.31–1.86 Å versus the reference NMR structure, and 1.38–1.97 Å versus the X-ray reference structure. Figure 3 shows the overlay of the backbone fold computed by RDC-CSA-ANALYTIC versus the NMR and X-ray reference structures.

These results indicate that RDC-CSA-ANALYTIC can be used to compute accurate global folds from a minimal amount of RDC and RCSA data. Protein backbone global folds of similar resolution have been used successfully in empirical high-resolution structure determinations [65], NOE assignment [22,66], and side-chain resonance assignment [67]. Therefore, our method will be useful in high-resolution protein structure determination. Furthermore, the use of RCSAs in the first stage of structure computation to compute accurate global folds is a novel concept, and our paper, being the first one to demonstrate this, can be a stepping stone to further research that exploits this new type of experimental data.

3.2   16-Fold Degeneracy of Peptide Plane Orientations

Our algorithm 3PLANAR was tested on the experimental RDC data for the protein ubiquitin (PDB id: 1D3Z) obtained from the BioMagResBank (BMRB) [54]. Using the singular value decomposition [29,56] module of RDC-CSA-ANALYTIC [52, 64,65], we computed the principal components of the alignment tensor for ubiquitin using its NMR structure. We used C^α-C', C'-N and N-H^N RDCs, measured in one alignment medium, for the peptide plane defined by the residues Ala28 and Lys29 of ubiquitin. 3PLANAR then computed the 16 oriented peptide planes (individual planes are shown in the online SI **Appendix D** [50]). In Figure 4 (a, b) and (c, d), the two sets of 8 oriented peptide planes corresponding to the two roots of the quadratic equation are shown. Figure 5 shows the 16 oriented peptide planes visualized together. A counterclockwise rotation of 90° about the x-axis elucidates the symmetry in the peptide plane orientations. Similar results were obtained when N-H^N RDC, ^{15}N-RCSA and ^{13}C$'$-RCSA data [12] was used, and the corresponding alignment tensor was computed by RDC-CSA-ANALYTIC.

3PLANAR is a $\Theta(1)$-time deterministic algorithm. During protein backbone structure determination, such as when using the RDC-ANALYTIC framework, the multiple possible peptide plane orientations consistent with RDC/RCSA are usually ruled out by the kinematic coupling between peptide planes along the polypeptide chain, standard biophysical and protein geometry assumptions, or using additional experimental restraints.

4   Conclusions

We described a novel algorithm, RDC-CSA-ANALYTIC, that uses a sparse set of RCSAs and RDCs to compute the protein backbone global fold accurately. Our algorithm is the first algorithm to demonstrate that the orientational restraints from RCSA can be used in the initial stage of structure computation. We hope that this breakthrough will shed new light on the information content of RCSA, and help NMR structural biologists use our new ways of using RCSA to solve protein structures. Our algorithm barely scratches the surface of this new area, and much work remains to be done. Computing loop conformations using RCSA is an immediate future extension. Ubiquitin is the only protein for which we were able to obtain experimental ^{15}N-RCSA and ^{13}C'-RCSA data, available in the public domain. In future, we would like to test our algorithms on other protein systems, when experimental data becomes available for those systems.

When using orientational restraints in structure determination, it is important to know all the possible degeneracies associated with them, and their implications for structure determination. We gave exact and tight bounds on the orientational degeneracy of peptide planes computed using RDCs and/or RCSAs, and described a $\Theta(1)$-time algorithm, 3PLANAR, to compute them.

Although RDCs have been regularly used in protein structure determination, RCSAs have been used only in a few cases for structure validation [14] and refinement [10, 26, 27, 53]. We envision that algorithms, such as RDC-CSA-ANALYTIC, that use RCSA data plus RDCs during the initial stages of protein structure determination will play a large role in future.

References

1. Berjanskii, M., Wishart, D.S.: Nat. Protoc. 1, 683–688 (2006)
2. Berman, H.M., et al.: Nucleic Acids Res. 28(1), 235–242 (2000)
3. Boehr, D.D., et al.: Science 313(5793), 1638–1642 (2006)
4. Bouvignies, G., et al.: Nature 477(7362), 111–114 (2011)
5. Brünger, A.T.: Yale University Press, New Haven (1992)
6. Burton, R.A., Tjandra, N.: J. Biomol. NMR 35(4), 249–259 (2006)
7. Burton, R.A., Tjandra, N.: J. Am. Chem. Soc. 129, 1321–1326 (2007)
8. Casey, J.: Proceedings of the Royal Society of London XIX, 495–497 (1871)
9. Cavalli, A., et al.: Proc. Natl. Acad. Sci. USA 104(23), 9615–9620 (2007)
10. Choy, W.-Y., et al.: J. Biomol. NMR 21, 31–40 (2001)
11. Clore, G., et al.: J. Magn. Reson. 133(1), 216–221 (1998)
12. Cornilescu, G., Bax, A.: J. Am. Chem. Soc. 122, 10143–10154 (2000)
13. Cornilescu, G., et al.: J. Biomol. NMR 13, 289–302 (1999)
14. Cornilescu, G., et al.: J. Am. Chem. Soc. 120, 6836–6837 (1998)
15. Das, B.B., et al.: J. Am. Chem. Soc. 134(4), 2047–2056 (2012)
16. Donald, B.R.: Algorithms in Structural Molecular Biology. The MIT Press (2011)
17. Donald, B.R., Martin, J.: Prog. NMR Spectrosc. 55(2), 101–127 (2009)
18. Dyson, H.J., Wright, P.E.: Nat. Rev. Mol. Cell. Biol. 6, 197–208 (2005)
19. Giesen, A.W., et al.: J. Biomol. NMR 25, 63–71 (2003)
20. Goldflam, M., et al.: Methods in Molecular Biology 831, 233–259 (2012)

21. Grishaev, A., et al.: J. Am. Chem. Soc. 131(27), 9490–9491 (2009)
22. Güntert, P.: Prog. Nucl. Magn. Reson. Spectrosc. 43, 105–125 (2003)
23. Hus, J.-C., et al.: J. Am. Chem. Soc. 130, 15927–15937 (2008)
24. Lange, O.F., et al.: Science 320, 1471–1475 (2008)
25. Leaver-Fay, A., et al.: Methods Enzymol. 487, 545–574 (2011)
26. Lipsitz, R.S., Tjandra, N.: J. Am. Chem. Soc. 123(44), 11065–11066 (2001)
27. Lipsitz, R.S., Tjandra, N.: J. Magn. Reson. 164(1), 171–176 (2003)
28. Liu, Y., Prestegard, J.H.: J. Biomol. NMR 47(4), 249–258 (2010)
29. Losonczi, J.A., et al.: J. Magn. Reson. 138, 334–342 (1999)
30. Loth, K., et al.: J. Am. Chem. Soc. 127(16), 6062–6068 (2005)
31. Lovell, S.C., et al.: Proteins: Struct., Funct., Genet. 40, 389–408 (2000)
32. Oas, T.G., et al.: J. Am. Chem. Soc. 109(20), 5962–5966 (1987)
33. Pervushin, K., et al.: Proc. Natl. Acad. Sci. USA 94(23), 12366–12371 (1997)
34. Prestegard, J.H., et al.: Chemical Reviews 104, 3519–3540 (2004)
35. Raman, S., et al.: Science 327, 1014–1018 (2010)
36. Religa, T.L., et al.: Nature 437(7061), 1053–1056 (2005)
37. Salmon, G.: Longmans. Green and Company, London (1912)
38. Salmon, L., et al.: J. Am. Chem. Soc. 132(24), 8407–8418 (2010) PMID: 20499903
39. Sanders II, C.R., Landis, G.C.: J. Am. Chem. Soc. 116(14), 6470–6471 (1994)
40. Saupe, A.: Angewandte Chemie 7(2), 97–112 (1968)
41. Schanda, P., et al.: J. Am. Chem. Soc. 132(45), 15957–15967 (2010)
42. Schwieters, C.D., et al.: J. Magn. Reson. 160, 65–73 (2003)
43. Seidel, K., et al.: ChemBioChem 6(9), 1638–1647 (2005)
44. Shen, Y., et al.: J. Biomol. NMR 44, 213–223 (2009)
45. Shen, Y., et al.: Proc. Natl. Acad. Sci. USA 105(12), 4685–4690 (2008)
46. Sitkoff, D., Case, D.A.: Prog. Nucl. Magn. Reson. Spectrosc. 32(2), 165–190 (1998)
47. Tian, F., et al.: J. Am. Chem. Soc. 123, 11791–11796 (2001)
48. Tjandra, N., Bax, A.: Science 278, 1111–1114 (1997)
49. Tolman, J.R., et al.: Proc. Natl. Acad. Sci. USA 92, 9279–9283 (1995)
50. Tripathy, C., Yan, A.K., Zhou, P., Donald, B.R.: Supporting information (2013), http://www.cs.duke.edu/donaldlab/Supplementary/recomb13/
51. Tripathy, C., Zeng, J., Zhou, P., Donald, B.R.: Protein Loop Closure Using Orientational Restraints from NMR Data. In: Bafna, V., Sahinalp, S.C. (eds.) RECOMB 2011. LNCS (LNBI), vol. 6577, pp. 483–498. Springer, Heidelberg (2011)
52. Tripathy, C., et al.: Proteins: Struct., Funct., Bioinf. 80(2), 433–453 (2012)
53. Tugarinov, V., et al.: Proc. Natl. Acad. Sci. USA 102(3), 622–627 (2005)
54. Ulrich, E.L., et al.: Nucleic. Acids. Res. 36(Database issue), D402–D408 (2008)
55. Vijay-kumar, S., et al.: J. Mol. Biol. 194, 531–544 (1987)
56. Wang, L., Donald, B.R.: J. Biomol. NMR 29(3), 223–242 (2004)
57. Wang, L., Donald, B.R.: Proceedings of CSB, pp. 189–202 (2005)
58. Wang, L., et al.: J. Comput. Biol. 13(7), 1276–1288 (2006)
59. Wishart, D.S., Case, D.A.: Methods Enzymol. 338, 3–34 (2002)
60. Wylie, B.J., et al.: J. Am. Chem. Soc. 127(34), 11946–11947 (2005)
61. Wylie, B.J., et al.: J. Am. Chem. Soc. 131(3), 985–992 (2009)
62. Yao, L., et al.: J. Am. Chem. Soc. 132(31), 10866–10875 (2010)
63. Yao, L., et al.: J. Am. Chem. Soc. 132(12), 4295–4309 (2010)
64. Yershova, A., et al.: Proceedings of WAFR, vol. 68, pp. 355–372 (2010)
65. Zeng, J., et al.: J. Biomol. NMR 45(3), 265–281 (2009)
66. Zeng, J., et al.: In: Proceedings of CSB, pp. 169–181 (2008) ISBN 1752-7791
67. Zeng, J., et al.: J. Biomol. NMR 50(4), 371–395 (2011)
68. Zeng, J., et al.: J. Biomol. NMR, 1–14 (2012)

Genome-Wide Survival Analysis of Somatic Mutations in Cancer[*]

Fabio Vandin[1,2,**], Alexandra Papoutsaki[1], Benjamin J. Raphael[1,2,***], and Eli Upfal[1,***]

[1] Department of Computer Science, Brown University, Providence, RI
[2] Center for Computational Molecular Biology, Brown University, Providence, RI
{vandinfa,alexpap,braphael,eli}@cs.brown.edu

Motivation. Next-generation DNA sequencing technologies now enable the measurement of exomes, genomes, and mRNA expression in many samples. The next challenge is to interpret these large quantities of DNA and RNA sequence data. In many human and cancer genomics studies, a major goal is to discover associations between an observed phenotype and a particular variable from genome-wide measurements of many such variables. In this work we consider the problem of testing the association between a DNA sequence variant and the *survival time*, or length of time that patients live following diagnosis or treatment. This problem is relevant to many cancer sequencing studies, in which one aims to discover somatic variants that distinguish patients with fast-growing tumors that require aggressive treatment from patients with better prognosis [1].

The most widely used statistical test of a difference in the survival time between two (or more) classes of samples is the nonparametric *log-rank* test. Nearly all implementations of the log-rank test rely on the asymptotic normality of the test statistic. However, this approximation gives poor results in many genomics applications where: (1) the populations are unbalanced; i.e. the population containing a given variant is significantly smaller than the population without that variant; (2) we test many possible variants and are interested in those variants with very small p-values that remain significant after multi-hypothesis correction. These issues do not arise in the traditional applications of the log-rank test, namely clinical trials and product reliability tests, where *pre-selected* groups that differ in one tested feature – e.g., receiving a particular drug or treatment – are compared.

Contributions. We show empirically that the asymptotic approximations used in most implementations of the log-rank test produce poor estimates of the true p-values when applied to unbalanced populations. An exact test, based on the exact distribution of the test statistic, is thus advantageous in such applications. Exact tests for comparing survival distributions have received scant attention in the literature.

---

[*] This work is supported by NSF grant IIS-1016648.
[**] Corresponding author.
[***] These authors contributed equally to the work.

M. Deng et al. (Eds.): RECOMB 2013, LNBI 7821, pp. 285–286, 2013.

We consider the two exact null distributions for the log-rank test, the conditional and permutational distributions [2,3]. We show empirically that in the range of parameters expected in genomic applications: (1) the p-value of the log-rank statistic is very sensitive to the choice of the null distribution; and (2) the permutational distribution produces p-values significantly closer to the true p-values than the p-values from the conditional distribution. We therefore prefer the permutational distribution that matches exactly the problem's parameters, and moreover has maximum local power among all order invariant tests [3]. No efficient algorithm for computing p-values for the permutational distribution has been proposed in the literature.

We introduce a novel fully polynomial time approximation scheme (FPTAS) for the p-value of the log-rank test under the permutational distribution. That is, we present an algorithm that computes for any $\varepsilon > 0$ a conservative estimate $\tilde{p}$ of the correct p-value such that: (1) $\tilde{p} \in [p, (1 + \varepsilon)p]$; and (2) the run-time of the algorithm is polynomial in ε^{-1} and the number of patients. The FTPAS uses a novel method to approximate the entire distribution of the log-rank statistic. The method is not only mathematically sound but also practical and efficient. In contrast to Monte-Carlo approaches for computing p-values, the run-time of the FPTAS is not a function of the p-value, which can be exponentially small.

Application to Cancer Data. We implement and test our algorithm on somatic mutation and survival data from The Cancer Genome Atlas (TCGA). In particular, we analyze data from four different cancer types, reporting in all cases substantial differences between the p-values obtained by our algorithm and the p-values obtained using either the exact conditional distribution, or the asymptotic approximation of the log-rank statistic – the latter as implemented in the survdiff package in R. We identify mutations with statistically significant association to survival time. Many of these are supported by the literature (e.g., BRCA2 in ovarian serous adenocarcinoma [4]), and are only identified using our FPTAS. Moreover, survdiff suspiciously reports a number of mutations of very small frequency as having statistically significant association with survival time, while our FPTAS does not. These results show that our algorithm is practical, efficient, and finds known associations in small and unbalanced populations with fewer false positives.

References

1. Jiao, Y., et al.: DAXX/ATRX, MEN1, and mTOR pathway genes are frequently altered in pancreatic neuroendocrine tumors. Science 331(6021), 1199–1203 (2011)
2. Mantel, N., Haenszel, W.: Statistical aspects of the analysis of data from retrospective studies of disease. Journal of the NCI 22(4), 719–748 (1959)
3. Peto, R., Peto, J.: Asymptotically efficient rank invariant test procedures. J. Roy. Stat. Soc. Ser. A 135, 185–207 (1972)
4. The Cancer Genome Atlas Research Network: Integrated genomic analyses of ovarian carcinoma. Nature 474(7353), 609–615 (2011)

Spectral Library Generating Function for Assessing Spectrum-Spectrum Match Significance

Mingxun Wang[1,2] and Nuno Bandeira[1,2,3]

[1] University of California, San Diego, Dept. of Computer Science and Engineering,
9500 Gilman Dr., La Jolla, CA, 92093, USA
{miw023,bandeira}@ucsd.edu
[2] Center for Computational Mass Spectrometry, CSE, UCSD
[3] Skaggs School of Pharmacy and Pharm. Sci., UCSD

Tandem mass spectrometry (MS/MS) continues to be the technology of choice for high-throughput analysis of complex proteomics samples. While MS/MS spectra are commonly identified by matching against a database of known protein sequences, the complementary approach of spectral library searching [1,2,3] against collections of reference spectra consistently outperforms sequence-based searches by resulting in significantly more identified spectra. But despite this demonstrated superior sensitivity, the development of methods to determine the statistical significance of Spectrum-Spectrum Matches (SSMs) in peptide spectral library searches is still in its early stages.

The most common approach to controlling the False Discovery Rate (FDR) in both database search [4] and spectral library search [5] is the Target-Decoy approach where one extends the database/library of true peptides with a complement of sequences/spectra from 'random' peptides and uses matches to the latter to estimate the number of false matches to true sequences/spectra. But while these FDR approaches continue to be very valuable in correcting for multiple hypothesis testing in large-scale experiments, they provide little to no insight on the statistical significance of individual SSMs or Peptide Spectrum Matches (PSMs).

The estimation of the significance of SSMs is currently hindered by difficulties in finding an appropriate definition of 'random' SSMs to use as a null model when estimating the significance of true SSMs. We propose to avoid this problem by changing the null hypothesis – instead of determining the probability of a random match with a score $\geq T$, our approach determines the probability that a *true* match has a score $\leq T$. To this end, we explicitly model the variation in instrument measurements of MS/MS peak intensities (using a reference spectral library and a set of matching experimental MS/MS spectra) and show how these models can be used to determine a theoretical distribution of SSM scores between reference and query spectra of the same molecule. While the proposed Spectral Library Generating Function (SLGF) approach can be used to calculate theoretical distributions for any additive SSM score (e.g., any dot product), we further show how it can be used to calculate the distribution of expected cosines

M. Deng et al. (Eds.): RECOMB 2013, LNBI 7821, pp. 287–288, 2013.
© Springer-Verlag Berlin Heidelberg 2013

between reference library and replicate query spectra. To assess the statistical significance of a score of a SSM between a reference library and unknown query spectrum, we used these SLGF calculated theoretical cosine score distributions to derive a p-value. In evaluating SLGF we explored both the accuracy of the theoretical distributions as well as SLGF's usefulness in the context of spectral library search. First we show that these expected cosine distributions did indeed approximate empirical score distributions and note that further work is necessary to enable more accurate theoretical distribution calculations. Second, using these p-values as scores, we demonstrate that these SLGF-based SSM p-value scores significantly outperform current state-of-the-art spectral library search tools such as SpectraST [1] in our test dataset. We additionally provide a detailed discussion of the multiple reasons behind the observed differences in the sets of identified MS/MS spectra.

Acknowledgements. This work was supported by the National Institutes of Health grant 3-P41-GM103484 from the National Institute of General Medical Sciences.

References

1. Lam, H., Deutsch, E.W., Eddes, J.S., Eng, J.K., King, N., Stein, S.E., Aebersold, R.: Development and validation of a spectral library searching method for peptide identification from ms/ms. Proteomics 7, 655–667 (2007)
2. Wang, J., Pérez-Santiago, J., Katz, J.E., Mallick, P., Bandeira, N.: Peptide identification from mixture tandem mass spectra. Mol. Cell. Proteomics 9, 1476–1485 (2010)
3. Dasari, S., Chambers, M.C., Martinez, M.A., Carpenter, K.L., Ham, A.J., Vega-Montoto, L.J., Tabb, D.L.: Pepitome: evaluating improved spectral library search for identification complementarity and quality assessment. J. Proteome Res. 11, 1686–1695 (2012)
4. Elias, J.E., Gygi, S.P.: Target-decoy search strategy for increased confidence in large-scale protein identifications by mass spectrometry. Nat. Methods 4, 207–214 (2007)
5. Lam, H., Deutsch, E.W., Aebersold, R.: Artificial decoy spectral libraries for false discovery rate estimation in spectral library searching in proteomics. J. Proteome Res. 9, 605–610 (2010)

SPARSE: Quadratic Time Simultaneous Alignment and Folding of RNAs without Sequence-Based Heuristics

Sebastian Will[1,2,*], Christina Schmiedl[1,*],
Milad Miladi[1], Mathias Möhl[1], and Rolf Backofen[1,3,4,5,**]

[1] Bioinformatics, Department of Computer Science, University of Freiburg, Germany
[2] Bioinformatics, Department of Computer Science, University of Leipzig, Germany
[3] Centre for Biological Signalling Studies (BIOSS), University of Freiburg, Germany
[4] Centre for Biological Systems Analysis (ZBSA), University of Freiburg, Germany
[5] Centre for Non-coding RNA in Technology and Health, Bagsvaerd, Denmark
backofen@informatik.uni-freiburg.de

Motivation: There is increasing evidence of pervasive transcription, resulting in hundreds of thousands of ncRNAs of unknown function. Standard computational analysis tasks for inferring functional annotations like clustering require fast and accurate RNA comparisons based on sequence and structure similarity. The gold standard for the latter is Sankoff's algorithm [3], which simultaneously aligns and folds RNAs. Because of its extreme time complexity of $O(n^6)$, numerous faster "Sankoff-style" approaches have been suggested. Several such approaches introduce heuristics based on sequence alignment, which compromises the alignment quality for RNAs with sequence identities below 60% [1]. Avoiding such heuristics, as e.g. in LocARNA [4], has been assumed to prohibit time complexities better than $O(n^4)$, which strongly limits large-scale applications.

Results: Breaking this barrier, we introduce SPARSE (Sparse Prediction and Alignment of RNAs using Structure Ensembles), a novel *quadratic time* Sankoff-style approach that does not rely on sequence-based heuristics but employs structural properties of RNA ensembles; its $O(n^2)$ complexity matches the one of sequence alignment. The approach is based on a novel lightweight Sankoff-style alignment model, for which we introduce the algorithm PARSE. For the first time it transfers the Sankoff-model completely to a lightweight energy model; thus, it is more expressive than all previous lightweight methods, which inherit the PMcomp model [2]. In comparison to LocARNA and similar approaches, the novel model enables much stronger sparsification based on the RNA structure ensemble; consequently, SPARSE aligns and folds RNAs with similar alignment and better folding quality in significantly less time. Finally, SPARSE aligns ncRNAs from the challenging low sequence identity region more accurately than tools relying on sequence-based heuristics.

---

* Joint first authors.
** Corresponding author.

M. Deng et al. (Eds.): RECOMB 2013, LNBI 7821, pp. 289–290, 2013.
© Springer-Verlag Berlin Heidelberg 2013

Conclusion: Our results indicate that a complete lightweight Sankoff-style model with stronger sparsification can increase the performance and accuracy of RNA alignment, where the potential of the model points far beyond the studied prototype. Not falling back on sequence comparison, SPARSE suggests itself for large scale similarity assessment of RNAs with moderate to very low sequence identity.

References

1. Gardner, P.P., Wilm, A., Washietl, S.: A benchmark of multiple sequence alignment programs upon structural RNAs. Nucleic Acids Res. 33(8), 2433–2439 (2005)
2. Hofacker, I.L., Bernhart, S.H., Stadler, P.F.: Alignment of RNA base pairing probability matrices. Bioinformatics 20(14), 2222–2227 (2004)
3. Sankoff, D.: Simultaneous solution of the RNA folding, alignment and protosequence problems. SIAM J. Appl. Math. 45(5), 810–825 (1985)
4. Will, S., Reiche, K., Hofacker, I.L., Stadler, P.F., Backofen, R.: Inferring non-coding RNA families and classes by means of genome-scale structure-based clustering. PLoS Comput. Biol. 3(4), e65 (2007)

An Algorithm for Constructing Parsimonious Hybridization Networks with Multiple Phylogenetic Trees

Yufeng Wu

Department of Computer Science and Engineering
University of Connecticut
Storrs, CT 06269, U.S.A.
ywu@engr.uconn.edu

Abstract. Phylogenetic network is a model for reticulate evolution. Hybridization network is one type of phylogenetic network for a set of discordant gene trees, and "displays" each gene tree. A central computational problem on hybridization networks is: given a set of gene trees, reconstruct the minimum (i.e. most parsimonious) hybridization network that displays each given gene tree. This problem is known to be NP-hard, and existing approaches for this problem are either heuristics or make simplifying assumptions (e.g. work with only two input trees or assume some topological properties). In this paper, we develop an exact algorithm (called $PIRN_C$) for inferring the minimum hybridization networks from multiple gene trees. The $PIRN_C$ algorithm does not rely on structural assumptions. To the best of our knowledge, $PIRN_C$ is the first exact algorithm for this formulation. When the number of reticulation events is relatively small (say four or fewer), $PIRN_C$ runs reasonably efficient even for moderately large datasets. For building more complex networks, we also develop a heuristic version of $PIRN_C$ called $PIRN_{CH}$. Simulation shows that $PIRN_{CH}$ usually produces networks with fewer reticulation events than those by an existing method.

1 Introduction

It is well known that reticulate evolution plays a significant role in shaping the evolutionary history of many species. There are several reticulate evolutionary processes, such as horizontal gene transfer and hybrid specification. To better model the effects of these reticulate evolutionary processes, a network-based model called phylogenetic network (rather than the traditional phylogenetic tree) is needed. Briefly, phylogenetic network is a directed acyclic graph, which has nodes (called reticulation nodes) with more than one incoming edges. See Figure 1 for an illustration of phylogenetic networks. The study of phylogenetic networks has received significant attention in recent years. Refer to the recent books [10,11] and also surveys (e.g. [12]) for background on phylogenetic networks.

Different models and formulations of phylogenetic networks with various modeling assumptions and different types of input have been proposed and studied.

M. Deng et al. (Eds.): RECOMB 2013, LNBI 7821, pp. 291–303, 2013.
© Springer-Verlag Berlin Heidelberg 2013

In this paper, we focus on one specific formulation of phylogenetic network, called "hybridization network" [14,10], which takes a set of gene trees as input. Here, a gene tree models the evolutionary history of some gene. Due to reticulate evolution, the gene trees may have different topologies. The goal is to construct a phylogenetic network that "displays" each of the gene trees. We provide more precise definitions in Section 2. Most current approaches for hybridization network inference are based on the parsimony principle [10,11]. That is, the goal is to find the hybridization networks with the smallest amount of reticulation events. In this paper, we also follow the parsimony principle.

It is often believed that hybridization networks may be useful in studying reticulate evolution. However, hybridization networks have not been widely used by biologists [11]. One obstacle is the computational challenge. Many existing computational formulations for inferring hybridization networks are known to be NP complete. Due to the computational difficulty, most existing approaches are heuristic. Moreover, existing approaches often impose simplifications on the hybridization network formulation. Simplification can be on the modeling of hybridization networks or the types of inputs allowed. For example, phylogenetic networks with structural assumptions such as galled networks [8] or the so-called level-k networks as in e.g. [15] have been previously studied. Another simplification often made in the study of hybridization networks is that only two input gene trees are allowed (e.g. [14,20,1]). Clearly, methods allowing multiple gene trees are likely to be more useful with the more available gene sequence data. Currently, there are only a few heuristic methods [18,13,5] on hybridization network construction or reticulation level estimation that allow multiple gene trees and do not rely on structural assumptions.

In this paper, we develop an algorithm (called $PIRN_C$) for inferring the parsimonious hybridization networks from multiple gene trees. To the best of our knowledge, $PIRN_C$ is the first exact algorithm for this formulation. $PIRN_C$ has the following features.

- $PIRN_C$ takes a set of rooted binary gene trees as input and constructs a hybridization network that displays each of the gene trees.
- $PIRN_C$ is an exact algorithm (i.e. it infers the most parsimonious networks).
- $PIRN_C$ allows any number of gene trees in principle, although longer running time and larger amount of memory may be needed for larger input. $PIRN_C$ also does not impose any structural constraints (for example, "gall"-like structures as in e.g. [9,6]) on phylogenetic networks.
- The running time of $PIRN_C$ is largely decided by the number of reticulation events in the inferred hybridization networks. When the number of reticulation events is relatively small (say five or fewer), $PIRN_C$ runs reasonably fast even for moderately large problem instance (say five gene trees with 30 taxa). On the other hand, for some larger dataset with say six or more reticulation events for five gene trees with 30 taxa, $PIRN_C$ becomes slow.

$PIRN_C$ may be best applied for inferring hybridization networks with relatively simple structure (i.e. the number of reticulation events is relatively small).

We note that real hybridization networks may indeed have relatively small number of reticulation events as suggested in [11]. Nevertheless, constructing parsimonious hybridization networks with larger number of reticulations is still an interesting problem from the computational perspective. In this paper, we also develop a heuristic version of $PIRN_C$ called $PIRN_{CH}$. $PIRN_{CH}$ does not always find the most parsimonious networks, but simulation shows that $PIRN_{CH}$ usually produces networks with fewer reticulations than an existing method.

2 Definitions and Background

Throughout this paper, we assume a phylogenetic tree is rooted, binary and leaf-labeled by a set of species (called taxa). In-degrees of all vertices (also called nodes) in a tree (except the root) are one. For convenience, for a tree node v, we often call the subtree rooted at v as the subtree v. Our definition of hybridization networks is similar to that in [14] with only some small changes. A hybridization network (sometimes simply network) is a directed acyclic graph with vertex set V and edge set E, where some nodes in V are labeled by taxa. V can be partitioned into V_T (called tree nodes) and V_R (called reticulation nodes). E can be partitioned into E_T (called tree edges) and E_R (called reticulation edges). Moreover,

1. Except the root, each node must have at least one incoming edge.
2. Reticulation nodes have in-degree two. Tree nodes have in-degree one.
3. E_R contains edges that go into some reticulation nodes. E_T contains edges that go into some tree nodes.
4. A node is labeled by some taxa iff its out-degree is zero (i.e. is a leaf).

In addition, we have one more restriction:

R_1 For a network $\mathcal{N}$, when only *one* of the incoming edges of each reticulation node is kept and the other is deleted, we always derive a tree T'.

In this paper, we assume the in-degree of reticulation nodes is two by noting that we can always convert a reticulation node with in-degree of three or more to several reticulation nodes with in-degree of two [18]. We call a branch of a hybridization network or a tree a "lineage". Intuitively, a lineage corresponds to some extant or ancestral species modeled in the phylogenetic network. There are two types of lineages: leaf lineages (those originated from the leaves of the network) and internal lineages (which correspond to ancestral species of the network). An internal lineage l_i in a network is created by either a reticulation or a coalescence.

We first consider the derived tree T' (that is embedded in $\mathcal{N}$) as stated in R_1. When we recursively remove non-labeled leaves and contract edges to remove degree-two nodes of T' (called cleanup), we obtain a phylogenetic tree T (for the same set of species as in $\mathcal{N}$). Now suppose we are given a phylogenetic tree T. We call T is *displayed* in $\mathcal{N}$ when we can obtain an induced tree T' from $\mathcal{N}$ by properly choosing a single edge to keep at each reticulation node so that T' is topologically *equivalent* to T after cleanup. We denote the induced T' (if exists) as $T_\mathcal{N}$. We call the choices of which reticulation edges to keep (and prune) the "display choices". In Figure 1, each of the three trees is displayed in the network. For example, one possible display choice for T_1 (the left most gene tree) is keeping lineages b and d (and pruning lineages a and e).

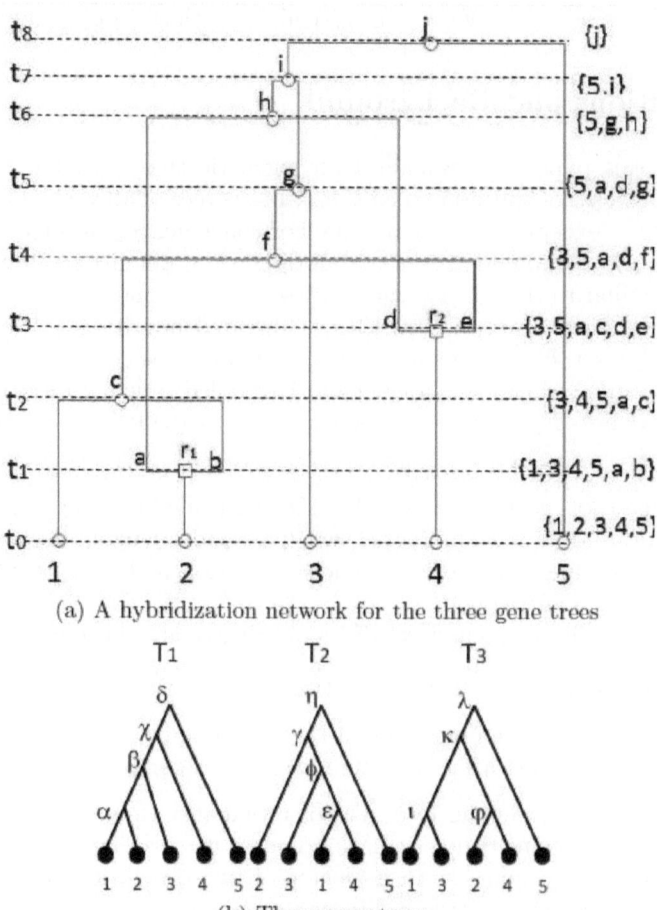

(a) A hybridization network for the three gene trees

(b) Three gene trees

Fig. 1. An illustration of a hybridization network with two reticulation events for three gene trees T_1, T_2 and T_3. Reticulation: square. Speciation (coalescence): oval. Dotted lines: time. Configurations are shown to the right, one for each time line. Leaf labels: numbers. Internal nodes (subtrees) of gene trees are labeled by Greek letters.

For a hybridization network $\mathcal{N}$, we define the hybridization number (denoted as $H_\mathcal{N}$) as the number of the reticulation nodes. Note that this is equivalent to using the summation of in-degree minus one of all reticulation nodes as in [14] since the in-degree of a reticulation node is assumed to be two. Sometimes $H_\mathcal{N}$ is also called the number of reticulation events in $\mathcal{N}$. Recall that the optimal hybridization network is the one with the smallest hybridization number. Now we formulate the central problem in this paper.

The Most Parsimonious Hybridization Network Problem. Given K rooted and binary gene trees $T_1, T_2, \ldots T_K$ (with the same n taxa), construct the most parsimonious hybridization network $\mathcal{N}_{min}$ such that (i) each gene tree T_i is displayed in $\mathcal{N}_{min}$ and (ii) $H_{\mathcal{N}_{min}}$ is *minimized* among all possible such networks. We call $H_{\mathcal{N}_{min}}$ the hybridization number of $T_1, \ldots, T_K$.

Constructing parsimonious hybridization networks for a set of K gene trees is a computationally challenging problem. Even the two-gene-tree (i.e. $K = 2$) case is known to be NP-complete [4]. This two-gene-tree case is closely related to computing the subtree prune and regraft (SPR) distance of two trees, a well-studied NP complete problem [7,3] in phylogenetics. Nonetheless, there are several practical algorithms for the SPR distance problem (e.g. [17,16]). For the two-gene-tree case of the hybridization network problem, there are also several exact methods [2,20,1]. Although the worst case running time of these practical methods are exponential, these methods may work reasonably well in practice. It becomes more computationally challenging when there are three or more gene trees. There are currently only a few heuristic methods for either estimating the hybridization number $H(T_1, \ldots, T_K)$ or reconstructing near optimal networks for trees $T_1, \ldots, T_K$ when $K \geq 3$ [18,13,5]. There are no existing methods for the exact computation of the hybridization number or reconstructing parsimonious networks with three or more trees.

3   Constructing Parsimonious Hybridization Networks

3.1   The Backward in Time View

The backward in time view is the foundation of our method. With a forward in time view, a tree node in a hybridization network refers to a speciation event where one lineage splits into two lineages; at a reticulation node a new lineage is created after two incoming lineages are merged. In this paper, we take a backward in time view instead. In this view of time, a tree node is called a coalescence: two lineages coalesce into a single lineage at a tree node when looking backward in time. Similarly, in this view, two new lineages are created by the reticulation of a lineage at a reticulation node. As an example, we consider the network shown in Figure 1. Lineages 1 and b coalesce at time t_2 to form the lineage c, and a reticulation occurs for the lineage 4 at t_3 and creates lineages d and e. It is important to note that a lineage created by a reticulation may "vanish" (i.e. be pruned) when we make the display choices for a tree. For example, to display T_2 (the middle tree in Figure 1), the lineage b vanishes. Displaying a tree T within a network can also be explained with this view of time. Imagine that we "cut" the network with the time line at time t and we only consider the portion of the network more recent than time t. We say a subtree T_s of T is displayed by time t if T_s can be obtained at the lineage l_i where l_i is cut by the time line t. That is, we can obtain T_s by following lineages backward in time to l_i at t.

In this case, we also say T_s is displayed in l_i or l_i displays T_s. When we start at the present time, only leaves (i.e. subtrees with singleton taxa) of T are displayed. As the time line moves backward, larger and larger subtrees are displayed. For example, in Figure 1, at time t_0, only singleton subtrees of T_1 are displayed by t_0. When we move the time back, the subtree α is also displayed by t_2 (and is displayed in the lineage c). In the end, we reach the root of the network where the entire T is displayed. This simple observation is important for the $PIRN_C$ algorithm described here.

3.2   The High Level Idea

Here is the high level idea of the $PIRN_C$ algorithm. We take a coalescent-style approach by going backward in time. At a particular time of phylogenetic history, there is a set of lineages that are present at that time. Let us call the snapshot of the phylogenetic history at a particular time the "ancestral configuration" (or simply configuration), which specifies the set of ancestral lineages alive at that time. At present time, there is a single fixed configuration, which contains all the n extant lineages in the given gene trees. When moving backward in time, configuration changes when some genealogical events (namely coalescence and reticulation) occur. Here, we assume there are no two genealogical events occurring at exactly the same time. For example, consider the example network in Figure 1. The initial configuration (denoted as C_0) contains lineages $1, 2, 3, 4$ and 5. The first event backward in time from the present time is the reticulation r_1 of lineage 2 at time t_1, which creates two new lineages a and b. So right before (i.e. more ancient than) t_1, the configuration contains $1, 3, 4, 5, a$ and b. When we continue tracing backward, the coalescence between lineage 1 and b happens at time t_2, which creates a new lineage c. Then the new configuration right before t_2 contains five lineages: $3, 4, 5, a$ and c. Eventually we reach the final configuration (denoted as C_f, which contains a single lineage j).

However, when only gene trees are given, we do not know what coalescent and reticulation events will occur nor the series of configurations at the time of genealogical events when tracing backward in time. In fact, if we knew, we would have already found the true hybridization network: configurations at all the genealogical events specify precisely the phylogenetic history. The key for our approach is finding configurations at genealogical events that correspond to the most parsimonious network. Suppose we start with one configuration C and consider what configurations can be reached from C by a single genealogical event backward in time. Here, each pair of lineages in C can coalesce and each lineage of C can have a reticulation. New configurations are generated with these genealogical events. If we trace backwards long enough, we will reach the final configuration C_f, where the hybridization network corresponding to C_f displays each given gene tree. If we also ensure C_f is the one that uses the fewest number of reticulations, we then know the minimum number of reticulations needed for the given input gene trees. Once such a C_f is found, we can then identify the series of genealogical events leading to C_f and this allows us to build the most parsimonious network.

The approach sketched above is a simple strategy. However, a moment of thoughts indicates that its naïve implementation will be too slow: the space of possible configurations is immense. Consider a configuration with n lineages from which we are to search for new configurations. If no restriction is imposed, there are $\binom{n}{2}$ possible coalescences and n reticulations among the n lineages. Suppose n is 30. Then there are up to 465 new configurations reachable from one configuration with one reticulation or one coalescence. The number of possible configurations to explore quickly becomes prohibitively large shortly after the start of the configuration search. In this paper, we show that the basic approach can be made much faster with additional techniques, which allows us to "cut corners" while still ensuring the finding of optimal hybridization networks. The key to our approach is that the search is guided by the given gene trees. That is, our algorithm is based on guided configuration search and configurations that do not lead to parsimonious networks for the given gene trees may be pruned early. We have also developed additional speedup techniques that further improve the efficiency. Together they turn the basic strategy into a practical approach.

3.3   The Guided Search for the Parsimonious Configurations

Ancestral configuration is the basic data structure used in our algorithm. An ancestral configuration $\mathcal{C}$ contains a set of lineages $l_1 \ldots l_m$. Recall that each subtree T_s of T is also displayed in some lineage l_i of the network. Initially, $\mathcal{C}_0$ only displays the singleton subtrees. As we explore the configuration space backward in time, we may find configurations where increasingly larger input subtrees are displayed within their lineages. The search stops when each whole gene tree is displayed in the single lineage of the final configuration $\mathcal{C}_f$. Therefore, the set of subtrees displayed in a configuration measures the progress made from $\mathcal{C}_0$ to the current configuration: the more large subtrees displayed in a configuration, the closer we are in finishing the construction of hybridization networks. For example, in Figure 1, the lineage 2 only displays singleton subtrees with taxon 2. And so do the lineages a and b. The lineage c is created by the coalescence of lineages 1 and b. Thus, the lineage c displays the subtree α. Note that b is created by a reticulation and thus b can vanish (i.e. b may be pruned in displaying a subtree). Thus, c also displays the singleton subtree with taxon 1 (in case b vanishes)

The above discussion suggests the set of displayed subtrees of a lineage is key to configuration search. We let a lineage l_i maintain the set of input subtrees, denoted as $T(l_i)$, that are displayed in l_i. For convenience, we sometimes use $T(l_i)$ to represent the lineage l_i (as in Figure 2). For a leaf lineage l_i that is labeled with taxon x, $T(l_i)$ contains the singleton subtrees labeled by x (which appears in each gene tree). When the lineage l_i is an internal lineage, $T(l_i)$ is determined when l_i is created by genealogical events as follows.

1. If l_i is created by a reticulation of the lineage l_i', then $T(l_i) = T(l_i')$.
2. If l_i is created by a coalescence of the lineages l_i^1 and l_i^2, then $T(l_i)$ contains new subtrees formed by coalescing one subtree displayed in l_i^1 and another

subtree displayed in l_i^2. More specifically, $T(l_i)$ contains $p(s^1, s^2)$ (if exists), where $s^1 \in T(l_i^1)$, $s^2 \in T(l_i^2)$. Here, $p(s^1, s^2)$ refers to the subtree in some input gene tree that has subtrees s^1 and s^2 as children; if s^1 and s^2 do not form a subtree in a gene tree, then $p(s^1, s^2)$ does not exist. For example, in Figure 1, $p(\alpha, 3) = \beta$, $p(\varphi, \iota) = \kappa$ but $p(\alpha, 4)$ does not exist.

As alluded before, when determining $T(l_i)$ formed by coalescing l_i^1 and l_i^2, we also need to consider whether l_i^1 or l_i^2 is vanishable. We say a lineage l_i is vanishable if either l_i is created by a reticulation or l_i is created by a coalescence between two lineages, where both of them are vanishable. Intuitively, a vanishable lineage means that the lineage may vanish and thus does not involve in forming new displayed subtrees with the other lineage if certain display choices are made. For example, in Figure 1, the lineages a, b, d and e (and also h since both of its children a and d are vanishable) are vanishable while the lineages $1, 2, 3, 4, 5, c, f, g, i$ and j are not. Suppose one coalescing lineage (say l_i^1) is vanishable. Then $T(l_i)$ also contains each $s \in T(l_i^2)$. For example, in Figure 1, the lineage f is created by the coalescence of c and e, where $T(c) = \{1, \alpha\}$ and $T(e) = \{4\}$. Then, subtrees 1 and 4 form the subtree ϵ of T_2, and thus $\epsilon \in T(f)$. Moreover, since e is vanishable, $1, \alpha \in T(f)$. Also, c is not vanishable. Thus, $T(f) = \{1, \alpha, \epsilon\}$.

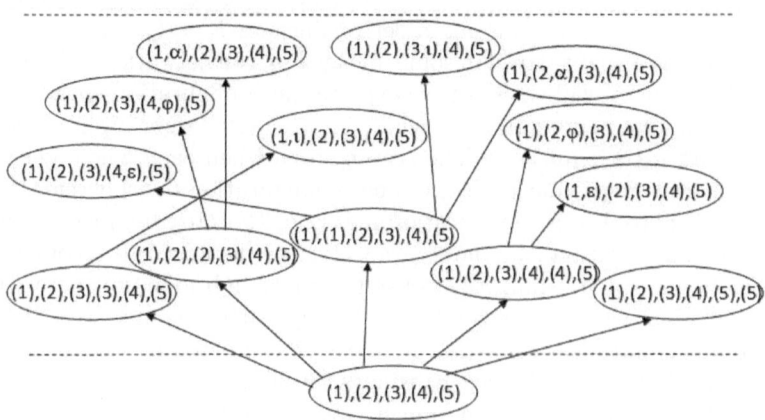

Fig. 2. The list of configurations of stages 0 and 1 for the example in Figure 1. A configuration (ellipse) contains a set of lineages, where each lineage is represented by its set of displayed subtrees (in numerical taxa form and Greek letters as in Figure 1).

There are some subtle issues about maintaining the displayed subtrees in configurations, which will be discussed in the full version of this paper.

The Configuration Search Algorithm. The basic algorithm for constructing parsimonious hybridization networks explores configurations in a breadth-first search style. The algorithm runs in stages, where at each stage the algorithm constructs a set of configurations in the following way. First, new configurations are added to this stage with one reticulation performed upon configurations found during the previous stage. Then, we perform as many coalescences on these newly formed configurations and obtain additional configurations for this stage. That is, configurations on one stage are obtained from the same number of reticulations from the initial configuration C_0. More specifically, in this algorithm, R refers to the breath-first search level, and is equal to the number of reticulations performed so far from C_0. $L_C(R)$ is the list of configurations found at level R. R_{max} is the user-defined maximum of reticulations allowed.

1. $R \leftarrow 0$. $L_C(0) \leftarrow \{C_0\}$.
2. While $R < R_{max}$
3.   For each $C \in L_C(R)$
4.   Perform one reticulation on each lineage of C and obtain new configurations C'.
5.   For each C', recursively try all ways of coalescences of two lineages in C' to create new configurations C''; then discard C'' if it is infeasible (see later this section); otherwise, $L_C(R+1) \leftarrow L_C(R+1) \cup \{C''\}$.
6.   If a final configuration is found, construct the optimal network by traceback and stop.
7. $R \leftarrow R + 1$
8. Report there is no solution with less than R_{max} reticulations.

See Figure 2 for an example of executing the configuration search algorithm on the trees shown in Figure 1 for the first two levels. At level 0, we start with a single configuration C_0. With proper preprocessing, we do not need to perform coalescences on C_0. This will be explained in the full version of this paper, At level 1, a single reticulation is performed on C_0 to obtain new configurations C'; then all possible coalescences are performed on each C'. We find thirteen configurations in total at level 1.

Optimality. The $PIRN_C$ algorithm examines configurations with non-decreasing reticulation distance from C_0. Since no configurations that lead to the final configuration are discarded, the found network is the most parsimonious hybridization network.

Infeasible Configurations. In principle, every pair of lineages in a configuration can coalesce to create a new configuration. However, some coalescence will lead to a configuration C that is infeasible: the final configuration C_f can not be obtained from C. Early removal of infeasible configurations can significantly speed up the search for optimal networks. Here is a simple test for finding infeasible configurations. Given a set of displayed subtrees S within a gene tree T, we say T is displayable from S if each leaf of T is "covered" by some subtree in S. Otherwise, we say T is not displayable from S. A leaf is covered by a subtree if the subtree contains the leaf. Intuitively, if subtrees in S can not cover each leaf of T, then T can not be displayed by S. Checking whether a tree T is displayable from S can be easily done by a traversal of T. A configuration C is infeasible if some input gene tree is not displayable from the set of displayed subtrees of all the lineages in C. For example, we consider the coalescence of lineages (1) and (2) in the configuration $\{(1),(2),(3),(3),(4),(5)\}$, which creates a new configuration C

$= \{(\alpha),(3),(3),(4),(5)\}$. For $\mathcal{C}$, the leaf lineages 1 and 2 in T_1 are covered by the subtree α; but the leaf lineages 1 and 2 in T_2 and T_3 are not covered by displayed subtrees. Thus, $\mathcal{C}$ is infeasible and should not be considered.

We have developed stronger infeasibility check techniques, which are more effective in pruning the search space of configurations. Due to the space limit, we will describe these techniques in the full version of this paper.

3.4 $PIRN_{CH}$: A Heuristic

$PIRN_C$ becomes slow when the number of reticulation events increases. To construct more complex networks, we develop a heuristic called $PIRN_{CH}$, which is based on the same principle of $PIRN_C$ but has more aggressive approaches to prune the search space of configurations. $PIRN_{CH}$ uses a scoring scheme to rank configurations. Intuitively, the score of a configuration $\mathcal{C}$ is based upon the progress made by $\mathcal{C}$ toward the final configuration. Then we keep the top N_c (chosen by the user) ranked configurations and prune the rest at each stage. There is a trade-off between accuracy and efficiency in choosing the value of N_c. Due to the space limit, details will be provided in the full version of this paper.

4 Results

We have implemented both $PIRN_C$ and $PIRN_{CH}$ for building the parsimonious network as part of the software package $PIRN$. It is available for download from: http://www.engr.uconn.edu/~ywu/. We test our new algorithms with simulated data on a 3192 MHz Intel Xeon workstation. We use the same simulation data generated by a two-stage approach as in [18]. Since $PIRN_C$ is designed to build networks with relatively small number of reticulation, we use the datasets generated in [18] with lower reticulation level. We test for several settings of n (the number of taxa) and K (the number of gene trees).

To test $PIRN_C$, we compare with the bounds computed by the program $PIRN$ [18]. $PIRN$ provides a lower bound (called the RH bound) and an upper bound (called the SIT bound). Note that when the RH bound matches the SIT bound, $PIRN$ finds the optimal network. When the two bounds do not match, we only know the range of hybridization number but not the true hybridization number, and this is a major weakness of the $PIRN$ approach [18]. It is known in [18] that the lower and upper bounds match often for lower reticulation level and smaller number of gene trees, but diverge more for higher reticulation level and larger number of gene trees. The reason for comparing with $PIRN$ is that $PIRN$ appears to infer networks that in practice are close to the optimum [18,13]. In our simulation, we restrict our attention to datasets whose hybridization number is at most 4 since $PIRN_C$ is designed for data with smaller hybridization number. For datasets with higher hybridization number, $PIRN_C$ simply reports that their hybridization number is larger than 4 and no network is constructed. Table 1 shows the results of our simulation. The "#Data ≤ 4" refers to the percentage of datasets that have hybridization number of 4 or less, and we only give results

Table 1. Average performance of $PIRN_C$ over 100 datasets for each setting on simulated data. Results are only for those datasets with hybridization number of 4 or less (i.e. datasets with hybridization number of 5 or more are excluded). #Data $\leq$ 4: percentage of datasets with the hybridization number of 4 or less (where $PIRN_C$ constructs the optimal networks). $PIRN_C$ = RH (resp. $PIRN_C$ > RH): among the datasets where $PIRN_C$ gives optimal results, percentage of datasets $PIRN_C$ gives the same (resp. larger) hybridization number as given by the RH lower bound. $PIRN_C$ < SIT (the other two are straightforward): percentage of datasets $PIRN_C$ gives the smaller hybridization number as given by the SIT upper bound. #Data not optimal by SIT: percentage of data where the RH bound and SIT bounds do not match (and thus the optimality is not determined by the two bounds) while $PIRN_C$ gives optimal solution. Time: average run time of $PIRN_C$ in seconds.

	n=10			n=20			n=30		
	K=3	K=4	K=5	K=3	K=4	K=5	K=3	K=4	K=5
#Data $\leq$ 4	98	98	93	88	77	65	84	76	65
$PIRN_C$ = RH	96	93	90	88	74	63	84	75	61
$PIRN_C$ > RH	2	5	3	0	3	2	0	1	4
$PIRN_C$ < SIT	0	1	0	0	1	0	0	1	0
$PIRN_C$ = SIT	98	97	93	88	76	65	84	75	65
$PIRN_C$ > SIT	0	0	0	0	0	0	0	0	0
#Data not optimal by SIT	2	6	3	0	4	2	0	1	4
Time	13.4	49.9	92.6	276.8	705.8	1686.6	606.7	2227.1	2811.5

for these datasets (i.e. $PIRN_C$ does not give results for some datasets). Table 1 shows that $PIRN_C$ can find optimal networks where $PIRN$ does not: for example, for $n = 10$ and $K = 4$ case, $PIRN_C$ finds the true optimum for 6 out of 98 datasets, where the bounds of $PIRN$ do not match (and thus $PIRN$ does not know whether its solutions are optimal or not) for these datasets. For some other settings, $PIRN_C$ gives the same results as $PIRN$ does. Still, it may be useful to have a method that always finds optimal solutions. The ability for finding optimal networks is the key advantage of $PIRN_C$ when compared with existing methods like $PIRN$ (and MURPAR [13]). The running time of $PIRN_C$ is more influenced by the hybridization number than by n or K. The case of hybridization number being 4 (or even 5) or smaller is usually practically solvable by $PIRN_C$.

For handling more complex networks, we also test our heuristic $PIRN_{CH}$ on datasets with higher hybridization number. Note that the choices of $PIRN_{CH}$ parameters (e.g. N_c, the maximum number of configurations kept at each search level) have a large impact on the accuracy and efficiency. For this simulation, we set N_c to be 100,000. Results are shown in Table 2. The coarse mode of the SIT bound is used for larger data (when $n = 40$ and 50) as in [18]. As shown in Table 2, $PIRN_{CH}$ performs well against $PIRN$: there is only one out of 900 datasets where $PIRN_{CH}$ constructs a network using more reticulation than $PIRN$; and $PIRN_{CH}$ finds optimal networks (when its reticulation number matches the RH bound) for 82% for data with 50 taxa and 5 gene trees, while the SIT bound can only do the same for 58%. Also the gap between the results by $PIRN_{CH}$ and the SIT bound increases for larger and more complex data.

Table 2. Performance of $PIRN_{CH}$ on 100 simulated datasets per settings. =RH (resp. SIT=RH): the number of datasets $PIRN_{CH}$ (resp. SIT bound) gives the same results as the RH lower bound (and thus optimal networks are found). *: coarse mode of the SIT bound is used for $n = 40$ and 50. Gap(RH): average gap between $PIRN_{CH}$ results and the RH bound. <SIT (the other two are straightforward): the number of datasets $PIRN_{CH}$ gives the smaller hybridization number as given by the SIT upper bound. Gap(SIT): average gap between $PIRN_{CH}$ results and the SIT bound. Gap of two values a and b is defined as $a - b$. Time: the time of $PIRN_{CH}$ in in seconds.

	n=30			n=40*			n=50*		
	K=3	K=4	K=5	K=3	K=4	K=5	K=3	K=4	K=5
=RH	98	93	77	97	90	83	98	89	82
SIT=RH	97	92	78	92	73	55	96	75	58
Gap(RH)	0.02	0.08	0.25	0.03	0.11	0.18	0.02	0.10	0.18
<SIT	1	3	3	5	22	37	2	16	34
=SIT	99	97	96	95	78	63	98	84	66
>SIT	0	0	1	0	0	0	0	0	0
Gap(SIT)	0.01	0.03	0.02	0.06	0.25	0.54	0.02	0.17	0.39
Time	850.6	3,321.3	6,453.6	2,942.7	5,299.8	16,384.3	2073.7	8,204.7	13,846.64

5   Discussion

Simulation shows that $PIRN_C$ and $PIRN_{CH}$ perform reasonably well comparing with $PIRN$ (previously the best approach for building hybridization networks of multiple trees), although constructing optimal hybridization networks is still challenging computationally. Our approach is based on the concept of ancestral configuration. A similar data structure has been used in studying the discordance of gene trees caused with the so-called incomplete lineage sorting (another important evolutionary process for the so-called gene tree and species tree problem) [19]. Ancestral configurations may be useful in developing new algorithms for studying multiple evolutionary processes together (e.g. reticulate evolution and incomplete lineage sorting) on a proper model.

Acknowledgments. This work is partly supported by U.S. National Science Foundation grants IIS-0803440 and CCF-1116175.

References

1. Albrecht, B., Scornavacca, C., Cenci, A., Huson, D.H.: Fast computation of minimum hybridization networks. Bioinformatics 28, 191–197 (2012)
2. Bordewich, M., Linz, S., John, K.S., Semple, C.: A reduction algorithm for computing the hybridization number of two trees. Evolutionary Bioinformatics 3, 86–98 (2007)
3. Bordewich, M., Semple, C.: On the computational complexity of the rooted subtree prune and regraft distance. Annals of Combinatorics 8, 409–423 (2004)
4. Bordewich, M., Semple, C.: Computing the minimum number of hybridization events for a consistent evolutionary history. Discrete Applied Mathematics 155, 914–928 (2007)

5. Chen, Z., Wang, L.: Algorithms for reticulate networks of multiple phylogenetic trees. IEEE/ACM Transactions on Computational Biology and Bioinformatics 9(2), 372–384 (2012)
6. Gusfield, D.: Optimal, efficient reconstruction of Root-Unknown phylogenetic networks with constrained and structured recombination. J. Comp. Sys. Sci. 70, 381–398 (2005)
7. Hein, J., Jiang, T., Wang, L., Zhang, K.: On the complexity of comparing evolutionary trees. Discrete Appl. Math. 71, 153–169 (1996)
8. Huson, D.H., Klöpper, T.H.: Beyond Galled Trees - Decomposition and Computation of Galled Networks. In: Speed, T., Huang, H. (eds.) RECOMB 2007. LNCS (LNBI), vol. 4453, pp. 211–225. Springer, Heidelberg (2007)
9. Huson, D., Rupp, R., Gambette, P., Paul, C.: Computing galled networks from real data. Bioinformatics 25, i85–i93 (2009); Bioinformatics Suppl., Proceedings of ISMB 2009
10. Huson, D.H., Rupp, R., Scornavacca, C.: Phylogenetic Networks: Concepts, Algorithms and Applications. Cambridge University Press, Cambridge (2010)
11. Morrison, D.A.: Introduction to Phylogenetic Networks. RJR Productions, Uppsala (2011)
12. Nakhleh, L.: Evolutionary phylogenetic networks: models and issues. In: Heath, L., Ramakrishnan, N. (eds.) The Problem Solving Handbook for Computational Biology and Bioinformatics, pp. 125–158. Springer (2010)
13. Park, H.J., Nakhleh, L.: MURPAR: A Fast Heuristic for Inferring Parsimonious Phylogenetic Networks from Multiple Gene Trees. In: Bleris, L., Măndoiu, I., Schwartz, R., Wang, J. (eds.) ISBRA 2012. LNCS, vol. 7292, pp. 213–224. Springer, Heidelberg (2012)
14. Semple, C.: Hybridization networks. In: Gascuel, O., Steel, M. (eds.) Reconstructing Evolution: New Mathematical and Computational Advances, Oxford, pp. 277–309 (2007)
15. van Iersel, L., Keijsper, J., Kelk, S., Stougie, L., Hagen, F., Boekhout, T.: Constructing Level-2 Phylogenetic Networks from Triplets. In: Vingron, M., Wong, L. (eds.) RECOMB 2008. LNCS (LNBI), vol. 4955, pp. 450–462. Springer, Heidelberg (2008)
16. Whidden, C., Zeh, N.: A Unifying View on Approximation and FPT of Agreement Forests. In: Salzberg, S.L., Warnow, T. (eds.) WABI 2009. LNCS, vol. 5724, pp. 390–402. Springer, Heidelberg (2009)
17. Wu, Y.: A practical method for exact computation of subtree prune and regraft distance. Bioinformatics 25, 190–196 (2009)
18. Wu, Y.: Close lower and upper bounds for the minimum reticulate network of multiple phylogenetic trees. Bioinformatics (Supplement Issue for ISMB 2010 Proceedings) 26, 140–148 (2010)
19. Wu, Y.: Coalescent-based species tree inference from gene tree topologies under incomplete lineage sorting by maximum likelihood. Evolution 66, 763–775 (2012)
20. Wu, Y., Wang, J.: Fast Computation of the Exact Hybridization Number of Two Phylogenetic Trees. In: Borodovsky, M., Gogarten, J.P., Przytycka, T.M., Rajasekaran, S. (eds.) ISBRA 2010. LNCS, vol. 6053, pp. 203–214. Springer, Heidelberg (2010)

Fast and Accurate Calculation of Protein Depth
by Euclidean Distance Transform

Dong Xu[1,2], Hua Li[3], and Yang Zhang[2]

[1] Bioinformatics & Systems Biology Program, Sanford-Burnham Medical Research Institute
dxu@sanfordburnham.org
[2] Department of Computational Medicine and Bioinformatics, University of Michigan
zhng@umich.edu
[3] Integration Application Center, Institute of Computing Technology,
Chinese Academy of Sciences
lihua@ict.ac.cn

Abstract. The depth of each atom/residue in a protein structure is a key attribution that has been widely used in protein structure modeling and function annotation. However, the accurate calculation of depth is time consuming. Here, we propose to use the Euclidean distance transform (EDT) to calculate the depth, which conveniently converts the protein structure to a 3D gray-scale image with each pixel labeling the minimum distance of the pixel to the surface of the molecule (i.e. the depth). We tested the proposed EDT method on a set of 261 non-redundant protein structures. The data show that the EDT method is 2.6 times faster than the widely used method by Chakravarty and Varadarajan. The depth value by EDT method is also highly accurate, which is almost identical to the depth calculated by exhaustive search (Pearson's correlation coefficient≈1). We believe the EDT-based depth calculation program can be used as an efficient tool to assist the studies of protein fold recognition and structure-based function annotation.

Keywords: Euclidean distance transform, fold recognition, molecular visualization, protein depth, protein tertiary structure, solvent accessibility.

1   Introduction

For a given protein tertiary structure, many residue level attributions can be extracted, such as the secondary structure type, dihedral angle and solvent accessibility. Those structural features help establish the properties of different amino acid types and categorize protein structure folds. For example, Ramachandran plot [1] revealed that the distribution of backbone dihedral angles (or the secondary structure) was highly regulated. Solvent accessibility (SA) evaluates the hydrophobicity of amino acids in different protein structures, which can be calculated accurately by EDTSurf [2] or approximately by DSSP [3].

However, SA usually specifies the residues in a binary form. For the residues that are completely buried in protein, it does not describe where the residues locate inside

M. Deng et al. (Eds.): RECOMB 2013, LNBI 7821, pp. 304–316, 2013.
© Springer-Verlag Berlin Heidelberg 2013

the molecule. Depth, which measures the distance of each atom/residue to the solvent accessible surface in a continuous form, greatly complements the missing information by SA. In fact, the depths of residues in a protein are highly related to their effects of mutations on protein stability and on protein-protein interactions [4]. The residue depth has also been widely used to specify protein folds in protein structure prediction [5-7] and assist structure-based protein function annotation [8].

Despite the importance, by far there are very few methods which can calculate the depth for protein structures efficiently at either an atom level or a residue level. In Ref. [4], Chakravarty and Varadarajan proposed to calculate the residue depth by rotating the protein in a box where the closest water molecule is identified for each atom in the protein. The accuracy of the method is compromised since the calculated depth value depends on the positions of the water molecules. One can also calculate the depth by first generating the explicit solvent accessibility surface (e.g. by EDT-Surf or MSMS [9]) and then identifying the vertex on the triangulated surface which is the closest one to the atom [10-11]. However, the computation of this kind of method is quite time-consuming since all the atoms in the protein need to be searched against the huge number of vertices on the surface.

In a recent study, we have established the relationships between the three kinds of macromolecular surfaces and Euclidean distance transform (EDT) theoretically and developed a fast algorithm for generating their triangulated surfaces precisely [2]. In this work, we apply the EDT technique to the calculation of protein atom depth and residue depth. The algorithm is fast since the explicit triangulated surface is not required. To investigate the efficiency and accuracy of this method, we compare the computational time and depth value with that by Chakravarty and Varadarajan (CV). We also analyze the relations of the depth with the commonly-used radius of gyration and solvent accessibility. The source code and executable program are freely available at http://zhanglab.ccmb.med.umich.edu/EDTSurf/.

2 Material and Method

2.1 Depth Definition

Atom depth is the shortest distance between the center of the atom and the outer solvent accessible surface (SAS) of the molecule, as illustrated in Fig. 1. SAS is the area traced out by the center of a probe sphere when it is rolled over the whole molecule [12]. When one atom is exposed (e.g. atom i in the figure), its depth will equal to the sum of the van de Waals radius and the radius of the probe sphere r_p which is often set to 1.4 Å. For atoms which are completely buried inside (e.g. atoms j and k in the figure), their solvent accessibilities are all equal to zero, but their depths may be different. Residue depth is the average value of the atom depths of all the atoms in a residue.

The definition of depth by Chakravarty and Varadarajan is a little different, which is the shortest distance to the explicit bulk water rather than the solvent accessible surface. Since water molecules don't have spherical shapes and may have different poses around the molecule, this difference will result in the slightly different depth values.

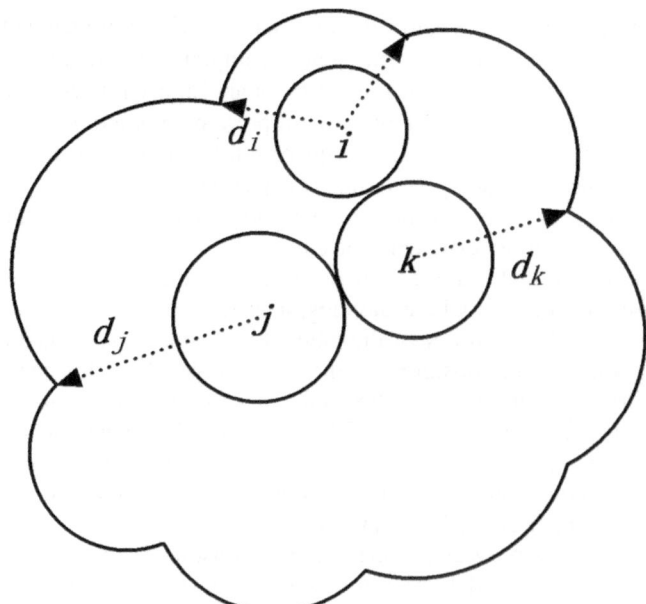

Fig. 1. Illustration of three atoms with different depth values in a 2D plane. The outside boundary stands for the solvent accessible surface

2.2   Euclidean Distance Transform

Euclidean distance transform (EDT) is the transformation that converts a digital binary image to another gray-scale image where the value of each pixel is the minimum Euclidean distance between that pixel and the boundary. We have developed a fast algorithm which can conduct EDT in arbitrary dimensional space [13]. EDT has been widely used in the fields of image processing and computer graphics, such as skeleton extraction [14], shortest path planning [15] and geometric shape description [16].

Given a protein structure, we suppose it has N atoms, each of which locates at p_i and has a van der Waals radius r_i. To calculate the atom depth in this protein, we first build the solvent accessible solid using equation (1), which is the union of all the spheres with radius equal to the sum of the van der Waals radius and the radius of the probe sphere. The union operation is conducted in the discrete 3D space using space-filling technique, with each sphere represented by a set of grid points.

$$O_{SA} = \bigcup_{i=1}^{N} sphere(p_i, r_i + r_p) \tag{1}$$

Then we can easily determine the outer shell of the solvent accessible solid, which is the discrete representation of solvent accessible surface. We do the EDT transform to the shell and can get the shortest Euclidean distance of each point to the shell, which happens to be the depth value of this point. Although there are other distance functions, such as City-block distance and Chessboard distance, only Euclidean distance has the direct relationship to the three macromolecular surfaces as well as the depth.

In the original CV method and its recent extension [17], non-bulk water molecules are removed in the regions of narrow cavities and internal voids. Otherwise, atoms around those regions will have small values of depths. Using equation (1), the solvent accessible solid has already filled most of the empty space in the same regions since the radius of each atom is enlarged by the radius of the probe sphere. Therefore, the two methods have consistent depth values in those special regions.

Fig. 2 shows an example of the EDT result to the same shape of the SAS in Figure 1, where the red curve stands for the SAS. After the EDT transform, every position has a shortest distance to the SAS, as represented by the gray-scale pixel value in the image. The lighter the point is, the longer distance to the surface it will have. Based on the definition of the depth, we can see that the gray-scale pixel value calculated by the EDT at each point exactly is the depth value of that point. In the figure, the centers of the three atoms, as represented by the blue dots, have different depth values.

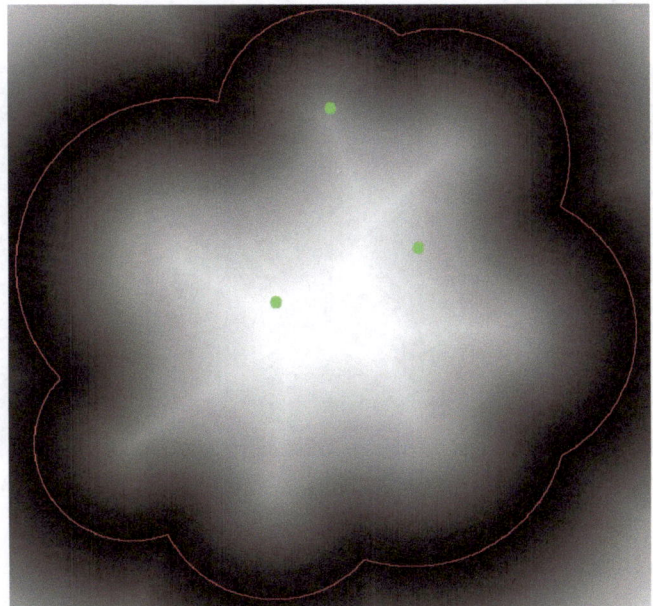

Fig. 2. Illustration of the EDT to the solvent accessible surface (in red) in a 2D plane. Centers of the three atoms are marked in blue.

Solvent accessibility of each residue is defined as the ratio of the total SAS area of all the atoms in the residue to the maximum SAS area of that residue type. Hence, we have to build the explicit triangulated surface from the discrete shell by surface triangulation algorithms such as the Marching Cube method [18]. Different to the solvent accessibility derivation, depth calculation doesn't require the generation of the explicit triangulated SAS.

Given the shell of the discrete SAS, we can also calculate the depth of each atom by exhaustive search (ES). That is to say, we search for the point on the shell which is the closest to the center of the atom.

3   Result and Discussion

3.1   Visualization of Depth

In order to visually check the depth information generated by the method described above, we have embedded the EDT-based depth calculation algorithm into our Macromolecular Visualization and Processing (MVP) program, which can be downloaded at http://zhanglab.ccmb.med.umich.edu/MVP/.

Fig. 3 shows two snapshots of the MVP visualization result of a *hypothetical protein from thermus thermophilus HB8* (PDB ID: 1whz, chain A), which contains 122 residues and 937 atoms. Atoms in the left figure are in the ball-stick style. Red color means high value of atom depth while blue means low. In the right image, we show the protein backbone structure where the color of each residue is also correlated with its residue depth. From both images, we can clearly see the layers of the protein structure, especially the hydrophobic core which is in red.

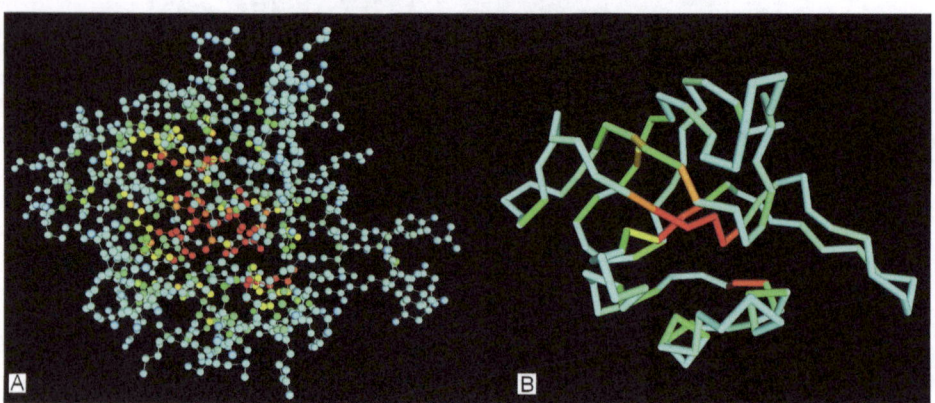

Fig. 3. (A) atom depth (B) residue depth of the protein 1whz chain A

3.2   Depth Distributions of Different Residue Types

Since different residue types have different hydrophobicities, their depth distribution should also be different. Therefore, we choose 36,556 protein domains used by our threading programs [6,19] for validation, which can be downloaded at http://zhanglab.ccmb.med.umich.edu/library/. Those structures are non-homologous to each other with sequence identity cutoff 70%. Protein chains which contain multiple domains are discarded from the list in this test, because multiple-domain proteins are often not well-packed.

The distributions of residue depths of the 20 residue types are summarized in Fig. 4, which are arranged in the order of their hydrophobicity scales [20]. Residue depths normally are in the range of 2.9 Å and 8.9 Å. Almost all the residue depths are less than 5 Å for the 8 hydrophillic residues on the top 2 rows. TRP and SER have similar hydrophobicities, but TRP has more depths which are deeper than 5 Å. This is probably because TRP has a longer side-chain and its depth can be large even part of the residue is exposed. For the 6 most hydrophobic residues on the bottom two rows, more depth values are around 6 Å than the hydrophillic residues. However, majority of the depths are still close to 3.1 Å, which means many hydrophobic residues still locate around the surface of the domain structures. It is understandable if the protein is stable only in the complex form instead of the monomeric form. Hydrophobic residues in the interface will have deep residue depths if we treat the complex as a whole.

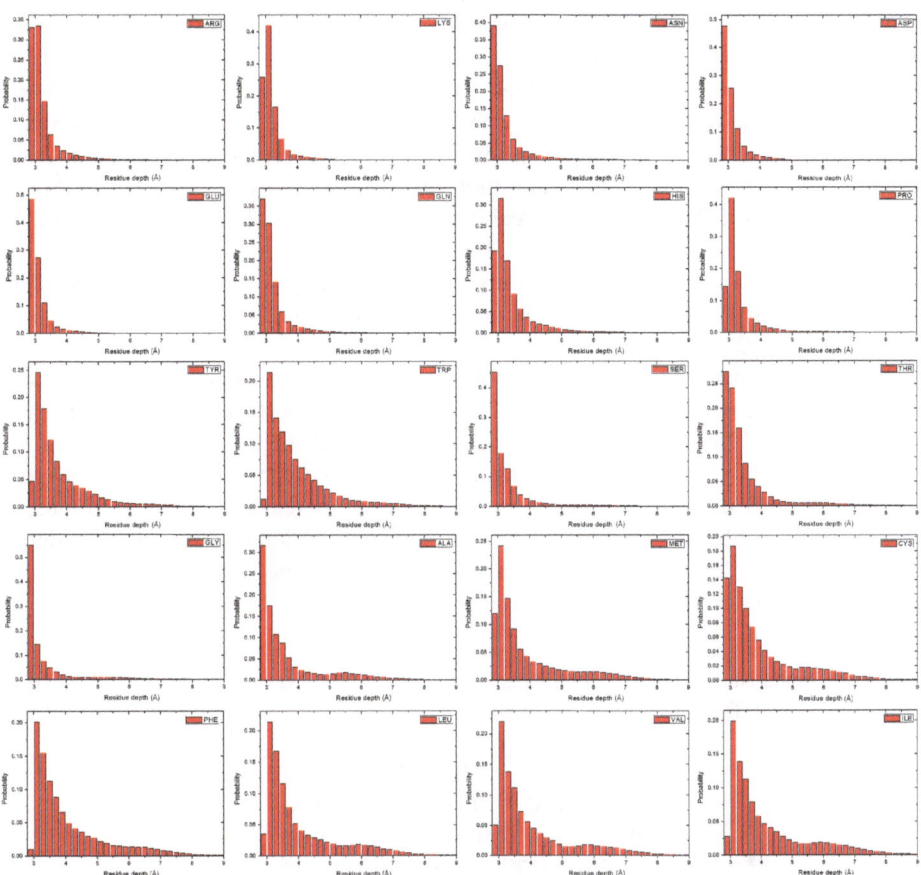

Fig. 4. distributions of residue depths for the 20 residue types

3.3   Comparison of Depth Generation Methods

We compare the depth results by the algorithm described by Chakravarty and Varadarajan (CV) and exhaustive search (ES) and EDT-based method (EDT) mentioned above. The test set here we choose contains 261 non-homologous protein chains randomly selected from the PISCES list [21]. We rotate each protein at 25 different orientations and find the shortest distance to the outer water molecule for each atom in the CV algorithm. For both ES and EDT, we first enlarge each protein 4 times and put it into a bounding box. Then we create the voxel shell which represents the solvent accessible surface. ES method directly searches the closest voxel for each atom without using EDT. EDT method only requires the EDT to the voxel shell to get the depth value for each atom.

We first compare the similarities of the depth values generated by the three methods. The Pearson's correlation coefficients (PCC) of the depth values by the three kinds of methods are shown in Table 1. Results by ES and EDT methods are highly close to each other. Although CV method is quite different to the other two, it still has a high correlation (>0.90) with them.

The difference of the depth values by CV and EDT mainly comes from two sources. First, the depth definitions by the two methods are slightly different, which have been described before. Second, since CV is an approximation method, depth value is highly dependent on the water molecules placed outside of the protein. Sometimes the depth value is close to the real depth if the water molecule happens to be the closest one while sometimes it doesn't. In contrast, depth values calculated by ES and EDT are close to the real one. The only error is caused by the discretization of the protein which makes the discrete shell not exactly the same as the actually continuous SAS.

We then compare the average computational time by the three methods, which is listed in the last row of the Table 1. The calculation is performed on a single node with a 2.27 GHZ Intel E5520 Xeon processor and 24 GB memory. EDT method is 2.6 times faster than CV and 1.9 times faster than ES. We can imagine the CPU time taken by the ES method will increase rapidly if we increase the scale factor to get more accurate SAS shell. We have also tried the new version of the DEPTH program using the CV method in [17], which takes even longer time (data not shown) due to the extensive search for the non-bulk water molecules.

Table 1. Comparison of the residue depths by methods of Chakravarty and Varadarajan (CV), exhaustive search (ES) and EDT-based method (EDT)

		CV	ES	EDT
	CV	1.00	0.91	0.90
PCC	ES	0.91	1.00	1.00
	EDT	0.90	1.00	1.00
Time(sec)		2.23	1.69	0.88

Compared with the accuracy, speed may be not an issue if we only calculate the depth once for a given protein structure. However, a lot of computational resources could be saved if depth information of thousands of structures has to be calculated. For example, in the application of protein fold recognition, the non-redundant template library often contains more than 30,000 protein chains/domains extracted from the Protein Data Bank (PDB) [22].

3.4   Depth vs. Radius of Gyration

The radius of gyration (RG) refers to the root mean square distance of the protein atoms from the center of gravity. Due to the simplicity of calculation, RG has been widely used to characterize the global shape and compactness of protein tertiary structures in protein structure prediction [23] and function annotation [24]. However, due to the high specificity of protein tertiary structure packing, the simple RG calculation cannot precisely reflect the shape and residue distribution related to the exposed surfaces on specific proteins. In this section, we examine the quantitative relation of RG and depth calculated from EDT technique which highlights the advantage of depth in characterizing the overall shape of protein tertiary structures.

We compare the radius of gyration with the maximum residue depth (MD) and the average residue depth (AD) in Fig. 5(A) and 5(B) separately. The data are acquired still based on the 36,556 domain structures in our threading template library. In the left figure, we can see that the two features have some correlation in most of the regions. Most times, when the radius of gyration is large, the maximum depth will also be high. Especially when the protein structure is compact and has a globular shape, its maximum depth will be highly correlated with its radius of gyration, such as the protein in Fig. 6(A). It is the chain A of the *Desulfovibrio vulgaris apoflavodoxin-riboflavin complex* (PDB ID: 1bu5), which has the radius of gyration around 14 Å. Since the five beta-strands and four helices are densely organized, the maximum depth is also very high and very close to the radius of gyration.

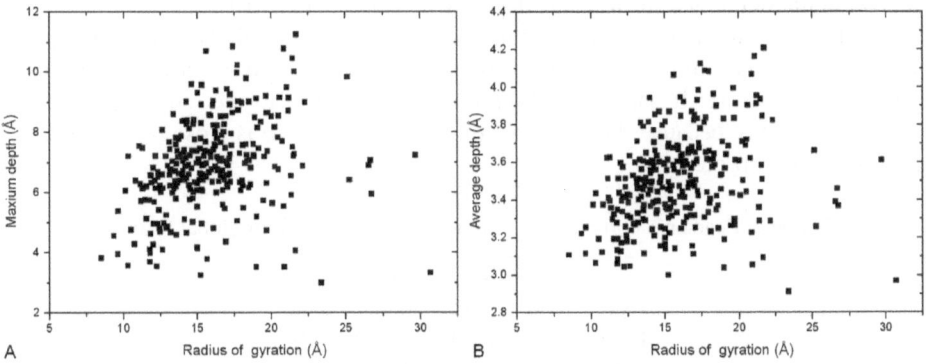

Fig. 5. Comparison of the radius of gyration with the maximum residue depth in (A) and the average residue depth in (B). Reduced number of points are shown in the figure by Origin

There are also exceptions where the radius of gyration is high but the maximum depth is extremely low. This is because some single-domain proteins (e.g. a super-long helix) have a loose shape which makes the depth values of most residues very low. Fig. 6(B) shows the chain L of the *Bacteriophage phi29 head-tail connector protein* (PDB ID: 1ijg). If we solely consider this chain, only one end is well-shaped. There are three other helices in the middle, which connect the other end with two short beta-strands and one short helix. This structure has an extremely large radius of gyration of 30 Å. However, since this chain is not compact and most of the residues are exposed, the maximum depth is only 7.603 Å.

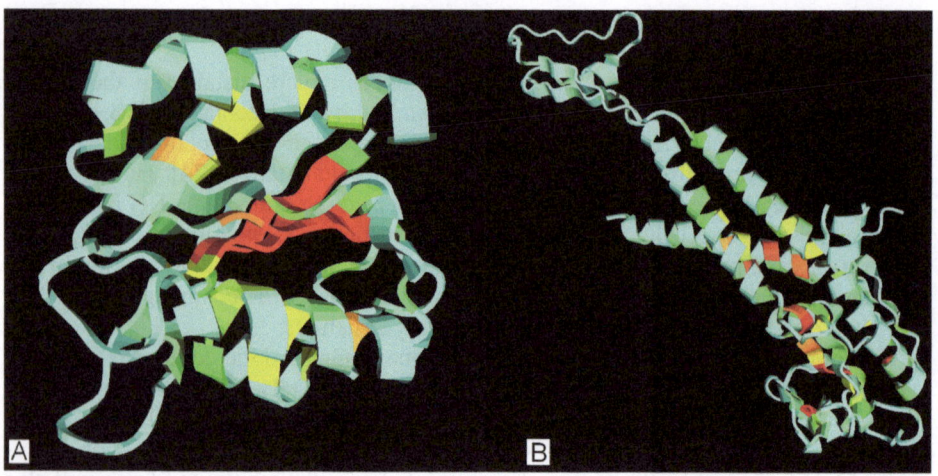

Fig. 6. Cartoon style of two protein chains with color representing the residue depth. (A) 1bu5 chain A, radius of gyration=13.840Å, maximum depth=10.459Å (B) 1ijg chain L, radius of gyration=30.428Å, maximum depth=7.603Å.

The scatter plot between the average depth and the radius of gyration in the Fig. 5(B) has the similar distribution to that between the maximum depth and the radius of gyration in Fig. 5(A). This is because the Pearson's correlation coefficient between the maximum depth and the average depth is very high (0.92 in Table 2).

Another measurement of the overall shape is the radius of the bounding sphere (RS), which is the minimum radius of the sphere which could cover all the atoms in the protein structure. It has a very high correlation (0.96) to the radius of gyration, which is probably because the center of the bounding sphere is close to the center of gravity for most proteins.

Table 2. Pearson's correlation coefficients between the four global features

	RG	RS	MD	AD
RG	1.00	0.96	0.07	0.02
RS	0.96	1.00	0.05	-0.02
MD	0.07	0.05	1.00	0.92
AD	0.02	-0.02	0.92	1.00

All the PCC values between the four global structural features are listed in Table 2. Due to the irregular shapes of some proteins, RG and RS have no obvious correlations with the maximum and average depths. For RG and RS, distance calculations are between the positions of all the residues and one fixed point (e.g. the center of gravity or the center of the bounding sphere). Those distances have no strong physical meaning when the protein has non-globular shape and residues are far away from this point. For depth, different atoms have different closest points on the SAS.

From the above analysis, we can draw the conclusion that RG and RS are very rough measurements of protein shapes. The maximum/average residue depth provides non-redundant information to RG/RS. They can help characterize the unique features of protein tertiary structures including the overall 3D shape and in particular the residue distribution relative to the surface exposition.

3.5   Depth vs. Solvent Accessibility

The range of solvent accessibility value is in [0, 1] after we normalize the SAS area by the maximum SAS area of each residue type. For the residues which are partially exposed to solvation, they may have the same solvent accessibility but different depth values due to the various sizes of the different residue types. By comparing the non-zero solvent accessibility and residue depth for each residue type based on the 36,556 protein domains, we find that SA and RD follow an exponential function:

$$RD = y + A \times e^{-SA/t} \tag{2}$$

In Table 3, we list values of the three parameters in equation (2) for all the 20 amino acids. As expected, those parameters are different for different residue types. The amplitude parameter A seems proportional to the size of each amino acid. For example, small amino acids GLY and AlA have small amplitudes while large amino acids ARG and TRP have big amplitudes. Hydrophillic residues tend to have a larger t parameter, such as ARG and LYS while more hydrophobic residues have a higher y parameter, e.g. PHE and ILE.

Table 3. Parameters of the exponential functions for the 20 amino acid types

	y	A	t		y	A	t
ARG	2.96	1.20	0.14	SER	2.89	0.84	0.11
LYS	2.96	1.18	0.16	THR	2.97	0.90	0.12
ASN	2.93	0.98	0.11	GLY	2.87	0.64	0.06
ASP	2.91	0.97	0.11	ALA	2.96	0.77	0.07
GLU	2.92	1.09	0.12	MET	3.08	1.17	0.08
GLN	2.95	1.09	0.12	CYS	2.93	0.98	0.11
HIS	3.00	1.20	0.11	PHE	3.17	1.29	0.07
PRO	3.04	0.83	0.11	LEU	3.16	1.05	0.06
TYR	3.08	1.35	0.10	VAL	3.13	0.95	0.07
TRP	3.12	1.44	0.10	ILE	3.18	1.05	0.06

In Fig. 7, we compare the solvent accessibility and residue depth for aspartic acid as an example. Generally, points in the scatter plot follow the exponential function, as illustrated by the black fitting curve. Depth difference is not significant when the solvent accessibility is high, which means that the majority of the residue is exposed. However, when SA is low (majority is buried), depth values can be quite different. Two reasons may cause the diversity of the depth values. The first one is the different relative positions and orientations of the residue to the solvent accessible surface while the other is the different side-chain conformations. The shape of each residue type is not unique due to the degrees of freedom of the side-chain torsion angles.

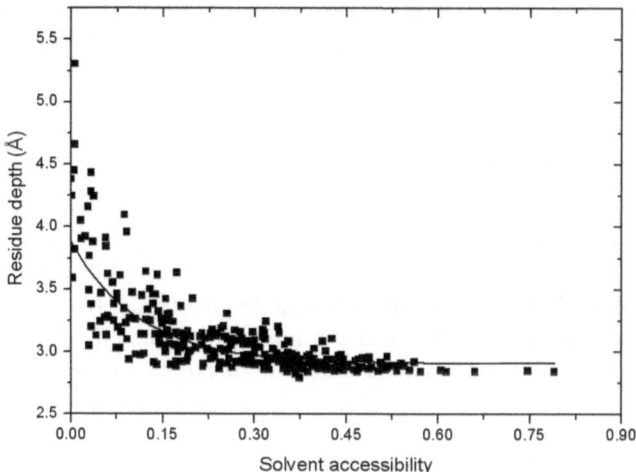

Fig. 7. Scatter plot of the solvent accessibility and residue depth for aspartic acid. Black curve is the fitting curve by an exponential function. Reduced number of points are shown in the figure by Origin.

4   Conclusions

We have developed a computational algorithm for the fast and accurate calculation of the atom/residue depth through Euclidean distance transform. The method was tested on a set of 261 non-redundant protein structures. It was shown that EDT-based method is 2.6 times faster than the widely-used method developed by Chakravarty and Varadarajan but the accuracy of the EDT-based method is higher than that of the latter compared to the actual depth from exhaustive search.

The depth data are systematically analyzed in the large-scale proteins that cover the entire PDB library at the sequence identity cutoff of 70%. It is found that the maximum/average residue depth has no obvious correlation with the commonly-used radius of gyration and radius of the bounding sphere. Hence, the maximum/average depth could be considered as a new geometric feature for describing the global shape of a protein tertiary structure. It is of potential use for protein fold classification and structure comparison.

When the residue is not completely buried inside of the protein molecule, solvent accessibility and residue depth follow an exponential relation. Different residue types have different parameters of the fitting functions and different distributions of residue depths even their hydrophobic scales are close to each other. The various sizes of the amino acids seem to be the major factor which causes the difference.

When the residue is completely buried inside of the protein, residue depth becomes a useful measurement as the solvent accessibility remains zero in this situation. It could be used as a complementary feature to the solvent accessibility for improving the fold recognition and the structure-based protein function annotation.

The source code and executable program for computing the atom depth and residue depth are freely available at http://zhanglab.ccmb.med.umich.edu/EDTSurf/. The associated software MVP (Macromolecular Visualization and Processing) for visualizing the depth information is at http://zhanglab.ccmb.med.umich.edu/MVP/.

Acknowledgements. The project is supported in part by the NSF Career Award (DBI 1027394), and the National Institute of General Medical Sciences (GM083107, GM084222).

References

1. Ramachandran, G.N., Sasisekharan, V.: Conformation of polypeptides and proteins. Adv. Protein Chem. 23, 283–438 (1968)
2. Xu, D., Zhang, Y.: Generating triangulated macromolecular surfaces by Euclidean distance transform. PLoS One 4(12), e8140 (2009)
3. Kabsch, W., Sander, C.: Dictionary of protein secondary structure: pattern recognition of hydrogen-bonded and geometrical features. Biopolymers 22(12), 2577–2637 (1983)
4. Chakravarty, S., Varadarajan, R.: Residue depth: a novel parameter for the analysis of protein structure and stability. Structure 7(7), 723–732 (1999)
5. Zhou, H., Zhou, Y.: Fold recognition by combining sequence profiles derived from evolution and from depth-dependent structural alignment of fragments. Proteins 58(2), 321–328 (2005)
6. Liu, S., Zhang, C., Liang, S., Zhou, Y.: Fold recognition by concurrent use of solvent accessibility and residue depth. Proteins 68(3), 636–645 (2007)
7. Wu, S., Zhang, Y.: MUSTER: Improving protein sequence profile-profile alignments by using multiple sources of structure information. Proteins 72(2), 547–556 (2008)
8. Roy, A., Yang, J., Zhang, Y.: COFACTOR: an accurate comparative algorithm for structure-based protein function annotation. Nucleic Acids Res. 40(Web Server issue), W471–W477 (2012)
9. Sanner, M.F., Olson, A.J., Spehner, J.C.: Reduced surface: an efficient way to compute molecular surfaces. Biopolymers 38(3), 305–320 (1996)
10. Zhang, H., Zhang, T., Chen, K., Shen, S., Ruan, J., Kurgan, L.: Sequence based residue depth prediction using evolutionary information and predicted secondary structure. BMC Bioinformatics 9, 388 (2008)
11. Yuan, Z., Wang, Z.X.: Quantifying the relationship of protein burying depth and sequence. Proteins 70(2), 509–516 (2008)
12. Lee, B., Richards, F.M.: The interpretation of protein structures: estimation of static accessibility. J. Mol. Biol. 55(3), 379–400 (1971)
13. Xu, D., Li, H.: Euclidean Distance Transform of Digital Images in Arbitrary Dimensions. In: Zhuang, Y.-T., Yang, S.-Q., Rui, Y., He, Q. (eds.) PCM 2006. LNCS, vol. 4261, pp. 72–79. Springer, Heidelberg (2006)
14. Choi, W.P., Lam, K.M., Siu, W.C.: Extraction of the Euclidean skeleton based on a connectivity criterion. Pattern Recognition 36(3), 721–729 (2003)
15. Shih, F.Y., Wu, Y.T.: Three-dimensional Euclidean distance transformation and its application to shortest path planning. Pattern Recognition 37(1), 79–92 (2004)

16. Xu, D., Li, H.: Shape analysis of volume models by Euclidean distance transform and moment invariants. In: 10th IEEE International Conference on Computer-Aided Design and Computer Graphics, pp. 437–440 (2007)
17. Tan, K.P., Varadarajan, R., Madhusudhan, M.S.: DEPTH: a web server to compute depth and predict small-molecule binding cavities in proteins. Nucleic Acids Res. 39(Web Server issue), W242–W248 (2011)
18. Lorensen, W.E., Cline, H.E.: Marching cubes: a high resolution 3d surface construction algorithm. Comput. Graph. 21(4), 163–169 (1987)
19. Wu, S., Zhang, Y.: LOMETS: a local meta-threading-server for protein structure prediction. Nucleic Acids Res. 35(10), 3375–3382 (2007)
20. Kyte, J., Doolittle, R.F.: A simple method for displaying the hydropathic character of a protein. J. Mol. Biol. 157(1), 105–132 (1982)
21. Wang, G., Dunbrack Jr., R.L.: PISCES: a protein sequence culling server. Bioinformatics 19(12), 1589–1591 (2003)
22. Berman, H.M., Westbrook, J., Feng, Z., Gilliland, G., Bhat, T.N., Weissig, H., Shindyalov, I.N., Bourne, P.E.: The Protein Data Bank. Nucleic Acids Res. 28(1), 235–242 (2000)
23. Zhang, Y., Kolinski, A., Skolnick, J.: TOUCHSTONE II: a new approach to ab initio protein structure prediction. Biophys. J. 85(2), 1145–1164 (2003)
24. Roy, A., Zhang, Y.: Recognizing protein-ligand binding sites by global structural alignment and local geometry refinement. Structure 20(6), 987–997 (2012)

Inference of Spatial Organizations of Chromosomes Using Semi-definite Embedding Approach and Hi-C Data

ZhiZhuo Zhang[1], Guoliang Li[2], Kim-Chuan Toh[3], and Wing-Kin Sung[1,2]

[1] School of Computing, National University of Singapore
[2] Genome Institute of Singapore
[3] Department of Mathematics, National University of Singapore
{zhizhuo,ksung}@comp.nus.edu.sg, ligl@gis.a-star.edu.sg,
mattohkc@nus.edu.sg

Abstract. For a long period of time, scientists studied genomes assuming they are linear. Recently, chromosome conformation capture (3C) based technologies, such as Hi-C, have been developed that provide the loci contact frequencies among loci pairs in a genome-wide scale. The technology unveiled that two far-apart loci can interact in the tested genome. It indicated that the tested genome forms a 3D chromsomal structure within the nucleus. With the available Hi-C data, our next challenge is to model the 3D chromosomal structure from the 3C-derived data computationally. This paper presents a deterministic method called ChromSDE, which applies semi-definite programming techniques to find the best structure fitting the observed data and uses golden section search to find the correct parameter for converting the contact frequency to spatial distance. To the best of our knowledge, ChromSDE is the only method which can guarantee recovering the correct structure in the noise-free case. In addition, we prove that the parameter of conversion from contact frequency to spatial distance will change under different resolutions theoretically and empirically. Using simulation data and real Hi-C data, we showed that ChromSDE is much more accurate and robust than existing methods. Finally, we demonstrated that interesting biological findings can be uncovered from our predicted 3D structure.

Keywords: Chromatin Interaction, 3D genome, Hi-C, Semi-definite Programming.

Program and Supplementary online
http://biogpu.ddns.comp.nus.edu.sg/~zzz/ChromSDE/

1 Introduction

Genome is usually assumed to be a set of linear chromosomes. This model, however, is over-simplified and it cannot explain the interactions among different genomic elements (e.g., enhancer, promoter, gene). Chromosome actually

M. Deng et al. (Eds.): RECOMB 2013, LNBI 7821, pp. 317–332, 2013.

forms a 3D structure within the nucleus and its spatial organization affects many chromosomal mechanisms such as gene regulation, DNA replication, epigenetic modification and maintenance of genome stability[3,8,15,18,19,20].

Generally, if two elements in the genome are close in the sequence level, they are also close in the structure level. But the converse statement is not necessarily true. For example, Li et al[15] showed that multiple related genes are located far away in linear model but are organized topologically close through long-range chromatin interactions and transcribed in a single "transcription factory". In the past, the 3D organization of chromosomes was usually studied by florescent in situ hybridization (FISH) which are low throughput and low resolution methods. Recently, several high throughput, high resolution methods [6,9,12,16,29] derived from the 3C method [4] have been proposed. These methods measure the contact frequencies for loci pairs. Two loci are expected to be spatially nearer if and only if the contact frequency of the loci pair is higher. 4C[29] and 5C[6] can measure the contact frequencies among a subset of loci while Hi-C[16] and its variant (TCC [12]) can capture the contact frequencies in a genome-wide manner.

Given the 3C-derived data, one interesting bioinformatics problem is to infer the 3D structure of the genome. A number of works have been proposed recently. All the current methods have two steps: (1) Converting the contact frequencies between loci to spatial distances and (2) Predicting the 3D chromosomal structure from the spatial distances. Duan et al. [7] converted the contact frequencies extracted from the 4C experiment on yeast to spatial distances and treated the 3D structure modeling problem as a constrained non-convex quadratic optimization problem using an optimization solver called IPOPT[26]. Bau et al. [1] translated the contact frequencies extracted from 5C experiments to spatial distances by inverting the Z-score of contact frequencies and treated the 3D structure modeling problem as finding an equilibrium state of a set of particles using Integrated Modeling Platform (IMP)[23] . With the same platform (IMP), Kalhor et al.[12] claimed that the Hi-C (or TCC) data can be better fitted by learning a set of 3D structures (since the sample has multiple cells where the chromatin structures in different cells are different) instead of one single structure. More recently, two Markov-chain Monte Carlo (MCMC)[21] sampling-based methods, MCMC5C[22] and BACH[10], were proposed to infer the 3D structures by maximizing the likelihood of the observed Hi-C data. Both methods assume that the expected contact frequencies and spatial distances among loci follow the power law distribution. MCMC5C[22] models the observed frequency with Gaussian distribution with respect to the expected frequency. BACH[10] models the observed frequency with Poisson distribution with respect to the expected frequency and takes the enzyme cutting site bias (e.g., CG content, mappability, fragment length) into account.

Although some works have been done, there are unsolved issues in both steps 1 and 2. For step 1, the conversion between the contact frequency and spatial distance has one parameter. Existing methods, except BACH, assume that the parameter is fixed or is known beforehand. We found that the parameter is actually different for different datasets. Thus it is important to have a method to estimate the parameter. For step 2, existing methods infer the 3D chromosomal structure by heuristics. They are not guaranteed to reconstruct the correct structure even in the noise-free case.

To fill in these gaps, we propose a novel chromosome structure modeling algorithm called ChromSDE (Chromosome Semi-Definite Embedding). ChromSDE models the problem as two parts:

1. Assuming that the parameter for the conversion from the contact frequency to the spatial distance is known, ChromSDE formulates the 3D structure modeling problem as a non-convex non-linear optimization problem similar to the previous works. Instead of directly solving the non-convex optimization which is NP-hard, ChromSDE relaxes it to a semi-definite programming(SDP) problem, whose global optimal solution can be computed in polynomial time. With this formulation, our approach is guaranteed to recover the correct 3D structure in the noise-free case when the structure is uniquely localizable[24].
2. For the parameter in our conversion function from the contact frequency to the spatial distance, ChromSDE formulates it as a univariate optimization problem and estimate the correct parameter by a modified version of the golden section search method.

This paper may have significant impact in three aspects. First, the SDP relaxation method in ChromSDE is a powerful relaxation technique, which is theoretically guaranteed to recover the correct structure in the uniquely localizable noise-free case[24]. The SDP approach has been successfully applied in other graph realization problems[2,14,27], but to our best knowledge, no one has introduced it in chromosome structure modeling. Second, we prove theoretically and empirically that the conversion parameter changes if we examine the data under different resolutions. Thus, it is inappropriate to assume that the conversion parameter is known. We developed an efficient algorithm to estimate the correct conversion parameter from the input data. Third, we proposed a measure called *Consensus Index* which can quantify if the input frequency data comes from a consensus structure or a mixture of different structures. It is arguable if Hi-C data is appropriate for modeling 3D structures, because the contact frequencies come from a population of cells instead of a single cell. Our simulation shows that if the data is from a consensus structure, the *Consensus Index* is high.

We evaluated our method with simulated data and real Hi-C data. Through simulation study, we showed that ChromSDE can perfectly recover different types of simulated structures in the noise-free setting while other tested programs fail in many cases. Even with noise, ChromSDE still significantly outperforms other tested programs. In addition, we also showed that ChromSDE can accurately estimate the conversion parameter and output the *Consensus Index* that can reflect the degree of mixture. Next, real Hi-C data replicates with different cutting enzymes are used to further validate the robustness and accuracy of ChromSDE comparing to other tested programs. The result indicates that ChromSDE can infer a more accurate and robust 3D model than existing methods. Finally, we show that ChromSDE can robustly handle different resolution data and the predicted high resolution 3D structure unveils interesting biological findings.

2   Method

The Hi-C and TCC technologies enable us to obtain paired-end reads from interacting loci in the genome. The interaction data can be summarized by a contact frequency matrix F, in which F_{ij} represents the number of contacts between loci i and j (loci i and j are genomic regions in a fixed bin size such as 1Mbp or 40kb). We expect two loci are spatially close if and only if the contact frequency between them is high. A further note is that the raw Hi-C or TCC interaction frequencies are affected by various biases (GC content, mappability and fragment length), and should be normalized [28].

The chromatin 3D modeling problem is defined as follows: Given a normalized interaction frequency matrix F, infer a 3D structure whose pairwise distances highly correlate with the interaction frequencies in F. This problem can be solved by 1) converting the frequency matrix F into a distance matrix D that describes the expected pairwise distance among the loci; 2) learning a 3D structure from the distance matrix D. Step 1 is based on the observation of Lieberman et al. [16] that the conversion between the frequency matrix F and the distance matrix D follows the power law distribution (Equation 1) where α is a parameter called the conversion factor and D_{ij} and F_{ij} are the distance and frequency between loci i and j.

$$D_{ij} = \begin{cases} (1/F_{ij})^{\alpha} & \text{if } F_{ij} > 0 \\ \infty & \text{otherwise} \end{cases} \tag{1}$$

There are two main challenges in this approach: 1) estimate α; and 2) convert the distance matrix D to the 3D model. In the following two sub-sections, we present ChromSDE that resolves these two challenges. First, assuming that the conversion factor α is known, we describe a method that estimates the 3D structure from the expected distance matrix D. Then, the next section explains how ChromSDE estimates the correct value of the conversion factor α. To note that, the scale between the converted distance and the real physical distance is not considered here, since the relative distance (without the scale) does not affect the predicted structure for visualization and further study.

2.1   From Distance Matrix to 3D Structure

Assuming the conversion factor $\alpha (> 0)$ is known, the interaction frequency matrix F can be converted to the expected distance matrix D by Equation 1.

The 3D chromatin structure modeling problem aims to compute a set of 3-dimensional coordinates $\{\overrightarrow{x_1}, ..., \overrightarrow{x_n}\}$ for the n loci, such that their distances can fit the distance matrix D well. In other words, we hope to ensure that $\|x_i - x_j\|$

(distance between loci i and j) is approximately the same as D_{ij} for all loci i and j.

Mathematically, this problem can be formulated as three alternative optimization models in Equations (2)-(4), where $\| \cdot \|$ denotes the Euclidean norm. Each equation has two terms. The first term aims to minimize the errors between the embedding distances and the expected distances. The three alternatives apply three different commonly used error functions in the literature: (a) sum of square errors of the distance differences [1,7], (b) sum of absolute errors of the distance square differences [2,14] and (c) sum of square errors of the distance square differences [2,17]. The second term is the same for the three alternatives. It is a regularization term that maximizes the pairwise distances for the loci without any interaction frequency data. It is based on the assumption that the spatial distances of loci pairs not captured by the experiment cannot be too short.

$$\min_{\vec{x}_1,\ldots,\vec{x}_n \in \mathbb{R}^3} \sum_{\{i,j|D_{ij}<\infty\}} \omega_{ij} \cdot \left(\|\vec{x}_i - \vec{x}_j)\| - D_{ij} \right)^2 - \lambda \sum_{\{i,j|D_{ij}=\infty\}} \|\vec{x}_i - \vec{x}_j\|^2 \tag{2}$$

$$\min_{\vec{x}_1,\ldots,\vec{x}_n \in \mathbb{R}^3} \sum_{\{i,j|D_{ij}<\infty\}} \omega_{ij} \cdot \left| \|\vec{x}_i - \vec{x}_j)\|^2 - D_{ij}^2 \right| - \lambda \sum_{\{i,j|D_{ij}=\infty\}} \|\vec{x}_i - \vec{x}_j\|^2 \tag{3}$$

$$\min_{\vec{x}_1,\ldots,\vec{x}_n \in \mathbb{R}^3} \sum_{\{i,j|D_{ij}<\infty\}} \omega_{ij} \cdot \left(\|\vec{x}_i - \vec{x}_j)\|^2 - D_{ij}^2 \right)^2 - \lambda \sum_{\{i,j|D_{ij}=\infty\}} \|\vec{x}_i - \vec{x}_j\|^2 \tag{4}$$

In the formulas, ω_{ij} represents the weight or confidence of the observed data D_{ij}. Since we expect the confidence of D_{ij} is higher when F_{ij} is large, this paper simply set $\omega_{ij} = 1/D_{ij}$. The parameter $\lambda > 0$ in the second term is the regularization coefficient to balance the error term and the regularization term. In practice, we found that the results are stable for $0.001 < \lambda < 0.1$ (Supp Figure 1) and we fix it to 0.01 in this paper.

All three formulations (2)–(4) are non-convex non-linear optimization problems, which are NP-hard to solve for their global minimizers. Existing methods solved them by heuristics like MCMC sampling [10,22] and local search [7,12,23]. Here, we show that, by relaxing the solution space of every $\vec{x}_i$ from R^3 to R^n (n is the number of loci), formulations (3) and (4) become convex semidefinite programming (SDP) problems for which we can compute their global minimizers to any given degree of accuracy in polynomial time. Furthermore, if the expected distance matrix is generated from a 3D object and is noise-free, the above relaxations can reconstruct the optimal R^3 solution by projecting the R^n points to certain R^3 subspace in theory [24]. In practice, even if the distance matrix is not noise-free, we still can find a good approximated solution in the R^3 subspace. The projecting technique to obtain a solution in R^3 will be introduced later.

Formulation of SDP Relaxation Problems. This section describes how to reformulate Equations (3) and (4) as linear and quadratic semidefinite programming (SDP) problems by relaxing the solution space of every $\vec{x}_i$ from R^3 to R^n. Let K be the kernel matrix for $X = [\vec{x}_1, \vec{x}_2, \ldots, \vec{x}_n]$ (i.e., $K_{ij} = \vec{x}_i \cdot \vec{x}_j = K_{ji}$), then every square distance can be expressed in term of K. Precisely, we have: $\|\vec{x}_i - \vec{x}_j\|^2 = K_{ii} + K_{jj} - 2K_{ij}$. In addition, we set the center of the points to be the origin, that is:

$$\sum_{i=1}^n \vec{x}_i = 0 \Rightarrow \|\sum_{i=1}^n \vec{x}_i\|^2 = 0 \Rightarrow \sum_{i,j} K_{ij} = 0. \tag{5}$$

By our definition of the kernel matrix, K must be symmetric positive semidefinite (i.e., $K \succeq 0$). We first describe the quadratic relaxation (Equation (4)), which is stated as below:

$$\min \sum_{\{i,j|D_{ij}<\infty\}} \omega_{ij}(K_{ii} + K_{jj} - 2K_{ij} - D_{ij}^2)^2 - \lambda \sum_{\{i,j|D_{ij}=\infty\}} (K_{ii} + K_{jj} - 2K_{ij})$$

$$\text{s.t. } \sum_{ij} K_{ij} = 0, \quad K \succeq 0.$$

$$\tag{6}$$

For Equation (3), the error term contains the absolute value operator $|\cdot|$, which cannot be handled directly by standard SDP solvers. Fortunately, without increasing the problem complexity, we can replace the absolute value operator $|\cdot|$ by adding two sets of slack variables. The linear SDP relaxation of Equation (3) is stated as below:

$$\min \sum_{\{i,j|D_{ij}<\infty\}} \omega_{ij}(\varepsilon_{ij}^+ + \varepsilon_{ij}^-) - \lambda \sum_{\{i,j|D_{ij}=\infty\}} (K_{ii} + K_{jj} - 2K_{ij})$$

$$\text{s.t. } K_{ii} + K_{jj} - 2K_{ij} + \varepsilon_{ij}^+ - \varepsilon_{ij}^- = D_{ij}^2 \tag{7}$$

$$\sum_{ij} K_{ij} = 0, \quad K \succeq 0, \quad \varepsilon_{ij}^+, \varepsilon_{ij}^- \geq 0.$$

Note that ε_{ij}^+ (and ε_{ij}^- respectively) represents the penalty when the embedding distance is shorter (and longer respectively) than the expected distance. Moreover, at least one of them must be zero in the final solution since they are non-negative and their summation is minimized.

A general purpose SDP solver, such as SDPT3 [25], can be used to solve the two SDP problems above. However, all the current general purpose SDP solvers (which are all based on interior-point methods) cannot handle large scale SDP problems. They can only comfortably handle distance matrix with around 40,000 expected distances (≈ 200 loci). Fortunately, for convex quadratic SDP such as the problem (6), recently developed advanced algorithm [11] based on partial proximal-point method (with semi-smooth Newton-CG method for solving the subproblems) can handle such a problem very efficiently even when the problem scale is large. In particular, it can handle 10,000,000 expected distances (≈ 3000 loci). In the result section, we present the results for both SDP relaxations in the small scale problems and the results for the quadratic SDP relaxation in the large scale problems (if not specially mentioned, the result is generated by quadratic SDP).

Obtaining 3D Coordinates from the Kernel Matrix. By solving the SDP problems (6) or (7), we obtain the solution as a positive semidefinite kernel matrix K. By computing the eigenvalue decomposition of K, the R^3 coordinates $X = [\vec{x}_1, \ldots, \vec{x}_n]$ can be recovered from K (i.e., $K \approx X^T X$). A 3-dimensional representation that approximately satisfies $K_{ij} \approx \vec{x}_i \cdot \vec{x}_j$ can be obtained from the top 3 eigenvalues $(\gamma_1, \gamma_2, \gamma_3)$ and eigenvectors $(\vec{\nu_1}, \vec{\nu_2}, \vec{\nu_3})$ of K. That is,

$$\vec{x}_i = [\sqrt{\gamma_1} \cdot \nu_{1,i} \quad \sqrt{\gamma_2} \cdot \nu_{2,i} \quad \sqrt{\gamma_3} \cdot \nu_{3,i}]^T. \tag{8}$$

In the ideal case where the input expected distance matrix is noise-free and dense enough (i.e., it has sufficient constraints to uniquely present a 3D structure), it can be shown that the approximation (8) is the exact solution and all other eigenvalues are equal to zero. This property is called unique localizability [24].

When the input expected distance matrix is noisy, ChromSDE performs further local refinement to the 3D coordinates obtained from the SDP relaxation problems [2]. Specifically, our ChromSDEx algorithm applies a local optimization method such as a quasi-Newton method or a gradient descent method to the original non-convex problem by using the 3D positions obtained from the SDP problems as the starting iteration point. Because the 3D positions produced by the SDP problems are generally close to a local minimizer, a local optimization method can generally converge to a good local minimizer for the original non-convex problems.

To measure if the input distance matrix can be represented as a single 3D structure, we propose a measure called *Consensus Index*, which includes two parts: the first part measures the degree of the input distance matrix D satisfying the triangle inequality, and is presented as the ratio between the embedded distance in R^n and the input distance; the second part measures how good the R^3 approximation is, and is presented as the ratio between the sum of top 3 eigenvalues and the sum of all eigenvalues of K. Precisely, Let $D'_{ij} = \sqrt{K_{ii} - 2K_{ij} + K_{jj}}$ be the embedded distance in R^n, then we have:

$$Consensus\ Index = \frac{\sum_{\{i,j | D_{ij} < \infty\}} min(D'_{ij}/D_{ij}, D_{ij}/D'_{ij})}{|\{i, j \mid D_{ij} < \infty\}|} \cdot \frac{\sum_{i=1}^{3} \gamma_i}{\sum_{i=1}^{n} \gamma_i} \tag{9}$$

Note that the *Consensus Index* is between 0 and 1. When the *Consensus Index* trends to 1, this means that the input distance matrix fits a single 3D structure well. The result section showed that the *Consensus Index* is a good indicator on whether the input data corresponds to a single 3D structure or a mixture of 3D structures.

2.2 Searching for the Correct Conversion Factor

In Section 2.1, the conversion factor $\alpha(> 0)$ is assumed to be known. However, the assumption is not valid . Even worse, Lemma 1 shows that the conversion factor changes with different resolutions.

Lemma 1. *Consider the frequency matrix F for loci $x_1, \ldots, x_{2n}$. Let the conversion factor of F be $\alpha > 0$, i.e., distance between loci x_i and x_j is $d_{ij} = (1/F_{ij})^\alpha$.*

Algorithm ChromSDE

Require: normalized frequency matrix F

Ensure: a set of 3D coordinates X, conversion factor α

1: $\alpha_{min} = 0.1$, $\alpha_{max} = 3$ # set search boundary for α

2: $\varphi = \frac{\sqrt{5}-1}{2}$ # golden section ratio

3: **repeat**

4:     $\eta = (\frac{\alpha_{max}}{\alpha_{min}})^{\varphi}$ # step size for updating α

5:     $x1 \leftarrow \alpha_{min} \cdot \eta$, $f1 \leftarrow goodness(x1, F)$

6:     $x2 \leftarrow \alpha_{max}/\eta$, $f2 \leftarrow goodness(x2, F)$

7:     **if** $f1 > f2$ **then**

8:        $\alpha_{min} \leftarrow x2$ # increase lower bound

9:     **else**

10:       $\alpha_{max} \leftarrow x1$ # decrease upper bound

11:     **end if**

12: **until** $(\alpha_{max} - \alpha_{min}) <$ tolerance

13: $\alpha \leftarrow \alpha_{min}$ # final value of α

14: $D \leftarrow (1/F)^{\alpha}$ # expected distance matrix

15: $X \leftarrow$ compute 3D structure using SDP method based on D

Function goodness(α, F)

1: $D \leftarrow (1/F)^{\alpha}$

2: $X \leftarrow$ compute 3D structure using SDP method based on D

3: $D' \leftarrow$ compute pair-wise distances from X

4: $F' \leftarrow (1/D')^{1/\alpha}$

5: Return $\sum_{\{(i,j)|F_{i,j}>0\}} -|F'_{i,j} - F_{i,j}|$

Fig. 1. Algorithm description for ChromSDE

Now, we reduce the resolution by merging adjacent loci, i.e., we generate the frequency matrix F' for the low resolution loci $y_1 \ldots, y_n$ where y_i is formed by merging adjacent loci x_{2i-1} and x_{2i}. Suppose $F'_{ij} = (F_{2i-1,2j-1} + F_{2i-1,2j} + F_{2i,2j-1} + F_{2i,2j})$ and d'_{ij} can be approximated as either arithmetic mean or geometry mean of $\{d_{2i-1,2j-1}, d_{2i-1,2j}, d_{2i,2j-1}, d_{2i,2j}\}$.

Then the conversion factor α' of F' is less than or equal to α .

Proof. Note that $\log F_{p,q} > 0$ and $\log d_{p,q} < 0$ since $F_{p,q} \geq 1$. Let $d_{\min} = \min_{p \in \{2i,2i-1\}, q \in \{2j,2j-1\}} d_{p,q}$. Since $d'_{ij} \geq d_{\min}$, we have $\log d'_{ij} \geq \log d_{\min}$. We also have

$$F'_{ij} = \sum_{p \in \{2i,2i-1\}, q \in \{2j,2j-1\}} \frac{1}{d_{p,q}^{1/\alpha}} \geq \frac{1}{d_{\min}^{1/\alpha}}.$$

Hence $\log F'_{ij} \geq -\frac{1}{\alpha} \log d_{\min}$. As $d'_{ij} = (1/F'_{ij})^{\alpha'}$, we have $\alpha' = \frac{-\log d'_{ij}}{\log F'_{ij}} \leq \frac{-\log d_{\min}}{-\frac{1}{\alpha} \log d_{\min}} = \alpha$. □

The Lemma 1 implies that the conversion factor of high-resolution Hi-C datasets is usually larger than that of low-resolution Hi-C datasets. Hence, we cannot assume that the conversion factor is a prior or is a fix value for different datasets. In fact, the predicted 3D structure is quite sensitive to the conversion factor. Given the same frequency matrix, different conversion factor leads to different expected distances and finally implies very different 3D structures (Supp Figure 2). Therefore, estimating the correct conversion factor for a frequency matrix F is important.

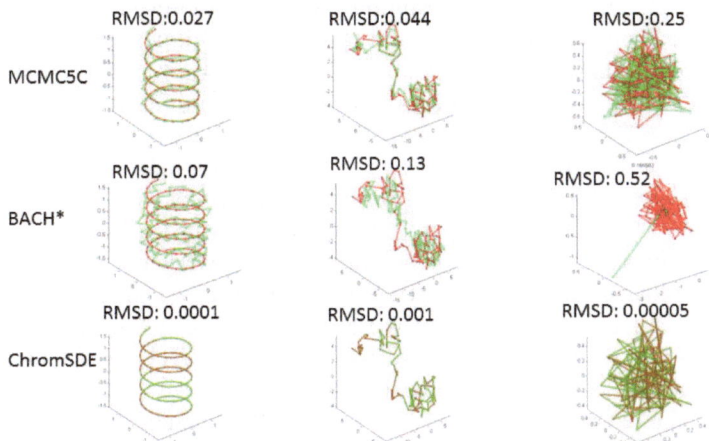

Fig. 2. Predicted 3D structures by different programs using simulated data.
Red curve is the true structure and green curve is the predicted structure. ChromSDE uses quadratic SDP here and the linear SDP has the same performance.

A correct conversion factor enables us to convert a frequency matrix to a correct 3D model, and vice versa. Based on this principle, for a frequency matrix F, the goodness of a conversion factor α ($goodness(\alpha, F)$) can be determined by comparing the predicted frequency matrix $\widehat{F}$ and the input frequency matrix F. Figure 1 details the function to compute $goodness(\alpha, F)$.

Our aim is to compute α that maximizes the goodness function. As there is no obvious well defined gradient for the goodness function, we cannot use methods such as gradient descent or Newton's method to optimize α. Instead, we perform the golden section search method to optimal α, assuming that the goodness function is unimodal in the search interval. Since $d_{ij} = (1/F_{ij})^\alpha$, we deduce that α cannot be too small; otherwise the spatial distance will be independent of the frequency (when $\alpha \to 0$). Also, α cannot be too large; otherwise, a small difference in frequencies will lead a very big difference in spatial distances, and small noise will seriously violation of the triangle inequality. In this paper, we assume that $0.1 \le \alpha \le 3$. Moreover, we observed that applying the standard golden section search on the logarithm domain of the interval is more efficient(see Supp Figure 3). The algorithm detail is in Figure 1.

3   Result

3.1   Simulation Study

To analyze the performance of ChromSDE, we generated three different types of 3D structures(Supp Figure 4): (1)Helix curve, (2)Brownian motion simulation of a single particle and (3)Uniform random points in a cube. Each structure is represented by 100 points. We assume that the Hi-C technique is sensitive enough to capture interactions with at most 50 nearest neighbours and the conversion factor α is 1, i.e., the contact frequency f of two given points can be computed as $f = (1/d)^{1/\alpha} = 1/d$, where d is the spatial distance between given points. We compared our algorithm with the existing methods MCMC5C[22] and BACH[10], which are the only publicly available standalone programs that are suitable for general Hi-C data. For MCMC5C, it cannot estimate the conversion factor by itself, so we supplied it with the correct value. For BACH, it can estimate the conversion factor with the default starting point equal to 1 (i.e., the correct value in our simulation study). Since there is no enzyme bias in our simulation, we also modify BACH to suppress this feature (called BACH*). For ChromSDE, we assume that the conversion factor is within the range (0.1,3), so we give advantages to the existing programs, but not our ChromSDE.

ChromSDE Guarantees Optimality in Noise-Free Case. Figure 2 shows the true simulated structures and the predicted structures by different programs. For the helix curve, all three programs can recover the structure correctly. For the Brownian motion curve, both ChromSDE and MCMC5C can almost perfectly recover the true structure and BACH* can only reproduce a not-so-accurate but similar structure. For the third case, MCMC5C produced a not-so-accurate structure and BACH* completely failed in this case, while our ChromSDE still can perfectly recover the true structure. The result is not surprising since SDP method is the only one that can guarantee perfect recovery of the true structure when the input data is noise-free and the structure is uniquely localizable. Based on the RMSD(root mean square deviation), ChromSDE also outperforms the other two methods in all the three simulated cases.

ChromSDE Outperforms the Existing Methods in Noisy-Data. The previous section showed that ChromSDE can recover the optimal chromatin structure in the noise-free case. Now, we test whether ChromSDE is robust in a noisy data setting. To study this, we simulated noisy contact frequency data in different noise level based on the Brownian curve structure. For any two loci i and j, the noisy frequency $\tilde{F}_{ij}$ is deviated from the true frequency $F_{ij} = 1/D_{ij}$ (D_{ij} is the spatial distance between loci i and j) by adding a uniform random noise δ within a given noise level. Precisely, $\tilde{F}_{ij} = F_{ij}(1 + \delta)$ where $|\delta|$ is smaller than the noise level.

Figure 3 shows the performance of the programs with different noise levels under different measurements. Figure 3(a) shows that, when the noise level increases, the Spearman correlation between the pairwise distances from the predicted structure and those from the true structure generally decreases. ChromSDE and MCMC5C perform similarly when the noise level < 0.6 and ChromSDE (both linear SDP and quadratic SDP) outperforms others when the noise level is higher than 0.6. Similar result is observed when we measure the RMSD between the predicted structure and the true structure(Supp Figure 5(a)). In Figure 3(c), we observed that ChromSDE can estimate the conversion factor quite accurately

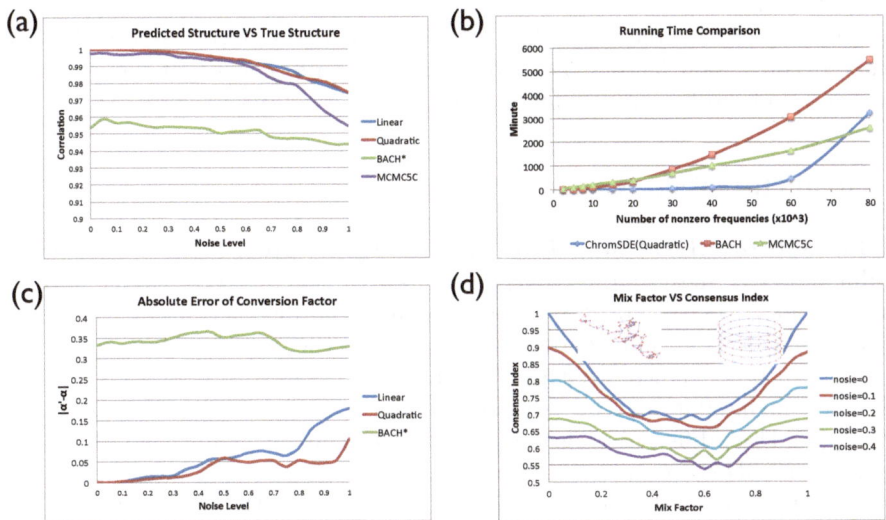

Fig. 3. Performance of different methods on simulated data. (a) Spearman correlation between the pair-wise distance matrices of the predicted structure and the true structure under different noise level. (b) Running times of tested programs given different number of pairs of observed frequency (test stop at 80000 pair-wise frequencies, 1600 points). (c) The absolute error of the estimated value of conversion factor under different noise levels. (d) The *Consensus Index* predicted by ChromSDE(quadratic model) under different degree of mixture of helix curve(right) and Brownian motion curve(left).

(deviation <0.1) when noise level <0.7. In contrast, the estimated conversion factor from BACH* tends to be incorrect (deviation around 0.35). This may be the reason why BACH* has worse performance comparing to others across different noise levels. Moreover, ChromSDE is faster than BACH and comparable to MCMC5C even though ChromSDE needs to search for the correct conversion factor but MCMC5C does not (Figure 3(b)). In summary, the result shows that the linear SDP and quadratic SDP models perform quite consistently and ChromSDE is more robust and accurate than existing methods.

Consensus Index **Indicates the Degree of Mixture of 3D Structures.** In Hi-C and TCC experiments, the data is from a population of cells, and each potentially has different 3D chromosomal structure. The method section proposed to use the *Consensus Index* to determine if the data is from a consensus 3D structure. To show that the *Consensus Index* is a good indicator of the degree of mixture, we generated a frequency matrix F_{merge} by merging the frequency matrix from the helix curve F_1 and the Brownian motion curve F_2 under different mix factor γ (i.e., $F_{merge} = \gamma \cdot F_1 + (1 - \gamma) \cdot F_2$). Figure 3(d) shows that the *Consensus Index* is affected by both the noise level and mix factor. For the same noise level, the *Consensus Index* approaches the minimum when the mix factor is close to 0.5. This indicates that the *Consensus Index* is the lowest when the two structures are highly mixed. For different noise levels, the *Consensus Index* decreases as the noise level increases. Also we note that the estimated conversion factors by ChromSDE are quite consistent with its true value even under different mix factors and noise levels(Supp Figure 5(c)).

3.2   Real Hi-C Data Study

Validate ChromSDE Using Two Enzyme Replicates. From the literature, two different enzymes(Hind3, NcoI) were used to generate Hi-C replicate data from the mouse ES cell(mESC)[5] and the human GM06990 cell(GM)[16]. Each enzyme replicate is an independent observation of the chromosome structure in the same cell type. Hence, we expect the result produced by a robust algorithm using one enzyme data can be validated using the other enzyme data.

We applied four different programs ChromSDE, BACH*, BACH and MCMC5C to predict the 3D structures of different chromosomes in the two cell lines using the Hi-C data from two replicates. For ChromSDE, BACH* and MCMC5C, the input is a normalized frequency matrix using the normalization pipeline by Yaffe and Tanay[28]. For BACH, we provide the raw Hi-C frequency and enzyme cutting point feature data. More detail can be found in the supplementary material.

We compute Spearman correlation between the normalized frequency of one enzyme data and the estimated frequency ($frequency \sim 1/distance$) of the predicted structure from the other enzyme data. (We use Spearman correlation instead of Pearson correlation since the Spearman correlation is independent to the conversion between frequency and distance, hence it is fair to every tested program.) Figure 4(a) shows that ChromSDE (both Linear SDP and Quadratic SDP) outperforms the other programs by at least 5% across all four tested Hi-C datasets. Especially, in the mESC dataset, ChromSDE obtains the average correlation of 0.9 across all chromosomes but other tested programs only obtain correlation at most 0.82. What's more, Figure 4(b),(c) and Supp Figure 6 showed the 3D structures of different chromosomes predicted by ChromSDE are highly reproducible and the conversion factors estimated by ChromSDE are more consistent than the ones estimated by BACH and BACH* across different chromosomes and different enzymes (Supp Table 1).

Besides, we observed that all the tested programs perform worse in GM than in mESC and the *Consensus Index* is around 0.9 in mESC and is only 0.7 in GM(Supp Figure 7). It indicates that mESC has a consensus 3D structure for its genome and GM is relatively diverse or has higher noise level due to the low sequencing depth.

ChromSDE Can Generate Consistent 3D Structures from Different Genomic Resolutions. We further tested ChromSDE on different genomic resolution data. Figure 5(a) showed that ChromSDE can predict similar structures of chromosome 13 under different resolutions using mESC Hind3 data (average Spearman correlation is 0.97, average RMSD is 0.08). In contrast, other existing programs cannot reproduce similar structures with different resolution

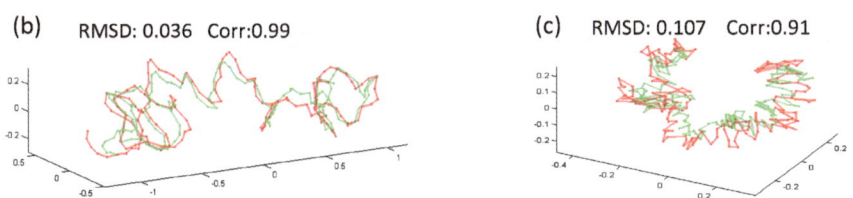

(a)

Validate Correlation	Quadratic SDP	Linear SDP	BACH	BACH*	MCMC5C
mESC_Ncol	**0.9096**	0.9095	0.7918	0.7871	0.8187
mESC_Hind3	**0.9080**	0.9070	0.8061	0.7999	0.7676
GM_Ncol	0.7674	**0.7899**	0.6740	0.7244	0.6942
GM_Hind3	**0.7832**	0.7763	0.7066	0.7358	0.7293

(b) RMSD: 0.036 Corr:0.99

(c) RMSD: 0.107 Corr:0.91

Fig. 4. Validate ChromSDE using mESC, GM Hi-C data with two different enzymes (Hind3, NcoI). (a) Average Spearman correlation across all chromosomes between inverse 3D distance and contact frequency from testing dataset. For each dataset, the best performer is highlighted. (b) Alignment between predicted structures of chromosome 1 of mESC Hind3(red) and mESC NcoI(green) by ChromSDE. (c) Alignment between predicted structures of chromosome 1 of GM Hind3(red) and GM NcoI(green) by ChromSDE.

data, especially for MCMC5C which cannot estimate the correct conversion factor(Supp Figure 8). We also showed that the conversion factor for each predicted structure in Figure 5(a). It demonstrated that the conversion factor increases as the resolution increases (also supported by BACH in Supp Figure 8). This further confirms the correctness of Lemma 1 even though the frequency has been normalized under different genomic resolutions.

To demonstrate the application of our predicted 3D structure, we generated a high resolution chromosome 3D structure for the region chr13:21Mb-25Mb (Figure 5(b)) using ChromSDE and mouse ES cell Hind3 data (40kbp resolution, estimated α is 0.83). Hist1h genes are highlighted with yellow color in the 3D structure, and we find that two groups of Hist1h genes are separated quite far away($\sim$1.5Mbp) in the linear genomic locations. In contrast, the promoters of two groups of Hist1h genes are spatially close to each other. To test if these two groups of genes interact each other for transcription, we checked the Pol2 ChIA-PET data available in our lab. We found that there are strong interactions(red dash line) between these two promoter regions mediated by Pol2, which indicates that the histone genes are co-regulated in the mouse ES cell.

Moreover, we found that the dense region and the loose region in the predicted 3D structure can be used to indicate the level of activity of those regions (from the snapshot of UCSC genome browser[13]). Dense regions (purple and blue color) correspond to repressive chromatin state in the cell, and there are few active histone modification and transcriptionx factor-binding events occurring in those regions. In contrast, loose regions(green and yellow color) correspond to active chromatin state in the cell, and there are a lot of histone modification and transcription factor-binding events occurring in those regions. Also, we found that loose regions usually containing more genes and are associated with early replication timing than the dense regions. It is also noted that the purple region is associated with LaminB1 binding and late replication timing, which suggests that Lamin may plays a part in the histone genes regulation and DNA replication.

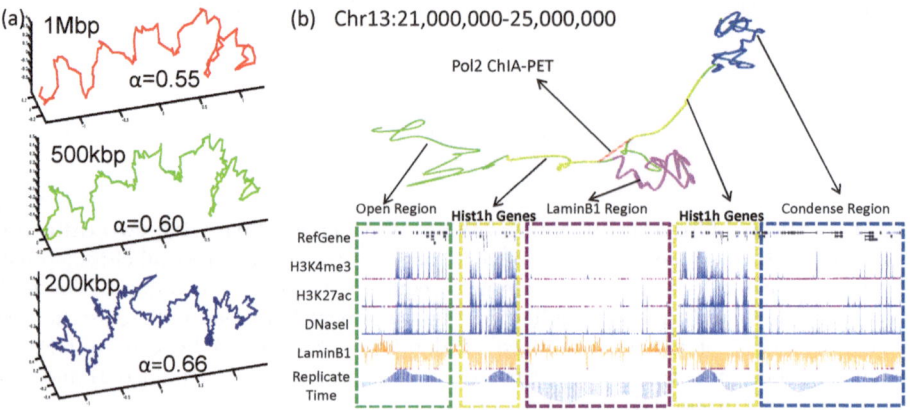

Fig. 5. Predicted structure of chromosome 13 from mESC Hind3 data. (a) The predicted structure of chromosome 13 under 1Mbp,500kbp,200kbp resolutions. (b) The predicted structure of the region chr13:21Mb-25Mb under 40kbp resolution and the different signal tracks of mESC from UCSC genome browser[13].

4   Discussion

In this study, we presented a method ChromSDE to reconstruct the consensus/dominate chromatin 3D structure of the given HiC data. To our best knowledge, ChromSDE is the only method which can guarantee recovering the correct structure in the noise-free case. In the noisy case, ChromSDE is much more accurate and robust than existing methods in both simulation and real data study. In addition, ChromSDE can automatically estimate the conversion factor, which is proved to change under different resolutions theoretically and empirically. Furthermore, we demonstrate that interesting biological findings can be uncovered from our predicted 3D structure.

We also developed the *Consensus Index* to determine how good the data can be explained by a single 3D structure. However, *Consensus Index* may not be informative when the noise level of the data is high or the mixing structures are similar. When the mixing structures are similar to each other then ChromSDE will learn the average structure. One future research is to recover all the mixing structures using Hi-C data.

Acknowledgements. The authors would like to thank Dr.Hugo Willy for the insightful discussions. This work was supported in part by the MOEs AcRF Tier 2 funding R-252-000-444-112.

References

1. Bau, D., Marti-Renom, M.A.: Genome structure determination via 3c-based data integration by the integrative modeling platform (2012)
2. Biswas, P., Liang, T.C., Toh, K.C., Ye, Y., Wang, T.C.: Semidefinite programming approaches for sensor network localization with noisy distance measurements. IEEE Transactions on Automation Science and Engineering 3(4), 360–371 (2006)
3. Dekker, J.: Gene regulation in the third dimension. Science Signalling 319(5871), 1793 (2008)
4. Dekker, J., Rippe, K., Dekker, M., Kleckner, N.: Capturing chromosome conformation. Science 295(5558), 1306–1311 (2002)
5. Dixon, J.R., Selvaraj, S., Yue, F., Kim, A., Li, Y., Shen, Y., Hu, M., Liu, J.S., Ren, B.: Topological domains in mammalian genomes identified by analysis of chromatin interactions. Nature 485(7398), 376–380 (2012)
6. Dostie, J., Dekker, J.: Mapping networks of physical interactions between genomic elements using 5c technology. Nat. Protoc. 2(4), 988–1002 (2007)
7. Duan, Z., Andronescu, M., Schutz, K., McIlwain, S., Kim, Y.J., Lee, C., Shendure, J., Fields, S., Blau, C.A., Noble, W.S.: A three-dimensional model of the yeast genome. Nature 465(7296), 363–367 (2010)
8. Fraser, P., Bickmore, W.: Nuclear organization of the genome and the potential for gene regulation. Nature 447(7143), 413–417 (2007)
9. Fullwood, M.J., Ruan, Y.: Chip-based methods for the identification of long-range chromatin interactions. Journal of Cellular Biochemistry 107(1), 30–39 (2009)
10. Hu, M., Deng, K., Qin, Z., Dixon, J., Selvaraj, S., et al.: Bayesian Inference of Spatial Organizations of Chromosomes. PLoS Comput. Biol. 9(1), e1002893 (2013), doi:10.1371/journal.pcbi.1002893
11. Jiang, K.F., Sun, D.F., Toh, K.C.: A partial proximal point algorithm for nuclear norm regularized matrix least squares problems. National University of Singapore (2012) (preprint)
12. Kalhor, R., Tjong, H., Jayathilaka, N., Alber, F., Chen, L.: Genome architectures revealed by tethered chromosome conformation capture and population-based modeling. Nat. Biotechnol. 30(1), 90–98 (2012)
13. Karolchik, D., Hinrichs, A.S., Kent, W.J.: The ucsc genome browser. Current Protocols in Bioinformatics, 1–4 (2009)
14. Leung, N.H.Z., Toh, K.C.: An sdp-based divide-and-conquer algorithm for large-scale noisy anchor-free graph realization. SIAM Journal on Scientific Computing 31(6), 4351–4372 (2009)
15. Li, G., Ruan, X., Auerbach, R.K., Sandhu, K.S., Zheng, M., Wang, P., Poh, H.M., Goh, Y., Lim, J., Zhang, J., Sim, H.S., Peh, S.Q., Mulawadi, F.H., Ong, C.T., Orlov, Y.L., Hong, S., Zhang, Z., Landt, S., Raha, D., Euskirchen, G., Wei, C.L., Ge, W., Wang, H., Davis, C., Fisher-Aylor, K.I., Mortazavi, A., Gerstein, M., Gingeras, T., Wold, B., Sun, Y., Fullwood, M.J., Cheung, E., Liu, E., Sung, W.K., Snyder, M., Ruan, Y.: Extensive promoter-centered chromatin interactions provide a topological basis for transcription regulation. Cell 148(1-2), 84–98 (2012)
16. Lieberman-Aiden, E., van Berkum, N.L., Williams, L., Imakaev, M., Ragoczy, T., Telling, A., Amit, I., Lajoie, B.R., Sabo, P.J., Dorschner, M.O., Sandstrom, R., Bernstein, B., Bender, M.A., Groudine, M., Gnirke, A., Stamatoyannopoulos, J., Mirny, L.A., Lander, E.S., Dekker, J.: Comprehensive mapping of long-range interactions reveals folding principles of the human genome. Science 326(5950), 289–293 (2009)

17. Liu, Y.J., Sun, D., Toh, K.C.: An implementable proximal point algorithmic framework for nuclear norm minimization. Mathematical Programming, 1–38 (2009)
18. Miele, A., Dekker, J.: Long-range chromosomal interactions and gene regulation. Mol. BioSyst. 4(11), 1046–1057 (2008)
19. Misteli, T.: Spatial positioning: A new dimension in genome function. Cell 119(2), 153–156 (2004)
20. Misteli, T., et al.: Beyond the sequence: cellular organization of genome function. Cell 128(4), 787 (2007)
21. Neal, R.M.: Probabilistic inference using markov chain monte carlo methods (1993)
22. Rousseau, M., Fraser, J., Ferraiuolo, M.A., Dostie, J., Blanchette, M.: Three-dimensional modeling of chromatin structure from interaction frequency data using markov chain monte carlo sampling. BMC Bioinformatics 12(1) (2011)
23. Russel, D., Lasker, K., Webb, B., Velazquez-Muriel, J., Tjioe, E., Schneidman-Duhovny, D., Peterson, B., Sali, A.: Putting the pieces together: integrative modeling platform software for structure determination of macromolecular assemblies. PLoS Biol. 10(1) (2012)
24. So, A.M.C., Ye, Y.: Theory of semidefinite programming for sensor network localization. Mathematical Programming 109(2), 367–384 (2007)
25. Toh, K.C., Todd, M.J., Tütüncü, R.H.: Sdpt3–matlab software package for semidefinite programming, version 1.3. Optimization Methods and Software 11(1-4), 545–581 (1999)
26. Wächter, A., Biegler, L.T.: On the implementation of an interior-point filter linesearch algorithm for large-scale nonlinear programming. Mathematical Programming 106(1), 25–57 (2006)
27. Weinberger, K.Q., Saul, L.K.: Unsupervised learning of image manifolds by semidefinite programming, vol. 2, pp. II-988–II-995. IEEE (2004)
28. Yaffe, E., Tanay, A.: Probabilistic modeling of hi-c contact maps eliminates systematic biases to characterize global chromosomal architecture. Nature Genetics (2011)
29. Zhao, Z., Tavoosidana, G., Sjolinder, M., Gondor, A., Mariano, P., Wang, S., Kanduri, C., Lezcano, M., Sandhu, K.S., Singh, U., Pant, V., Tiwari, V., Kurukuti, S., Ohlsson, R.: Circular chromosome conformation capture (4c) uncovers extensive networks of epigenetically regulated intra- and interchromosomal interactions. Nat. Genet. 38(11), 1341–1347 (2006)

Boosting Prediction Performance
of Protein-Protein Interaction Hot Spots
by Using Structural Neighborhood Properties
(Extended Abstract)

Lei Deng[1,3], Jihong Guan[1,*], Xiaoming Wei[1],
Yuan Yi[1], and Shuigeng Zhou[2,*]

[1] Department of Computer science and technology,
Tongji University, Shanghai, China
[2] Shanghai Key Lab of Intelligent Information Processing,
and School of Computer Science, Fudan University, Shanghai, China
[3] School of Software, Central South University, Changsha, China
jhguan@tongji.edu.cn, sgzhou@fudan.edu.cn

Abstract. Binding of one protein to another in a highly specific manner to form stable complexes is critical in most biological processes, yet the mechanisms involved in the interaction of proteins are not fully clear. The identification of hot spots, a small subset of binding interfaces that account for the majority of binding free energy, is becoming increasingly important in understanding the principles of protein interactions. Despite experiments like alanine scanning mutagenesis and a variety of computational methods have been applied to this problem, comparative studies suggest that the development of accurate and reliable solutions is still in its infant stage.

We developed *PredHS* (Prediction of Hot Spots), a computational method that can effectively identify hot spots on protein binding interfaces by using 38 optimally chosen properties. The optimal combination of features was selected from a set of 324 novel structural neighborhood properties by a two-step feature selection method consisting of a random forest algorithm and a sequential backward elimination method. We evaluated the performance of PredHS using a benchmark of 265 alanine-mutated interface residues (Dataset I) and a trimmed subset (Dataset II) with 10-fold cross validation. Compared with the state of the art approaches, PredHS achieves a significant improvement on the prediction quality, which stems from the new structural neighborhood properties, the novel way of feature generation as well as the selection power of the proposed two-step method. We further validated the capability of our method by an independent test and obtained promising results.

The PredHS web server and supplementary data are available at http://admis.tongji.edu.cn/predhs.

---

* Corresponding authors.

M. Deng et al. (Eds.): RECOMB 2013, LNBI 7821, pp. 333–344, 2013.
© Springer-Verlag Berlin Heidelberg 2013

1 Introduction

Protein-protein interactions play an important role in nearly all aspects of cellular function ranging from cell differentiation to apoptosis [1]. Studies of protein binding interfaces have revealed that only a small subset of critical residues called *hot spots* makes dominant contributions to the binding free energy [2], and provides useful targets within these interfaces. Identifying and understanding hot spots and their mechanisms on a large scale would have significant implications for practical applications, such as protein engineering and drug design. Since alanine-scanning mutagenesis to identify binding hot spots is currently expensive and time-consuming, the number of hot spots recognized by wet-experiments is quite limited. Therefore, there is a need for developing computational prediction methods to complement the mutagenesis experiments.

Efforts have been made to explain the rules between binding hot spots and protein structure and sequence information. Analysis of hot spots has shown that some residues are more favorable rather than a random composition. The fundamental ones, Tyr (21%), Arg (13.3%) and Trp (12.3%), are critical due to their sizes and conformations in hot spots [3, 4]. Also, it reveals that hot spots are usually located at the center of the interface and surrounded by energetically less important residues that are shaped like an O-ring to occlude bulk water molecules from the hot spots [2]. To refine the influential O-ring theory, a "double water exclusion" hypothesis [5] was proposed to characterize the topological organization of residues in a hot spot and their neighboring residues. Although these rules make sense to analyze specific interfaces, there are no simple patterns of features, such as hydrophobicity, shape or charge, can be used for predicting hot spots from a larger set of protein-protein complexes [6].

Current methods of hot spots prediction can be divided essentially into three main types: molecular simulation techniques, knowledge-based methods and machine learning methods. Molecular dynamics (MD) simulations were first introduced to simulate alanine substitutions and estimate the induced changes in binding free energy ($\Delta\Delta G$). Although some molecular simulation methods [7–10] are successful to identify hot spots from protein complexes, they are not applicable for large-scale hot spot predictions due to their enormous computational cost. On the other hand, empirical functions or simple physical methods, such as FOLDEF [11] and Robetta [12], which use experimentally calibrated knowledge-based simplified models to evaluate the binding free energy, provide an alternative way to probe hot spots with much less computation. Recently, considerable interest has focused on applying machine learning methods to predict hot spots, such as neural networks [13], decision tree [14], support vector machine [15, 17, 18], Bayesian networks [19], minimum cut trees [20] and random forest [30].

In this paper, we report a novel structure-based computational method, *PredHS* (Prediction of Hot Spots), that combines three main sources of information, namely site, Euclidean and Voronoi features describing the properties of either the target residue or the target residue's structural neighborhood. PredHS integrates a set of 38 optimal features selected from 324 site, Euclidean and Voronoi properties by a two-step feature selection method. We have benchmarked PredHS using a set of experimentally verified hot spot residues and an independent dataset. Results show that PredHS significantly outperforms the state of the art methods, and indicate that structural neighborhood properties are important determinants of hot spots. The framework of PredHS is shown in Figure 1.

2 Methods

2.1 Datasets

The complete benchmark dataset (called Dataset I), the same as that in the work of Cho et al [15], was obtained from ASEdb [22] and the published data of [12]. The interface residues in Dataset I are divided into 65 hot spots and 200 energetically unimportant residues. To evaluate the proposed method and compare it with the existing methods more comprehensively and fairly, a trimmed dataset (called Dataset II) was generated. Positive samples (hot spots) in Dataset II are the same as that in Dataset I, the only difference is the way to select negative samples (non-hot spots). In Dataset II, the interface residues with $\Delta\Delta G<0.4$ kcal/mol are labeled as non-hot spots and the other residues with $\Delta\Delta G$ between 0.4 and 2.0 are eliminated for the purpose of increasing discrimination as described in [21] and [17]. Details of the two datasets are presented in the Supplementary Material[1].

An independent test dataset was extracted from the BID database [23] to further assess the performance of our proposed method. This test dataset consists of 18 complexes containing 127 alanine-mutated data, of which 39 interface residues are hot spots.

2.2 Site Features

A wide variety of 108 sequence, structural and energy attributes are used to characterize potential hot spot residues, including conventional ones and new ones exploited in this kind of study. Detailed descriptions of other features are available in the Supplementary Material.

2.3 Structural Neighborhood Properties

Most of the conventional features such as physicochemical features, evolutionary conservation and solvent accessible area, describe only the properties of the current binding site itself, cannot represent the real situation well, and thus are insufficient to predict hot spots with high accuracy. Here, we develop a new way to calculate two types of structural neighborhood properties using Euclidean distance and Voronoi diagram.

The Euclidean neighborhood is a group of residues located within a sphere of $5\mathring{A}$ defined by the minimum Euclidean distances between any heavy atoms of the surrounding residues and any heavy atoms from the central residue. The value

---

[1] Available at http://admis.tongji.edu.cn/predhs/supplementary-material.pdf

of a specific residue-based feature f for neighbor j with regard to the target residue i is defined as

$$P_f(i,j) = \begin{cases} \text{the value of feature } f \text{ for residue } j & \text{if } |i - j| \geq 1 \text{ } and \text{ } d_{i,j} \leq 5\text{Å}, \\ 0 & \text{otherwise.} \end{cases}$$
(1)

Above, $d_{i,j}$ is the minimum Euclidean distance between any heavy atoms of residue i and any heavy atoms of residue j. The Euclidean neighborhood property of target residue i is defined as follows:

$$ENP_f(i) = \sum_{j=1}^{n} P_f(i,j)$$
(2)

where n is the total number of Euclidean neighbors.

We also use Voronoi diagram/Delaunay triangulation to define neighbor residues in 3D protein structures. For a protein structure, Voronoi tessellation partitions the 3D space into Voronoi polyhedra around individual atoms. Delaunay triangulation is the dual graph of Voronoi diagram, a group of four atoms whose Voronoi polyhedra meet at a common vertex form a unique Delaunay tetrahedra. In the context of Voronoi diagram (Delaunay triangulation), a pair of residues are said to be neighbors when at least one pair of heavy atoms of each residue have a Voronoi facet in common (in the same Delaunay tetrahedra). The definition of neighbors is based on geometric partitioning other than the use of an absolute distance cutoff, and hence is considered to be more robust. Voronoi/Delaunay polyhedra are calculated using the Qhull package that implements the Quickhull algorithm developed by Barber et al [16].

Give the target residue i and its neighbors $\{j = 1, ..., n\}$, for each site feature f, a Voronoi/Delaunay neighborhood property is defined as follows:

$$VDP_f = \sum_{j=1}^{n} P_f(j)$$
(3)

where $P_f(j)$ is the value of the site feature f for residue j.

2.4   Two-Step Feature Selection

Feature selection is performed in order to eliminate uninformative properties, which in turn improves model performance and provides faster and more cost-effective models. In this paper we propose a two-step feature selection method to select a subset of features that contribute the most in the classification.

In the first step, we assess the feature vector elements using the mean decrease Gini index (MDGI) calculated by the RF package in R [29]. MDGI represents the importance of individual feature vector element for correctly classifying an interface residue into hot spots and non-hot spots. The mean MDGI Z-Score of each vector element is defined as

$$MDGI \text{ } Z\text{-}Score = \frac{x_i - \bar{x}}{\sigma}$$
(4)

where x_i is the mean MDGI of the *i-th* feature, $\overline{x}$ is the mean value of all elements of the feature x, and σ is the standard deviation (SD). Here, we select the top 77 features with MDGI Z-Score larger than 2.5.

The second step is performed using a wrapper-based feature selection where features are evaluated by 10-fold cross-validation performance with the SVM algorithm, and redundant features are removed by sequential backward elimination (SBE). The SBE scheme sequentially removes features from the whole feature set till an optimal feature subset is obtained. Each removed feature is the one whose removal maximizes the performance of the predictor. The ranking criterion $R_c(i)$ represents the prediction performance of the predictor, which is built on a subset features exclusive of feature i, and is defined as follows:

$$R_c(i) = \frac{1}{k} \sum_{j=1}^{k} \{AUC_j + Accu_j + Sen_j + Spe_j\} \tag{5}$$

where k is the repeat times of 10-fold cross validation; AUC_j, $Accu_j$, Sen_j and Spe_j represent the values of AUC score, accuracy, sensitivity and specificity of the j-th 10-fold cross validation, respectively.

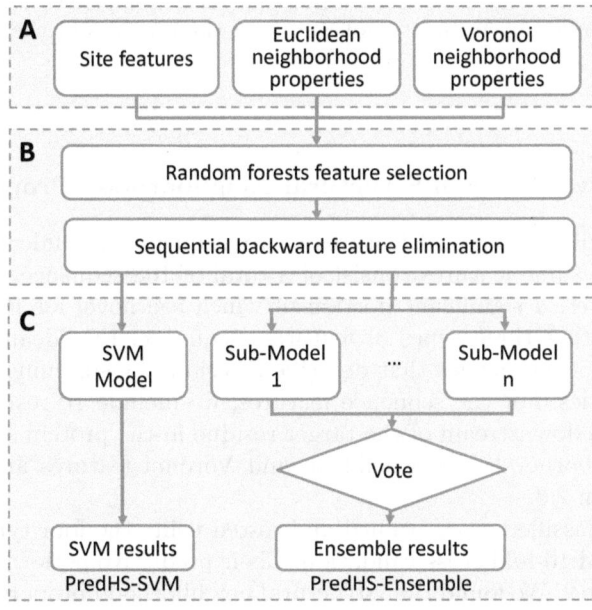

Fig. 1. The framework of PredHS. (A) Feature representation; (B) Two-step feature selection;(C) Prediction models: PredHS-SVM and PredHS-Ensemble.

2.5   The Classifiers

In this paper, two predictors were implemented under the PredHS framework shown in Figure 1. One is PredHS-SVM, another is PredHS-Ensemble, all are based on the 38 optimal features. The former is a support vector machine, the latter is an ensemble classifier built to handle the imbalanced classification problem. In what follows, we describe the implementation details of PredHS-Ensemble.

PredHS-Ensemble uses an ensemble of n classifiers and decision fusion technique on the training datasets. An asymmetric bootstrap resampling approach is adopted to generate subsets. It performs random sampling with replacement only on the majority class so that its size is equal to the number of minority samples, and keeps the entire minority samples in all subsets.

First, the majority class of non-hot spots is under-sampled and split into n groups by random sampling with replacement, where each group has the same or similar size as the minority class of interaction sites. After the sampling procedure, we obtain n new datasets from the set of non-hot spots. Each of the new dataset and the set of hot spots are combined into n new training datasets. Then, we train n sub-models by using the n new training datasets as input. Each of these classifiers is a Support Vector Machine (SVM). Here the LIBSVM package 2.8[2] is used with radial basis function (RBF) as the kernel. Finally, a simple majority voting method is adopted in the fusion procedure, and the final result is determined by majority votes among the outputs of the n classifiers.

3   Results

3.1   Predictive Power of Structural Neighborhood Properties

We investigated four types of features - site, sequence, Euclidean and Voronoi features. The residue features consist of a total of 108 sequence, structural and energy attributes, a significant portion of which are novel for hot spot identification. The other three types of features (sequence, Euclidean and Voronoi) are neighborhood properties that describe a residue by summing its neighbors' residue properties. For the sequence features, we include 10 residues upstream and 10 residues downstream of the target residue in the protein sequence as the sequence neighborhood. The Euclidean and Voronoi features are described in detail in Section 2.3.

Four SVM classifiers were trained and tested using the four types of features in Dataset I and 10-fold cross-validation. Their predictive performances are presented in Figure 2. We found that structural neighborhood properties (Euclidean and Voronoi) achieve the best performance, suggesting that structural neighborhood properties are more predictive than site properties in determining hot spots. We also observed that the classifier with linear sequence neighborhood properties is the worst performer, whose area under ROC curve is significantly smaller than that of the classifier with site features.

---

[2] Available at http://www.csie.ntu.edu.tw/~cjlin/libsvm/

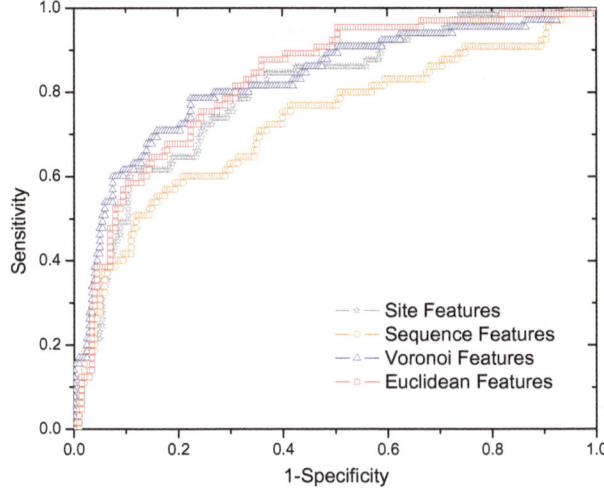

Fig. 2. ROC curves of classifiers with four types of features (site, sequence, Euclidean and Voronoi)

3.2   Selection of Optimal Features

The main goal of this study is to build effective and accurate models to predict hot spots. To this end, identification of a set of informative features is critical for performance boosting, and subsequently will enhance our understanding in the molecular basis of hot spots. We combine 324 site, Euclidean and Voronoi features for further feature selection. The 108 sequence features are not included in the combination since they perform significantly worse in the comparison study of Section 3.1. To assess the feature importance of the 324 features in predicting hot spots, we applied a two-step feature selection method on the Dataset I. As a result, a set of 38 optimal features are obtained and listed in Supplementary Material Table 4. We found that structural neighborhood properties (Euclidean and Voronoi properties) dominate the top-38 list, suggesting that structural neighborhood properties are more predictive than site properties in determining hot spot residues.

To quantitatively assess the performance of the two-step feature selection algorithm in PredHS, we compare it with four widely-used feature selection methods: Random Forests, Information Gain, Chi-squared and F-score. Figure 3 shows the ROC plots of the five feature selection methods based on Dataset I and 10-fold cross-validation. As can be seen from Figure 3, our two-step feature selection algorithm achieves the best performance. The proposed two-step feature selection algorithm, which is a hybrid approach integrating the merits of both filter methods and wrapper methods, can effectively improve the prediction performance with less computational cost and reduce the risk of overfitting.

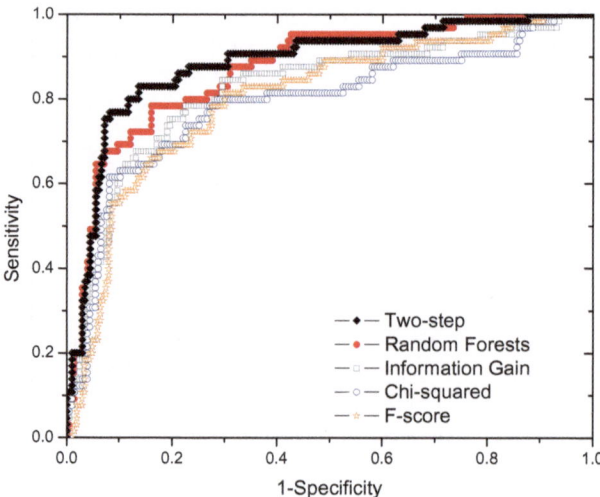

Fig. 3. ROC curves of our two-step algorithm and 4 existing feature selection methods

3.3 Performance Comparison with the State of the Art Approaches

To evaluate the performance of the proposed PredHS, eight existing hot spot prediction methods, Robetta [12], FOLDEF [11], KFC [14], MINERVA2 [15], HotPoint [21], APIS [17], KFC2a and KFC2b [18] are implemented and evaluated on both Dataset I and the Dataset II with 10-fold cross-validation. The performance of each model is measured by six metrics: accuracy (Accu), sensitivity (Sen), specificity (Spe), precision (Pre), CC and F1 score. F1-score is the harmonic mean of the *precision* and *recall* (equivalent to *sensitivity*), which is widely used to handle unbalanced data such as hot spot data.

Table 1 shows the detailed results of comparing our method with the existing methods. On Dataset I, our approach (PredHS-SVM and PredHS-Ensemble) show dominant advantage over the existing methods in five metrics: accuracy, sensitivity, precision, CC and F1-score. Only in specificity, FOLDEF and MINERVA2 perform as good as PreHS-SVM, all have the highest specificity value 0.93. Concretely, PredHS-Ensemble predicts the most actual hot spots as hot spots among these methods (with sensitivity = 0.85), while PredHS-SVM identifies the second most hot spots (with sensitivity = 0.75). Especially, PreHS-Ensemble's sensitivity is 47% higher than that of MINERVA2, which has the highest sensitivity among the existing methods. This suggests that our PredHS model is superior for predicting hot spot residues. Furthermore, PredHS-SVM's CC and F1 score are 25.5% and 19% respectively higher than that of MINERVA2 (still is the best in these two measures among the existing methods). Compared with PredHS-SVM, PredHS-Ensemble is much higher in sensitivity but relatively lower in specificity, however PredHS-Ensemble has much better balance of prediction accuracy between positive examples and negative examples.

Table 1. Performance comparison on Dataset I and Dataset II. Six performance measures are used: accuracy (Accu), sensitivity (Sen), specificity (Spe), precision (Pre), CC and F1 score. The highest values are highlighted.

Methods	Dataset I						Dataset II					
	Accu	Sen	Spe	Pre	CC	F1	Accu	Sen	Spe	Pre	CC	F1
PredHS-SVM	**0.88**	0.75	**0.93**	**0.79**	**0.69**	**0.76**	**0.87**	0.86	0.87	0.84	**0.74**	0.84
PredHS-Ensemble	0.82	**0.85**	0.81	0.60	0.60	0.69	**0.87**	**0.91**	0.84	0.80	**0.74**	**0.85**
Robetta	0.80	0.51	0.9	0.62	0.44	0.56	0.77	0.51	0.96	0.89	0.54	0.65
FOLDEF	0.78	0.31	**0.93**	0.59	0.31	0.40	0.70	0.31	**0.98**	0.91	0.41	0.46
KFC	0.79	0.55	0.87	0.58	0.43	0.57	0.74	0.55	0.87	0.75	0.45	0.64
MINERVA2	0.84	0.58	**0.93**	0.72	0.55	0.64	0.81	0.58	0.97	**0.93**	0.62	0.72
HotPoint	0.72	0.54	0.78	0.44	0.29	0.48	0.73	0.54	0.86	0.73	0.43	0.62
APIS	0.76	0.43	0.87	0.54	0.33	0.46	0.75	0.74	0.75	0.70	0.50	0.71
KFC2a	0.82	0.55	0.91	0.66	0.49	0.59	0.83	0.84	0.82	0.78	0.67	0.80
KFC2b	0.78	0.47	0.88	0.60	0.38	0.51	0.81	0.74	0.86	0.82	0.62	0.75

Table 2. Performance comparison on the independent test dataset

Methods	Accu	Sen	Spe	Pre	CC	F1
PredHS-SVM	**0.83**	0.59	**0.93**	**0.79**	**0.57**	**0.68**
PredHS-Ensemble	0.79	**0.74**	0.80	0.63	0.53	**0.68**
Robetta	0.70	0.33	0.86	0.52	0.23	0.41
FOLDEF	0.68	0.26	0.87	0.48	0.16	0.33
KFC	0.68	0.31	0.85	0.48	0.18	0.38
MINERVA2	0.77	0.46	0.91	0.69	0.42	0.55
HotPoint	0.69	0.59	0.74	0.5	0.31	0.54
APIS	0.71	0.56	0.77	0.52	0.33	0.54
KFC2a	0.74	**0.74**	0.74	0.56	0.41	0.64
KFC2b	0.79	0.59	0.87	0.68	0.47	0.63

As for Dataset II, PredHS still performs best in four performance metrics (accuracy, sensitivity, CC and F1-score). Again, this shows that PreHS can predict correctly more hot spots and has better balance in precision and recall than the existing methods. For almost all compared predictors, results on Dataset II are better than that on Dataset I, this is because Dataset II is a trimmed dataset where residues with $\Delta\Delta G$ between 0.4 and 2.0 are eliminated, which makes the prediction task not so tough.

3.4 Performance Evaluation by Independent Test

We further validate the performance of the proposed model (PredHS-SVM and PredHS-Ensemble) on the independent test dataset. Results of the independent test are presented in Table 2. We can see that our PreHS approach substantially outperforms the existing methods in five performance metrics (accuracy, specificity, precision, CC and F1 score), only KFC2a has a similar sensitivity value to that of PreHS-Ensemble, that is 0.74, the highest among the 10 compared predictors. Furthermore, the F1-scores of PredHS-SVM and PredHS-Ensemble are 0.68 and 0.68 respectively, while those of the existing methods fall in the range of 0.33-0.64. The findings from the independent test also indicate that the proposed PredHS model performs significantly better than the state of the art approaches.

4   Conclusion

Protein-protein interaction hot spots at the interfaces comprise a small fraction of the interface residues that make a dominant contribution to the free energy of binding. Alanine scanning mutagenesis experiments to identify hot spot residues are expensive and time-consuming, and computational methods can thus be helpful in suggesting residues for possible experimentation. In this study, we proposed a novel method, PredHS, including PredHS-SVM and PredHS-Ensemble, to predict hot spot residues in protein interfaces. Two key factors are responsible for our success. First, the wide exploitation of heterogeneous information, i.e. sequence-based, structure-based and energetic features, together with two types of structural neighborhood (Euclidian and Voronoi), provides more important clues for hot spot identification. A total of 324 features, including 108 site properties, 108 Euclidian neighborhood properties and 108 Voronoi neighborhood properties, have been investigated. Second, our two-step feature selection approach, which combines random forest and a sequential backward elimination, provides a ideal way for selecting an optimal subset of features within a reasonable computational cost. Also, the two-step method can significantly improve the prediction performance and reduce the risk of overfitting.

Our results highlight the advantages of basing hot spot prediction method on structural neighborhood properties. Compared with other computational hot spot prediction models, PredHS offers significant performance improvement both in terms of precision and recall as well as F1 score that measures the balance between precision and recall. PredHS-Ensemble has the highest sensitivity compared to other methods, but it has a lower specificity than PredHS-SVM. This is because that PredHS-Ensemble incorporates bootstrap resampling technique and SVM-based fusion classifiers to balance sensitivity and specificity.

As for the future work, major existing hot spots prediction methods, including MINERVA2 and KFC2a/b, are considered to be integrated into the PredHS web server to further improve the prediction performance by using Bayesian Networks.

Acknowledgement. This work was supported by China 863 Program under grant No. 2012AA020403, and National Natural Science Foundation of China under grants No. 61173118 and No. 61272380.

References

1. Alberts, B.D., et al.: Molecular Biology of the Cell. Garland, New York (1989)
2. Clackson, T., Wells, J.A.: A hot spot of binding energy in a hormone-receptor interface. Science 267, 383–386 (1995)

3. Bogan, A.A., Thorn, K.S.: Anatomy of hot spots in protein interfaces. J. Mol. Biol. 280, 1–9 (1998)
4. Moreira, I.S., et al.: Hot spots-A review of the protein-protein interface determinant amino-acid residues. Proteins 68, 803–812 (2007)
5. Li, J., Liu, Q.: 'Double water exclusion': a hypothesis refining the O-ring theory for the hot spots at protein interfaces. Bioinformatics 25, 743–750 (2009)
6. DeLano, W.L.: Unraveling hot spots in binding interfaces: progress and challenges. Current Opinion in Structural Biology 12, 14–20 (2002)
7. Massova, I., Kollman, P.A.: Computational Alanine Scanning To Probe Protein-Protein Interactions: A Novel Approach To Evaluate Binding Free Energies. J. Am. Chem. Soc. 120, 9401–9409 (1998)
8. Huo, S., et al.: Computational Alanine Scanning of the 1:1 Human Growth Hormone-Receptor Complex. J. Comput. Chem. 23, 15–27 (2002)
9. Grosdidier, S., Fernández-Recio, J.: Identification of hot-spot residues in protein-protein interactions by computational docking. BMC Bioinformatics 9, 447 (2008)
10. Brenke, R., et al.: Fragment-based identification of druggable 'hot spots' of proteins using Fourier domain correlation techniques. Bioinformatics 25(5), 621–627 (2009)
11. Guerois, R., et al.: Predicting Changes in the Stability of Proteins and Protein Complexes: A Study of More Than 1000 Mutations. J. Mol. Biol. 320, 369–387 (2002)
12. Kortemme, T., Baker, D.: A simple physical model for binding energy hot spots in protein-protein complexes. Proc. Natl. Acad. Sci. 99(22), 14116–14121 (2002)
13. Ofran, Y., Rost, B.: Protein-Protein Interaction Hotspots Carved into Sequences. PLoS Comput. Biol. 3(7), e119 (2007)
14. Darnell, S.J., et al.: An automated decision-tree approach to predicting protein interaction hot spots. Proteins 68, 813–823 (2007)
15. Cho, K., et al.: A feature-based approach to modeling protein-protein interaction hot spots. Nucleic Acids Research 37(8), 2672–2687 (2009)
16. Barber, C.B., et al.: The Quickhull algorithm for convex hulls. ACM Ttransactions on Mathematical Software 22(4), 469–483 (1996)
17. Xia, J., et al.: APIS: accurate prediction of hot spots in protein interfaces by combining protrusion index with solvent accessibility. BMC Bioinformatics 11, 174 (2010)
18. Zhu, X., Mitchell, J.C.: KFC2: A knowledge-based hot spot prediction method based on interface solvation, atomic density, and plasticity features. Proteins 79, 2671–2683 (2011)
19. Assi, S.A., et al.: PCRPi: Presaging Critical Residues in Protein interfaces, a new computational tool to chart hot spots in protein interfaces. Nucleic Acids Research 38(6), e86 (2009)
20. Tuncbag, N., et al.: Analysis and network representation of hotspots in protein interfaces using minimum cut trees. Proteins 78, 2283–2294 (2010)
21. Tuncbag, N., et al.: Identification of computational hot spots in protein interfaces: combining solvent accessibility and inter-residue potentials improves the accuracy. Bioinformatics 25(12), 1513–1520 (2009)
22. Thorn, K.S., Bogan, A.A.: ASEdb: a database of alanine mutations and their effects on the free energy of binding in protein interactions. Bioinformatics 17, 284–285 (2001)
23. Fischer, T., et al.: The binding interface database (BID): a compilation of amino acid hot spots in protein interfaces. Bioinformatics 19, 1453–1454 (2003)
24. Chan, C.H., et al.: Relationship between local structural entropy and protein thermostability. Proteins 57, 684–691 (2004)

25. Kabsch, W., Sander, C.: Dictionary of protein secondary structure: pattern recognition of hydrogen-bonded and geometrical features. Biopolymers 22, 2577–2637 (1983)
26. Liang, S., Grishin, N.V.: Effective scoring function for protein sequence design. Proteins 54, 271–281 (2004)
27. Liang, S., et al.: Consensus scoring for enriching near-native structures from protein-rotein docking decoys. Proteins 75, 397–403 (2009)
28. Hartshorn, M.J.: AstexViewer: a visualisation aid for structure-based drug design. J. Comput. Aided Mol. Des. 16, 871–881 (2002)
29. Liaw, A., Wiener, M.: Classification and Regression by randomForest. R News 2, 18–22 (2002)
30. Wang, L., et al.: Prediction of hot spots in protein interfaces using a random forest model with hybrid features. Protein Engineering, Design & Selection 25(3), 119–126 (2012)
31. Kvansakul, M., et al.: Structural basis for the high-affinity interaction of nidogen-1 with immunoglobulin-like domain 3 of perlecan. EMBO J. 20(19), 5342–5346 (2001)

Author Index